T0214199

Lecture Notes
in Business Information Processing　　　**416**

More information about this series at http://www.springer.com/series/7911

Mohamed Fakir · Mohamed Baslam ·
Rachid El Ayachi (Eds.)

Business Intelligence

6th International Conference, CBI 2021
Beni Mellal, Morocco, May 27–29, 2021
Proceedings

 Springer

Editors
Mohamed Fakir ⓘ
Université Sultan Moulay Slimane
Beni-Mellal, Morocco

Mohamed Baslam ⓘ
Université Sultan Moulay Slimane
Beni-Mellal, Morocco

Rachid El Ayachi ⓘ
Université Sultan Moulay Slimane
Beni-Mellal, Morocco

ISSN 1865-1348   ISSN 1865-1356 (electronic)
Lecture Notes in Business Information Processing
ISBN 978-3-030-76507-1   ISBN 978-3-030-76508-8 (eBook)
https://doi.org/10.1007/978-3-030-76508-8

This Springer imprint is published by the registered company Springer Nature Switzerland AG
The registered company address is: Gewerbestrasse 11, 6330 Cham, Switzerland

Preface

This book of proceedings collects papers and posters accepted for presentation at the 6th International Conference on Business Intelligence (CBI 2021). CBI 2021 was organized by the Faculty of Sciences and Techniques (FST), and the laboratory of Information Processing and Decision Support (TIAD) at Sultan Moulay Slimane University along with Association of Business Intelligence (AMID), and held during May 27–29, 2021, in Beni Mellal, Morocco.

CBI 2021 received 60 paper submissions from authors at various universities, of which 43% were included in this book. The papers were selected by the event chairs and their selection was based on a number of criteria including treated topic, originality of the contribution, and comments provided by scientific committee members; each paper was reviewed by at least three reviewers. The authors of selected papers were then invited to submit revised versions.

The goal of the CBI conference is to provide an international forum for scientists, engineers, and managers in academia, industry, and government to address recent research results and to present, discuss, and share their ideas, theories, technologies, systems, tools, applications, and experiences on all theoretical and practical issues. CBI 2021 covered all topics related to business intelligence, optimization and decision support and database and web environment.

We would like to thank the president of Sultan Moulay Slimane University, and the dean of the Faculty of Sciences and Techniques for their support to the conference, and everyone who contributed to the success of this conference.

Special thanks to the members of the different committees for their support and collaboration. Also, we would like to thank the local Organizing Committee, reviewers, speakers, authors, and all conference attendees. Finally, we want to thank Springer for their support of this publication.

April 2021

Mohamed Fakir
Mohamed Baslam
Rachid El Ayachi

Organization

General Chair

Mohamed Fakir — Sultan Moulay Slimane University, Morocco

Program Chairs

Mohamed Baslam — Sultan Moulay Slimane University, Morocco
Rachid El Ayachi — Sultan Moulay Slimane University, Morocco

Steering Committee

Lekbir Afraites — Sultan Moulay Slimane University, Morocco
Ebad Banissi — London South Bank University, UK
Karim Benouaret — Université Claude Bernard Lyon 1, France
Belaid Bouikhalene — Sultan Moulay Slimane University, Morocco
Charki Daoui — Sultan Moulay Slimane University, Morocco
A. l. Manuel De Oliveira — University of Aveiro, Portugal
Mohammed Elammari — Garyounis University, Libya
Halima El Biaze — University of Quebec at Montreal, Canada
Youssef El Mourabit — Sultan Moulay Slimane University, Morocco
Mohamed Erritali — Sultan Moulay Slimane University, Morocco
Hammou Fadili — National Conservatory of Arts and Crafts, France
Majed Haddad — University of Avignon, France
Suliman Hawamdeh — University of North Texas, USA
Najlae Idrissi — Sultan Moulay Slimane University, Morocco
Mostafa Jourhmane — Sultan Moulay Slimane University, Morocco
Brahim Minaoui — Sultan Moulay Slimane University, Morocco
Hicham Mouncif — Sultan Moulay Slimane University, Morocco
Mourad Nachaoui — Sultan Moulay Slimane University, Morocco
Muhammad Sarfraz — Kuwait University, Kuwait
Nora Tigziri — Université Mouloud Mammeri de Tizi-Ouzou, Algeria
Hicham Zougagh — Sultan Moulay Slimane University, Morocco

Program Committee

Nora Tigziri — Université Mouloud Mammeri de Tizi-Ouzou, Algeria
Pedro Teixeira Isaias — University of Queensland, Australia
Weitong Chen — University of Queensland, Australia
Adnan Mahmood — Macquarie University, Australia
Seng Loke — Deakin University, Australia

Contents

Networking, Cloud Computing and Networking Architectures in Cloud (Full Papers)

Big Data, Datamining, Web Services and Web Semantics (Poster Papers)

Signal, Image and Vision Computing (Poster Papers)

Decision Support, Information Systems and NLP (Full Papers)

Part-of-Speech Tagging Using Long Short Term Memory (LSTM): Amazigh Text Written in Tifinaghe Characters

Otman Maarouf[✉] and Rachid El Ayachi

Laboratory TIAD, Department of Computer Science, Faculty of Science and Technology, Sultan Moulay Slimane University, Beni Mellal, Morocco
rachid.elayachi@usms.ma

Abstract. Long short term memory (LSTM) networks have been gaining popularity in modeling sequential data such as phoneme recognition, speech translation, language modeling, speech synthesis, chatbot-like dialog systems, and others. This paper investigates the attention-based encoder-decoder LSTM networks in TIFINAGH part-of-speech (POS) tagging when it is compared to Conditional Random Fields (CRF) and Decision Tree. The attractiveness of LSTM networks is its strength in modeling long-distance dependencies. The experiment results show that Long short-term memory (LSTM) networks perform better than CRF and Decision Tree that have a near performance.

Keywords: Tifinagh · Part-of-speech · Conditional Random Fields (CRF) · Decision tree · Recurrence neural network (RNN); Long short term memory (LSTM) networks · Sequence-to-sequence learning

1 Introduction

Part-of-speech tagging is the process of assigning a part-of-speech marker to each word in an input text. The input to a tagging algorithm is a sequence of (tokenized) words and a tag set, and the output is a sequence of tags, one per token.

Recently, neural networks have been gaining popularity in the field of artificial intelligence. The advancements are due to the breakthrough in the algorithms that learn and recognize very complex patterns using deep layers of neural networks or commonly known as the deep neural networks (DNN) [1], and the introduction of different types of neural network such as convolutional neural network and recurrent neural network (RNN). For instance, convolutional neural networks, which are special type of feed-forward neural networks with two dimensions networks, have shown tremendous accuracy in classifying images through local receptive fields, shared weights, pooling, from simple handwritten digit recognition to more complex face recognition. In the modeling of sequential patterns, such as phoneme recognition, automatic speech recognition [2], speech synthesis, speech translation [3], chatbot and many others, RNN or the more specialized type of RNN, the long short term memory (LSTM) networks have shown

© Springer Nature Switzerland AG 2021
M. Fakir et al. (Eds.): CBI 2021, LNBIP 416, pp. 3–17, 2021.
https://doi.org/10.1007/978-3-030-76508-8_1

to be better than many of the traditional approaches. This paper presents a comparative study of three methods to solve the problem of Amazigh part-of-speech (POS) tagging. These methods are LSTM networks, Conditional Random Fields (CRF) and Decision tree. The objective is to examine the performance of the current state of LSTM networks while compared to CRF and Decision tree in POS tagging. POS tagging is a language-processing task that assigned a POS tag (e.g., noun, verb, adjective, etc.) to each word in a sentence [1]. This paper content 4 section firstly we talk about the different approaches used in the Amazighe part of speech, in the second section we present the Amazighe background knowledge related to this research, in the third section we dispute the tree approaches used in this paper, in the last section we discus the result obtained by the different approaches, finally we end with a conclusion.

2 Works About Amazigh POS Tagging

In the field of Part of Speech tagging, several researchers find the attack use different approaches, for example, support vector machines [SVM], or Conditional random fields [CRF]. Concerning stochastic models we find Decision Tree, Hidden Markov Models (HMM) are very prominent. Dynamite is a generally utilized stochastic trigram HMM tagger which utilizes a suffix examination system to assess lexical probabilities for obscure tokens dependent on properties of the words in the preparation corpus which share the equivalent suffix. In this case, S. Amri implement a new tool for the Amazigh part of speech tagging using Markov Models and decision trees in manually annotated approximately 75000 tokens, collected from the written texts with a POS tagset of 28 tags defined for the Amazigh language written with Latin characters, the accuracy founnd by CRFs is 89.18%, [4] and by SVM is accurate 88.02% [4]. Amri et al. used a TreeTagger technique and they found 93.19% accuracy, specifically, 94.10% on known words and 70.29% on unknown words [5].

3 Linguistic Background

3.1 Amazigh Language

The Amazigh language has a place with the Hamito-Semitic/"Afro-Asiatic" dialects, with rich morphology. In phonetic terms, the language is described by the expansion of lingos because of authentic, topographical and sociolinguistic variables. It is spoken in Morocco, Algeria, Tunisia, Libya, and Siwa (an Egyptian Oasis); it is likewise spoken by numerous different networks in parts of Niger and Mali. It is used by a huge number of individuals in North Africa essentially for oral correspondence and has been presented in broad communications and in the instructive framework as a team with a few services in Morocco. In Morocco, the Amazigh language uses various vernaculars in its normalization (Tachelhit, Tarifit, and Tamazight) [6]. Amazigh NLP (Natural Language Processing) presents numerous difficulties for specialists. Its significant highlights are:

- It has its own content: the Tifinagh, which is composed from left to right.
- It doesn't contain capitalized.

- Like other normal dialects, Amazigh presents for NLP ambiguities in sentence structure classes, named elements, which means, and so forth For instance, linguistically; the word (illi) contingent upon the setting can mean a thing in this sentence (tfulki illi: my girl is wonderful) or an action word in this sentence (ur illi walou: there isn't anything).
- As most dialects whose exploration in NLP is new, Amazigh isn't supplied with semantic assets and NLP apparatuses.
- Amazigh, as the greater part of the dialects which have as of late began being examined for NLP, actually experience the ill effects of the shortage of language preparing apparatuses and assets.

3.2 Amazigh Script

Like any language that passes through oral to writing mode, Amazigh language needed a graphics system.

The Tifinagh is the name of the amazigh language lettre set. The name Tifinagh conceivably signifies 'the Phoenician letters', or potentially from the expression tifin negh, which signifies 'our innovation'.

Variants of Tifinagh are used to compose Berber dialects in Morocco, Algeria, Mali and Niger. The Arabic and Latin letter sets are additionally utilized. The cutting edge Tifinagh content is otherwise called Tuareg, Berber or Neo-Tifinagh, to recognize it from the old Berber Script.

In 2003 Tifinagh turned into the official content for the Tamazight language in Morocco. It is likewise utilized by the Tuareg, especially the ladies, for private notes, love letters and in enrichment. For public purposes, the Arabic letter set is regularly utilized.

Tifinagh is written from left to right and contains 33 graphemes which correspond to [5]:

- 27 consonants including the labials (ⵓ,ⵀ,ⵞ), dental (+,ⵐ,Ⴓ,Ⴇ,ⵏ,ⵔ,Q,ⵕ), the alveolar (ⵔ,ⵟ,ⵝ,ⵜ), the palatals (ⵞ,ⵔ), the velar (ⵕ,ⵜ), the labiovelars (ⵕ̈,ⵜ̈), the uvular (ⵍ,ⵜ,ⵝ), the pharyngeal (ⵏ,ⵖ) and the laryngeal (ⵔ);
- 2 semi-consonants: ⵢ and ⵓ;
- vowels: three full vowels ₒ, ⵜ, ₒand neutral vowel (or schwa) ₒ̃.

3.3 Amazigh Morphology

The Amazigh language is a morphological rich language which is agglutinative. The most utilized syntactic classes are Noun, Verb, Adjective or Adverb. Basically, things and action words are the base of the Amazigh morphology and the more significant classes to zero in on, as others can be gotten from them. We will introduce underneath these two linguistic Amazigh classes [6]:

Nouns

There are a various types of nouns in Amazigh: common nouns, noun adjectives, Arabic loans, proper nouns, and kinship terms. Amazigh common nouns carry gender and number affixes. Noun in Amazigh characterized by gender, number, and status. The noun is either masculine or feminine. It is plural or singular: plural starts from two. The noun is free or annexed.

Nouns beginning with a vowel are generally masculine, and those beginning with the consonant /ⵜ/ are generally feminine:

    a.   ⴰⴳⴰⵣ   « man »,   ⴰⵇⵇⴰ   « boy »

    b.   ⵜⴰⴻⴻⴰⵜ « woman »,   ⵜⴰⵇⵇⴰⵜ   « girl »

Noun adjectives or nominal adjectives are similar to common nouns; they are morphologically a word class that has special syntactic patterns.

Arabic loans used in Amazigh usually take the Arabic article al its phonological variants, which mark definiteness. This article obeys the phonotactic rules of Arabic and may realize a geminate initial consonant in the latter is a tongue blade articulation, just as in Arabic.

Verbs

The morphological aspect of the verb in Amazigh relies fundamentally upon the appendage and arrangement. A few action words are inductions by appendage (prefixes, postfixes) and different action words are fundamentally gotten from things, either from an action word and a thing or either from two action words.

3.4 Amazigh Tagset

3.4.1 Corpus

A corpus is a set of documents, artistic or not (texts, images, videos, etc.), grouped together for a specific purpose. We can use corpora in several fields: literary, linguistic, scientific studies, philosophy, etc. The branch of linguistics which is more specifically concerned with corpora is logically called corpus linguistics. It is linked to the development of computer systems, in particular to the constitution of textual databases. Since 2009, the university journal CORPUS has been dedicated to this field. We speak of a corpus to designate the normative aspect of the language: its structure and its code in particular. "Corpus" is generally opposed to "status", which corresponds to the terms of use of the language. This opposition is common in the study of language policies. Generally, a corpus contains up a few millions of words and can be lemmatized and annotated with information about the parts of speech. Among the corpus, there is the British National Corpus (100 million words) and the American National Corpus (20 million words). A balanced corpus would provide a wide selection of different types of texts and from various sources such as newspapers, books, encyclopedias or the web. For the Moroccan Amazigh language, it was difficult to find ready-made resources. We can just mention the manually annotated corpus of Maarouf et al. after the modification applied on the corpus of Outahajala et al. [7]. This corpus contains 20k words and we collected more than 40k words from MAP news so we have in global corpus more than 60k that are why we decided to use in this study.

3.5 Amazigh Tagset

Defining the sufficient tag set is a center undertaking in building a programmed POS tagger. It targets characterizing a calculable tag set with the suitable degree of granularity, for example not very fine-grained nor excessively shallow for the potential unites frameworks that will use it. The pre-owned corpus comprises of a rundown of writings separated from an assortment of sources, for example, a few books, just as certain writings from IRCAM's site. We had the option to arrive at an all out number of words better than 60k tokens. This corpus is commented on morphologically using the label set presented in [8].

Table 1. AMTS tag set.

N°	POS	Designation
1	NN	Common noun
2	NNK	Kinship noun
3	NNP	Proper noun
4	VB	Verb, base form
5	S	Preposition
6	FW	Foreign word
7	NUM	Numeral
8	ROT	Residual, other
9	PUNC	Punctuation
10	SYM	Symbol
11	VBP	Verb, participle
12	ADJ	Adjective
13	ADV	Adverb
14	C	Conjunction
15	DT	Determiner
16	FO	Focalize
17	IN	Interjection
18	NEG	Particle, negative
19	VOC	Vocative
20	PRED	Particle, predicate
21	PROR	Particle, orientation
22	PRPR	Particle, preverbal
23	PROT	Particle, other
24	PDEM	Demonstrative pronoun
25	PP	Personal pronoun
26	PPOS	Possessive pronoun
27	INT	Interrogative
28	REL	Relative

The Table 1 indicate the different tagset used in the Amazigh part of speech that contains 28 POS in different classes example nouns, verbs, prepositions...

4 Description of the Three Methods

4.1 Conditional Random Fields

CRFs are non-directed graphical models that seek to represent the distribution of annotation probabilities (or labels) and conditionally to observation x from labeled examples (examples with expected labels) [9]. They are thus models obtained by supervised learning, very used especially in the problems of labeling of sequences. We used this algorithm in our case of tagging words written in Tifinagh character.

In the sequential case, that is to say, the labeling of observations x_i by labels y_i, the potential function at the heart of the CRFs is written:

$$p(x|y) = \frac{1}{Z(x)} exp(\sum_{k=1}^{k_1} \sum_{i=1}^{n} \lambda_k f_k(y_i, x) \sum_{k=1}^{k_2} \sum_{i=1}^{n} \mu_k g_k(y_{i-1}, y_i, x)) \tag{1}$$

- $Z(x)$ a normalization factor;
- The local and global characteristic functions f and g: the functions f characterize the local relations between the current label in position i and the observations; the functions g characterize the transition between the nodes of the graph, that is to say between each pair of labels i and $i - 1$, and the sequence of observations.
- The values $k1, k2$ and n are respectively the number of functions f, the number of functions g, and the size of the sequence of labels to be predicted. The functions f and g are generally binary functions that verify a certain combination of labels and attributes describing the observations and applied to each position of the sequence.

These functions are defined by the user; they reflect his knowledge of the application. They are weighted by the λk and μk which estimate the importance of the information they bring to determine the class.

The learning of the CRFs consists in estimating the vector of parameters $\theta = \lambda 1, \lambda 2, ..., \lambda k1, \mu 1, \mu 2, ..., \mu k2$ (weight of the functions f and g) from training data, i.e. N labeled sequences $(x(i), y(i))i = Ni = 1$. In practice, this problem is reduced to an optimization problem, usually using methods of the type similar to Newton, such as the L-BFGS algorithm). After this learning step, the application of CRFs to new data is to find the most likely sequence of labels given a sequence of non-sighted observations. As for other stochastic methods, this one is generally obtained with a Viterbi algorithm [10].

The (unnormalized) CRF model for a sentence $x = (x1, ..., x|x|)$ and a POS sequence $y = (y1, ..., y|x|)$ is defined as:

$$p(y|x; w) = \prod_{i=n}^{|x|} \exp(w.\phi(y_{i-n}, ..., y_i, x, i)) \tag{2}$$

Where n denotes the model order, w the model parameter vector, and ϕ the feature extraction function. We denote the tag set as Y, that is, $yi \in Y$ for $i \in 1...|x|$ [11].

4.2 Decision Tree Tagger

Decision trees have long been considered as one of the most practical and straightforward approaches to classification. Strictly speaking, induction of decision trees is a method that generates approximations to discrete-valued functions and has been shown, experimentally, to provide robust performance in the presence of noise. Moreover, decision trees can be easily transformed to rules that are comprehensible by people [12].

There is a couple of very good reasons why decision trees are good candidates for NLP problems, from the classification point of view and especially for POS tagging [13]:

- Decision trees are ideally suited for symbolic values, which is the case for NLP problems.
- Disjunctive expressions are usually employed to capture POS tagging rules. By using decision trees such expressions can still be "discovered" and be associated with relevant linguistic features (note, that the linguistic bias inherent in the representation may also serve as an encoding of produced rules).

Decision trees are built top-down. One selects a particular attribute of the instances available at a node, and splits those instances to children nodes according to the value each instance has for the specific attribute. This process continues recursively until no more splitting along any path is possible, or until some splitting termination criteria are met. After splitting has ceased, it is sometimes an option to prune the decision tree (by turning some internal nodes to leaves) to hopefully increase its expected accuracy.

The underlying phase of building the choice tree occurs during the preparation stage. It will parse through the content and break down trigrams, embedding each unigram into the tree. Probabilities for which tag to utilize are resolved for a given hub of the tree depends on the data got from the two past hubs (trigram). When the tree is made, its hubs are pruned. On the off chance that the data addition of a specific hub is beneath a defined limit, its kids' hubs are evacuated [14].

4.3 Recurrence Neural Network (RNN)

A recurrent neural network (RNN) is a class of artificial neural organizations where associations between nodes structure a coordinated diagram along a transient arrangement. This permits it to show fleeting unique conduct. Gotten from feedforward neural network, RNNs can utilize their inner state (memory) to handle variable length groupings of information sources. This makes them material to undertakings, for example, unsegmented, connected handwriting recognition or speech recognition [15].

The expression "recurrent neural network" is utilized aimlessly to allude to two wide classes of organizations with a comparable general structure, where one is limited drive and the other is boundless motivation. The two classes of organizations display transient powerful conduct. A limited motivation intermittent organization is a coordinated non-cyclic diagram that can be unrolled and supplanted with a carefully feedforward neural organization, while a boundless drive repetitive organization is a coordinated cyclic chart that can not be unrolled.

Both limited motivation and endless drive intermittent organizations can have extra put away states, and the capacity can be under direct control by the neural organization. The capacity can likewise be supplanted by another organization or chart, if that joins time delays or has input circles. Such controlled states are alluded to as gated state or gated memory, and are important for long transient memory organizations (LSTMs) [16] and gated repetitive units (Fig. 1).

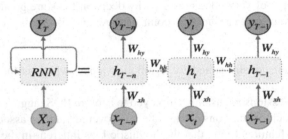

Fig. 1. A recurrent neural network. Note: X_i is input, Y_i is output for a RNN

The restriction of the essential RNN is that it isn't doing great in displaying the information that are a distance away practically speaking.

LSTM networks are acquainted with take care of the drawn out conditions issue in RNN. In a LSTM organization, there are entryways that permit data to be failed to remember and refreshed contingent upon the helpfulness the data is.

LSTMs can frame various sorts of organizations. The encoder decoder networks have been shown to be awesome in succession to grouping demonstrating.

Given a source arrangement (for example words), the encoder will encode the contribution as a vector and passes it to the decoder. The decoder will produce yield from the vector passed from the encoder until an exceptional finish of sentence tag is reached. A consideration based encoder-decoder LSTM organization will permit certain piece of the source arrangement to join in or center around certain piece of the objective grouping during preparing and disentangling, rather than the entire sentence encode as a solitary vector. In another word, the consideration esteems tell the strength of the arrangement between a mix of info words and yield words, permitting more setting explicit encoding and translating [2] (Fig. 2).

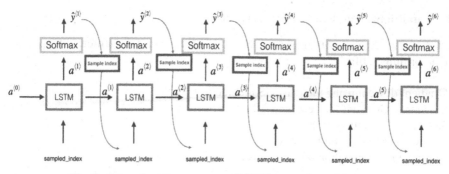

Fig. 2. Long short term memory (LSTM) networks architecture

5 Experiments and Results

The Amazighe POS annotated text used in the experiments consists of 60K words, which were tagged using 28 POS tags. Contrary to the usual norm where each word is assigned a POS tag, he word in the tagged sentences is not assign a tag explicitly. Instead, the tags are implicitly assigned in sequence to one or more words, just like in a parallel text in machine translation. Thus, the POS training algorithm must learn the word(s)/POS alignments from the "word/POS parallel text". From the total tagged words, about 80% words were used for training, and the remaining 20% sentences was used for testing and development respectively.

The performances are estimated by calculating the average of the precision (P), recall (R) and F1-score (F1) on the total number of tests [17]. These latter are calculated for each label l as follows:

$$P_l = \frac{number\ of\ items\ correctly\ labeled\ by\ the\ model\ with\ l}{number\ of\ items\ labeled\ by\ the\ model\ with\ l} \tag{3}$$

$$R_l = \frac{number\ of\ items\ correctly\ labeled\ by\ the\ model\ with\ l}{number\ of\ items\ manually\ labeled\ with\ l} \tag{4}$$

$$F(1)_l = 2 \times \frac{P_l \times R_l}{P_l + R_l} \tag{5}$$

5.1   Conditional Random Fields Approach

For testing the CRF algorithm we using Scikit-Leran Library (sklearn-crfsuite) provides five algorithms: 'lbfgs'

- Gradient descent using the L-BFGS method – 'l2sgd'
- Stochastic Gradient Descent with L2 regularization term – 'ap'
- Averaged Perceptron – 'pa'
- Passive Aggressive (PA) – 'arow'
- Adaptive Regularization of Weight Vector (AROW)

Choice of algorithm wasn't based on implementation details. Playing with different values of different parameters and different features.

The following table contains the different basic training results.

Table 2. CRF Classification matrix.

Tagset	Precision	Recall	F1-score
S	0,993	0,99	0,997
ADJ	0,851	0,731	0,786
PUNC	1	1	1
VB	0,948	0,949	0,948
NN	0,922	0,966	0,944
ADV	0,806	0,701	0,75
DT	0,853	0,905	0,878
VBP	0,842	0,734	0,784
C	0,796	0,912	0,85
REL	0,934	0,782	0,851
PROR	0,934	0,799	0,861
PP	0,984	0,96	0,981
NNP	0,868	0,799	0,832
PRPR	0,914	0,943	0,929
SYM	1	1	1
INT	0,65	0,915	0,759
NEG	0,997	1,02	0,988
PDEM	0,645	0,395	0,475
PRED	1	0,52	0,687
IN	1	0,734	0,853
PROT	0,52	0,395	0,449
NUM	0,993	0,993	0,993
VOC	0,913	0,982	0,946
NNK	0,967	0,877	0,92
PPOS	0,82	0,62	0,706
FOC	0,742	0,704	0,723
ROT	0,353	0,12	0,174
Weighted avg	0,934	0,935	0,933

According to the results indicated in Table 2, we found the weighted average .934, this result is justified by the lack of the sentences which containing ROT, FOC and ADV that is why we ask the people of linguistics to create sentences which contains these types.

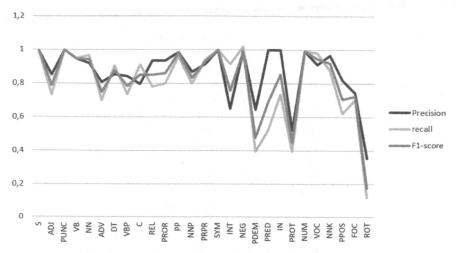

Fig. 3. Precision, recall and F1-score using CRF algorithm

Figure 3 indicates the variations of precision, recall, and F1-score after training the model using a conditional random fields algorithm that gives the same variation for the three values.

5.2 Decision Tree Tagger Approach

Next, we evaluated the Decision Tree approach by using Scikit-Leran Library, for each attribute in the dataset; the decision tree algorithm forms a node, where the most important attribute is placed at the root node. For evaluation, we start at the root node and work our way down the tree by following the corresponding node that meets our condition. This process continues until a leaf node is reached, which contains the prediction.

The following table contains the different basic training results:

According to Tables 2 and 3, we observe that the CRF algorithm is better than Decision Tree algorithm that gives the weighted average .876.

Figure 4 indicates the variations of precision, recall, and F1-score after training the model using a decision tree algorithm that gives the same variation for the three values.

5.3 Long Short Term Memory Approach

In this evaluation, we used the multi-layered LSTM networks using Google's Tensorflow framework (https://www.tensorflow.org/) to test the Amazighe POS tagging. The LSTM models were trained using the training set and configure with the development set. We

Table 3. Decision tree classification matrix.

Tagset	Precision	Recall	F1-score
S	0,893	0,99	0,897
ADJ	0,751	0,781	0,686
PUNC	1	1	1
VB	0,848	0,899	0,848
NN	0,822	0,916	0,844
ADV	0,706	0,701	0,65
DT	0,793	0,805	0,778
VBP	0,772	0,804	0,684
C	0,716	0,812	0,75
REL	0,854	0,902	0,771
PROR	0,804	0,809	0,811
PP	0,954	0,98	0,931
NNP	0,821	0,899	0,802
PRPR	0,874	0,913	0,879
SYM	1	1	1
INT	0,65	0,815	0,729
NEG	0,997	1	0,988
PDEM	0,545	0,675	0,375
PRED	0,9	0,67	0,587
IN	0,9	0,834	0,753
PROT	0,47	0,475	0,409
NUM	0,993	0,993	0,993
VOC	0,873	0,882	0,846
NNK	0,867	0,907	0,82
PPOS	0,82	0,77	0,606
FOC	0,642	0,704	0,623
ROT	0,253	0,22	0,124
Weighted avg	0,876	0.842	0,832

tested different sizes of word embedding vectors, and we found that 64 is the most optimum size for Amazighe POS tagging. In this evaluation we used 10 epochs in a different size batch and size cells. Then we found the results shown in the table below:

In Table 4, the results show that the best result of LSTM networks is using 64 size cells, where the accuracy given is 97.81% with 32 and 64 in size batch and 97.87% with 128 in size batch.

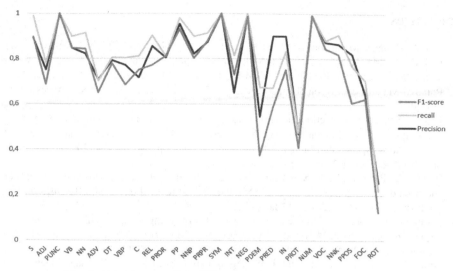

Fig. 4. Precision, recall and F1-score using decision tree algorithm

Table 4. POS tagging accuracy using LSTM networks.

Size cells	Size batch	Accuracy
32	32	97.41%
	64	97.68%
	128	97.74%
64	32	97.81%
	64	97.81%
	128	97.87%
128	32	97.79%
	64	97.79%
	128	97.74%

6 Conclusion

The CRF produces a more accurate POS tagging compared to Decision Tree. However, when we used the LSTM networks, we see that the results is best to what we get with CRF an Decision Tree. This shows that LSTM networks can capture the knowledge for a language. This has tremendous benefit especially to be used on languages that we do not have much linguistic studies on. However, in term of training the models, LSTM networks is the fastest to train and run. Qualitatively, a LSTM networks takes only few minutes to run about 48k training. This is followed by CRF, and subsequently followed by Decision Tree. In the future works, we will ameliorate the obtained results and treat other tasks concerning Amazighe language.

References

1. Tan, T., Ranaivo-Malançon, B., Besacier, L., Yeong, Y., Gan, K.H., Tang, E.K.: Evaluating LSTM networks, HMM and WFST in Malay part-of-speech tagging (2017). https://www.sem anticscholar.org//paper/Evaluating-LSTM-Networks%2C-HMM-and-WFST-in-Malay-Tan-Ranaivo-Malan%C3%A7on/5085ceeaa3c22b3a72b7482785251c13cd6855dc. Accessed 29 Oct 2020
2. Zen, H., Sak, H.: Unidirectional long short-term memory recurrent neural network with recurrent output layer for low-latency speech synthesis. In: 2015 IEEE International Conference on Acoustics, Speech and Signal Processing (ICASSP), pp. 4470–4474 (2015). https://doi.org/10.1109/ICASSP.2015.7178816
3. Berard, A., Pietquin, O., Servan, C., Besacier, L.: Listen and translate: a proof of concept for end-to-end speech-to-text translation. arXiv:161201744 [cs], 6 December 2016. https://arxiv.org/abs/1612.01744. Accessed 17 Jan 2021
4. Samir, A., Lahbib, Z., Mohamed, O.: Amazigh PoS tagging using machine learning techniques. In: Ben Ahmed, M., Boudhir, A.A. (eds.) SCAMS 2017. LNNS, vol. 37, pp. 551–562. Springer, Cham (2018). https://doi.org/10.1007/978-3-319-74500-8_51
5. Amri, S., Zenkouar, L.: Amazigh POS tagging using TreeTagger: a language independent model. In: Ezziyyani, M. (ed.) AI2SD 2018. AISC, vol. 915, pp. 622–632. Springer, Cham (2019). https://doi.org/10.1007/978-3-030-11928-7_56
6. Nejme, F., Boulaknadel, S., Aboutajdine, D.: Finite state morphology for Amazigh language. In: Gelbukh, A. (ed.) CICLing 2013. LNCS, vol. 7816, pp. 189–200. Springer, Heidelberg (2013). https://doi.org/10.1007/978-3-642-37247-6_16
7. Outahajala, M., Rosso, P., Zenkouar, L.: Building an annotated corpus for Amazigh. In: Proceedings of 4th International Conference on Amazigh and ICT (2011)
8. Outahajala, M., Benajiba, Y., Rosso, P., Zenkouar, L.: POS tagging in Amazighe using support vector machines and conditional random fields. In: Muñoz, R., Montoyo, A., Métais, E. (eds.) NLDB 2011. LNCS, vol. 6716, pp. 238–241. Springer, Heidelberg (2011). https://doi.org/10.1007/978-3-642-22327-3_28
9. Zheng, S., Jayasumana, S., Romera-Paredes, B., et al.: Conditional random fields as recurrent neural networks, pp. 1529–1537 (2015). https://www.cv-foundation.org/openaccess/content_iccv_2015/html/Zheng_Conditional_Random_Fields_ICCV_2015_paper.html. Accessed 7 Apr 2021
10. Claveau, V., Ncibi, A.: Découverte de connaissances dans les séquences par CRF non-supervisés. In: 20ème Conférence Sur Le Traitement Automatique Des Langues Naturelles, TALN, vol. 1 (2013). https://hal.archives-ouvertes.fr/hal-00912314
11. Silfverberg, M., Ruokolainen, T., Lindén, K., Kurimo, M.: Part-of-speech tagging using conditional random fields: exploiting sub-label dependencies for improved accuracy. In: Proceedings of the 52nd Annual Meeting of the Association for Computational Linguistics (Volume 2: Short Papers), pp. 259–264. Association for Computational Linguistics (2014). https://doi.org/10.3115/v1/P14-2043
12. Prokhorov, S., Safronov, V.: AI for AI: what NLP techniques help researchers find the right articles on NLP. In: 2019 International Conference on Artificial Intelligence: Applications and Innovations (IC-AIAI), pp. 76–765. IEEE (2019). https://doi.org/10.1109/IC-AIAI48757.2019.00023
13. Samir, A., Lahbib, Z.: Training and evaluation of TreeTagger on Amazigh corpus. IJIE 6(2/3/4), 230 (2019). https://doi.org/10.1504/IJIE.2019.101130
14. Research Institute for Artificial Intelligence, Romanian Academy, Boros, T., Dumitrescu, S.D., Pipa, S.: Fast and accurate decision trees for natural language processing tasks. In: Recent Advances in Natural Language Processing Meet Deep Learning, RANLP 2017, pp. 103–110. Incoma Ltd. Shoumen, Bulgaria (2017). https://doi.org/10.26615/978-954-452-049-6_016

15. Perez-Ortiz, J.A., Forcada, M.L.: Part-of-speech tagging with recurrent neural networks. In: Proceedings of the International Joint Conference on Neural Networks, IJCNN 2001 (Cat. No. 01CH37222), vol. 3, pp. 1588–1592. IEEE (2001). https://doi.org/10.1109/IJCNN.2001. 938396
16. Plank, B., Søgaard, A., Goldberg, Y.: Multilingual part-of-speech tagging with bidirectional long short-term memory models and auxiliary loss. arXiv:160405529 [cs], 21 July 2016. https://arxiv.org/abs/1604.05529. Accessed 29 Oct 2020
17. Das, D., Kolya, A., Ekbal, A., Bandyopadhyay, S.: Temporal analysis of sentiment events – a visual realization and tracking. In: Gelbukh, A.F. (ed.) CICLing 2011. LNCS, vol. 6608, pp. 417–428. Springer, Heidelberg (2011). https://doi.org/10.1007/978-3-642-19400-9_33

Contribution to Arabic Text Classification Using Machine Learning Techniques

Imad Jamaleddyn[1(✉)] and Mohamed Biniz[2]

[1] Laboratory TIAD, Department of Computer Science, Faculty of Sciences and Technics of Beni Mellal, Sultan Moulay Slimane University, Beni Mellal, Morocco
[2] Laboratory LIMATI, Department of Mathematics and Computer Science, Polydisciplinary Faculty of Beni Mellal, Sultan Moulay Slimane University, Beni Mellal, Morocco

Abstract. With the increase of text stored in electronic format, it is no longer possible for humans to understand all the incoming data or even categorize it. We need an automatic text classification system in order to classify them into predefined classes and quickly retrieve information. Text classification can be achieved by machine learning, it requires a set of approaches for vectorization and classification. In vectorization phase, this work proposes two approaches (BOW and TF-IDF), but in the classification phase, the algorithms of machine learning used are: RL, SVM and ANN. At the end, a comparison study is given.

Keywords: Machine learning · Natural language processing · Text representation · Text vectorization · Arabic text classification

1 Introduction

The availability of electronic text in companies and universities, as well as on the Internet, has increased exponentially in recent years. This makes it necessary to develop automatic classification tools that allow fast content exploration and analysis. The most important of these tools is that they are based on the concepts of artificial intelligence methods, machine learning and natural language processing.

Machine learning (ML) is a data analysis method that automates the creation of analytical models. It is a branch of artificial intelligence that allows systems to learn from data, identify patterns and make decisions with minimal human intervention.

Natural Language Processing (NLP) is a subfield of artificial intelligence that enables computers to understand and analyze human language. It is applied now in several areas such as document indexing, automatic summarization of documents, plagiarism detection and document classification.

Text classification (TC) is the process of grouping texts into categories regarding their content. Text classification is an important learning issue that is at the core of much information management and retrieval tasks. Until now a lot of works focused on text classification such as the classification of English, French and Chinese texts […], but, for Arabic texts, there are few studies relating to the classification of Arabic texts. This is explained by the complexity of the Arabic texts.

© Springer Nature Switzerland AG 2021
M. Fakir et al. (Eds.): CBI 2021, LNBIP 416, pp. 18–32, 2021.
https://doi.org/10.1007/978-3-030-76508-8_2

Arabic is one of the commonly spoken languages with more than 420 million speakers around the world. The Arabic alphabet is a collection of 28 letters:

أ ب ت ث ج ح خ د ذ ر ز س ش ص ض ط ظ ع غ ف ق ك ل م ن ه و ي

In the Arabic linguistics, there is also hamza (ء) considered as a letter, The vowels are (ا و ى), it also differs by diacritics that represent a small vowel like (fatha, kasra, damma, sukun, shadda, and tanween). The orthography system of the Arabic language is based on the diacritical effect [1].

The objective of this paper is to contribute to the classification of Arabic texts using machine learning algorithms. A set of approaches for vectorization and classification are needed. The methods used in the vectorization step are "BOW" and "TF-IDF". On the other hand, in the classification, the adopted approaches are Logistic Regression (LR), Support Vector Machine (SVM) and Artificial Neural Networks (ANN).

The rest of the paper is organized as follows. In Sect. 2 an overview of related works in text classification, Sect. 3 presents a definition of machine learning and ML algorithms for Text classification phase, Sect. 4 shows an overview of text classification and text representation, Sect. 5 presents our proposed system, in Sect. 6 results and evaluation methods are presented and finally in Sect. 7 conclusion is made.

2   Related Works

Mesleh presents an implementation of a Support Vector Machine (SVM) based text classification system for Arabic-language articles. This classifier uses the CHI square method as the feature selection method in the preprocessing step of the text classification system design procedure. In comparison with other classification methods, our classification system shows high classification efficiency for Arabic articles in terms of Macro-mean F1 = 88.11 and Micro-mean F1 = 90.57 [2].

The text classification task is basically treated as a supervised machine learning problem, where the documents are projected into the so-called Vector Space Model (SVM), basically using the words as features. Various classification methods have been successfully applied, Bayesian classifiers, decision trees, k-Nearest Neighbors (kNN), rule-learning algorithms, neural networks, fuzzy logic-based algorithms, maximum entropy, and support vector machines. However, the task suffers from the issue that the feature space in VSM is highly dimensional which negatively affects the performance of the classifier [3].

Abu-Errub proposes a new Arabic text classification method in which a document is compared to predefined document categories based on its content using the TF.IDF (Term Frequency times Inverse Document Frequency) measure, and then the document is classified into the appropriate subcategory using the chi-square measure [4].

Bahassine et al. propose an improved method for Arabic text classification that employs the Chi-square feature selection (referred to, hereafter, as ImpCHI) to enhance the classification performance. Besides, they have also compared this improved chi-square with three traditional features selection metrics namely mutual information, information gain, and Chi-square, they extend the current work to assess the method in terms of other evaluation methods using an SVM classifier. For this purpose, a dataset of 5070 Arabic documents are classified into six independent classes. In terms of performance, the experimental findings show that combining the ImpCHI method and SVM classifier outperforms other combinations in terms of precision, recall, and f-measures. This combination significantly improves the performance of the Arabic text classification model. The best f-measures obtained for this model is 90.50% when the number of features is 900 [5].

Boukil and al introduce an innovative method for Arabic text classification. It uses an Arabic stemming algorithm to extract, select, and reduce the features. After that, use the Term Frequency-Inverse Document Frequency technique as feature weighting technique. And finally, for the classification step, use one of the deep-learning algorithms that are very powerful in another field such as the image processing and pattern recognition, but still rarely used in text mining, this algorithm is the Convolutional Neural Networks. And obtain excellent results on multiple reference points with this combination and some adjustment of the hyperparameters in the convolutional neural network algorithm [6].

3   Machine Learning Algorithms

The Machine Learning (ML) is a field of artificial intelligence. It can be defined as covering the study, design, and development of algorithms giving the possibility to machines to learn without having been explicitly programmed. Instead of writing a program by hand, a ML algorithm will analyze many instances of examples to produce a program or model to perform the task illustrated by the examples.

A Machine Learning task can learn to: classify, predict, recommend, optimize, detect patterns or anomalies, filter... Machine learning tasks can obviously be combined with other ML algorithms or not to produce more advanced algorithms (definition of Arthur Samuel in 1959) [7] (Fig. 1).

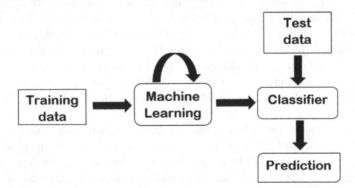

Fig. 1. Machine Learning Model

In Machine Learning, we basically try to create a model to predict on the test data. So, we use the training data to fit the model and testing data to test it. The models generated are to predict the results unknown which is named as the test set. As you pointed out, the dataset is divided into the train and test set to check accuracy, precision by training and testing it on it.

The approaches used in this document are:

3.1 Logistic Regression

Logistic regression is primarily used to model a binary variable (0,1) as a function of one or more other variables, called predictors. The binary variable modeled is usually called the response variable, or dependent variable. I will use the term "response" for the modeled variable because it has become the preferred way to refer to it [8].

Logistic regression can be considered a special case of a generalized linear model and therefore analogous to linear regression. However, the logistic regression model is based on very different assumptions (regarding the relationship between the dependent and independent variables) than the linear regression. The main differences between these two models can be seen in the following paragraphs. Two features of logistic regression. First, the conditional distribution P (y | x) is a Bernoulli distribution rather than a Gaussian distribution because the dependent variable is binary. Second, the estimated probabilities are constrained to [0,1] by the logistic distribution function because logistic regression predicts the probability that the instance is positive.

Logistic regression can be either binomial or multinomial. Binomial or binary logistic regression deals with situations in which the observed outcome for a dependent variable can have only two possible types (e.g., "dead" vs. "alive" or "winning" vs. "losing"). The Multinomial logistic regression (MLR) deals with situations in which the outcome can have three or more possible types (for example, "class A" vs. "class B" vs. "class C"). In binary logistic regression, the outcome is usually coded as "0" or "1", as this leads to the simplest interpretation [9]. If a particular observed outcome for the dependent variable is the outstanding possible outcome (called "success" or "case"), it is usually coded as "1" and the opposite outcome (called "failure" or "no case") as "0".

Multinomial logistic regression uses a one-to-one classifier to build decision boundaries for different class member types. We introduce an essential function in machine learning called the softmax function. The softmax function is used to compute the probability that an instance belongs to one of K classes when K > 2 [10].

For building a classification model with k classes, the multinomial logistic model is formally defined as follows:

$$\hat{y}(k) = \theta_0^k + \theta_1^k x_1 + \theta_2^k x_2 + \ldots + \theta_n^k x_n \tag{1}$$

The previous model takes in consideration the parameters of the k different classes. The softmax function is formally expressed as follows:

$$p(k) = \sigma\left(\hat{y}(k)\right)_i = \frac{e^{\hat{y}(k)_i}}{\sum_j^k e^{\hat{y}(k)_j}} \tag{2}$$

With:

$i = \{1, \ldots, k\}$ classes.

$\sigma\left(\hat{y}(k)\right)_i$ outputs the probability estimates that an example in the training dataset belongs to one of the K classes.

The cost function for learning the class labels in a multinomial logistic regression model is called the cross-entropy cost function. Gradient descent is applied to find the optimal values of the parameter θ that will minimize the cost function to predict the class with the greatest probability estimate accurately (Fig. 2).

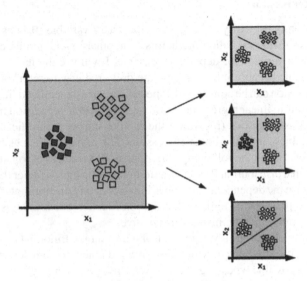

Fig. 2. An example of multinomial regression.

3.2 Support Vector Machine

SVM is a set of related supervised learning method used for classification problems. SVM simultaneously minimizes the empirical classification error and maximize the geometric margin. Designed to solve classification and regression problems that generate a lot of interest for its good performance in a wide range of practical applications [11].

The SVM represents a generalization of linear classifiers. It has been designed to separate two sets of data into linearly separable and non-linearly separable data.

SVM is known as a binary classifier and presents itself as a new method resulting from the formulation of the theory of statistical learning, largely due to Vapnik's 1995 book "the nature of learning statistical theory" [12].

SVM is extendable to classify k classes in a data set, where k > 2. There are two standard approaches to solve this problem. The first is the one-vs.-one (OVO) multi-class classification, and the other is the one-vs.-all (OVA) or one-vs.rest (OVR) multi-class classification technique [10].

One-vs.-One (OVO). In the one-to-one approach, when k, is greater than 2, the algorithm builds "k combination 2", $\left(\frac{k}{2}\right)$ classifiers, where each classifier is for a pair of classes. So, if we have 6 classes in our dataset, a total of 15 classifiers are built or trained for each pair of classes. This is illustrated with four classes in Fig. 3.

After training, the classifiers are rated by comparing the examples in the test set with each of the $\left(\frac{k}{2}\right)$ classifiers. The predicted class is then identified by selecting the largest number of times an example is assigned to a particular class.

The one-vs-one multi-class approach can typically lead to a large number of classifiers being built and thus slower processing time. Conversely, classifiers are more robust to class imbalances when the training of each classifier.

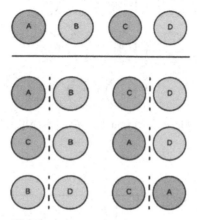

Fig. 3. An example of one-vs.-One

One-vs.-All (OVA). The OVA technique for adapting an SVM to a multi-classification problem where the number of classes k is greater than 2 is to fit each class k to the remaining k − 1 classes. Suppose we have ten classes, each of the classes will be classified against the remaining nine classes. This example is illustrated with four classes in Fig. 4.

3.3 Artificial Neural Network

Artificial neural networks are usually used for classification tasks. By analogy with biology, these units are called formal neurons. Neural networks consist of a set of neurons (nodes) connected by links that propagate signals from neuron to neuron. Neural networks make it possible to discover complex non-linear relationships between many variables without external intervention [13].

The input layer receives information from the features of the dataset, after which some computation takes place, and the information that captures the learned models of the data is propagated through the hidden layer(s) in hopes of improving the learned models. The hidden layer(s) is where the core horse of deep learning occurs. The hidden layer(s) may consist of multiple neural modules, as shown in Fig. 5.

Each hidden network layer learns a more sophisticated set of entity representations. The decision about the number of neurons in a layer (network width) and the number of hidden layers (network depth) that form the network topology is a design choice when training deep learning networks [10].

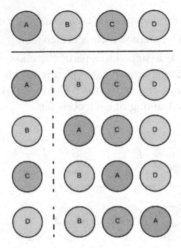

Fig. 4. An example of One-vs.-All

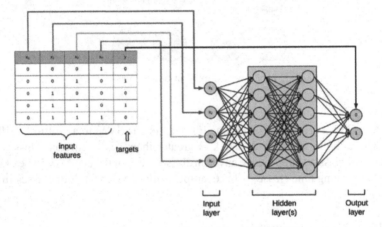

Fig. 5. An example of artificial neural networks

4 Text Classification and Text Representation

4.1 Text Classification

Text classification is a natural language processing method that allows us to classify texts into predefined classes. It is a very popular technique that has a variety of use cases. Text classification consists in assigning a Text to one or more classes to index the Text in a predefined set of categories, originally designed to assist in the documentary classification of books or articles in technical or scientific domains. The objective of this process is to be able to automatically perform the categories of a set of new texts. Text classification consists in learning, from examples characterizing thematic classes, a set of discriminating descriptors to classify a given Text into the class (or classes) corresponding to its content [14].

4.2 Text Representation

The vectorization of texts is one of the most important tasks in document classification. In fact, it is necessary to use an efficient representation technique that allows texts to be represented in a machine-readable form. The most popular representation is the vector model in which each text is represented by a vector of n weighted terms. There are many approaches to do this.

Bag of Words. The idea is to transform texts into vectors, each component of which represents a word. Words have the advantage of having an explicit meaning. However, there are several problems. Firstly, it is necessary to define what "a word" is in order to be able to process it automatically. It can be considered as a sequence of characters belonging to a dictionary, or, more practically, as a sequence of non-delimiting characters framed by delimiting characters (punctuation characters); it is then necessary to manage the acronyms, as well as the compound words; this requires linguistic preprocessing [15].

TF-IDF. TF × IDF coding has been introduced as part of the vector model and uses a function of the occurrence multiplied by a function of the inverse of the number of different documents in which a term appears. This acronym comes from English and Means "'term frequency' × 'inverse document frequency'". The terms characterizing a class appear several times in the documents of this class, and less, or not at all, in the others. That is why coding TF-IDF is defined as follows:

$$TFIDF(T, D) = TF(T, D) \times \log\left(\frac{N}{DF(T)}\right) \tag{3}$$

5 Proposed System

This part is dedicated to the description of the developed system, it aims to set the objectives and needs behind the realization of the classification system, as well as the steps to follow to build a text classifier based on vectorization approaches and machine learning algorithms.

The system can be schematized as follows (Fig. 6):

The classification system developed is a textual classification system capable of automating the classification process from two phases: a training phase and a prediction phase.

For the training phase, we have a collection of texts divided into a set of classes. This collection will be cleaned in the pre-processing step, then it will be transformed into a vector representation in the vectorization step, finally it will be used in the classification step to generate a classifier capable of classifying new unlabeled texts, this step will adopt Machine Learning algorithms.

For the prediction phase, we have a set of unlabeled texts, this set will follow the same steps as the learning phase (Pre-processing, Vectorization and Classification) in order to predict a class of each text in the set.

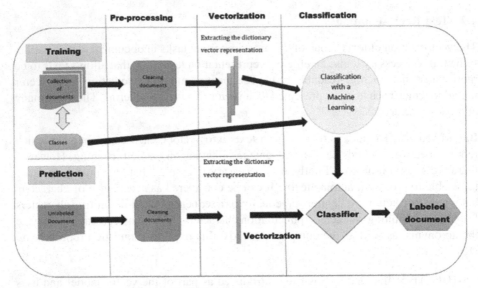

Fig. 6. An illustration of our proposed system.

5.1 Pre-processing

This step involves implementing several operations on classified and unclassified texts to extract essential information and make it available in an exploitable form to the machine. So, we must do a sequence of operations on the text like (Removing Undesirable Characters- Elimination of the stop words- Tokenization and Lemmatization/Stemming). The process of "cleaning" the data may vary depending on the data source.

5.2 Vectorization

The vectorization is a method allowing the conversion of the representation of a document into a vector containing a set of terms. There are several approaches that can be used, namely "word bag" and "TF-IDF". These two approaches are already detailed in Sect. 4.

5.3 Classification

We will now focus on the part of the classification: we use labeled data to predict which class a text belongs to. We will especially talk about binary classification, where it is a question of distinguishing whether or not a text belongs to a class.

For example, say if a text represents a subject or not. If so, we can say that this text is part of the category (sport or health or general culture…).

After the vectorization step using BOW or TFIDF, we receive a matrix characteristic of our dataset.

Before starting the classification, the dataset should be split into two parts: one for training and one for testing.

5.4 Training Techniques

Train – Test Split. The first idea (train/test split) is to split the dataset into two parts: a training set, and a test set. Afterwards, use the training set to train the model. This will allow us to calculate the performance on the test set, and the result will be a good approximation of the performance on unknown data.

The data set is separated into a training set and a test set. The test set is not used to train the model, but only to evaluate it.

If the division we make is not random, for example, a subset of our data included only text of a certain class. This will result in an "overfitting" overlearning is a statistical analysis that corresponds too closely or exactly to a particular set of data. Thus, this analysis may not correspond to additional data or reliably predict future observations. We are trying to avoid that with cross validation.

Cross Validation. Cross validation is a technique that allows you to use the dataset for training and validation, it works as follows:

The dataset is split into k parts or folds. Each of the k folds is used in turn as a test set. The rest (the union of the k-1 other parts) is used for training. At the end, each point (or observation) was used once in a test set (k-1) was used in a training set.

6 Results and Discussion

6.1 Dataset

The training and testing of the developed system are applied to datasets described below:

DATASET. CNN Arabic corpus: collected from cnnarabic.com, includes 5,070 text documents. Each text document belongs to 1 of 6 categories (Business 836, Entertainment 474, Middle East News 1462, Science & Technology 526, Sports 762, World News 1010). The corpus contains 2,241,348 (2.2M) words and 144,460 district keywords after the stop words removal (Table 1).

Table 1. Categories of dataset used.

Categories	Number of Text
Middle East	1462
Business	836
World	1010
Entertainment	474
Sport	762
SciTech	526
All	5070

6.2 Performance Evaluations

Confusion Matrix. In fact, when working with binary prediction models, new factors may come in for consideration. We are not only interested in the number of correct predictions, but we also want to know if a prediction was made positive when the example actually showed a negative result (we talk about false positive) or the opposite (we talk about false negative), this factor called the Confusion Matrix (Fig. 7).

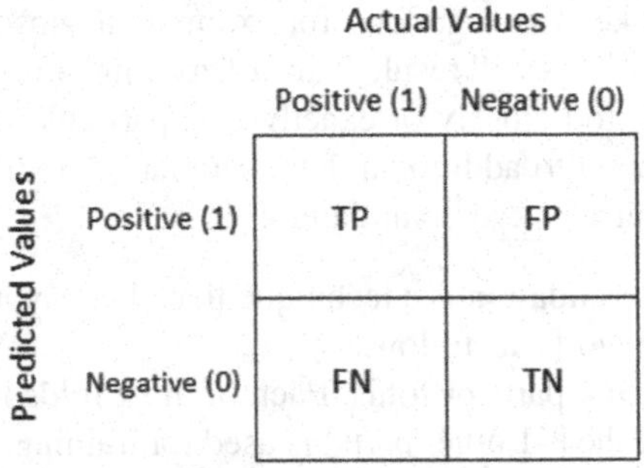

Actual Values

	Positive (1)	Negative (0)
Positive (1)	TP	FP
Negative (0)	FN	TN

Predicted Values

Fig. 7. Confusion Matrix representation.

Accuracy. Is one of the criteria for evaluating classification models. It refers to the proportion of correct text predictions made by the model. Formally, accuracy is defined as follows:

$$Accuracy = \frac{Number\ of\ correct\ predictions}{Total\ number\ of\ predictions} = \frac{TP + TN}{TP + TN + FP + FN} \tag{4}$$

Recall. Or sensitivity is the rate of true positives, i.e., the proportion of positives that have been correctly identified. It is the ability of our model to detect the class of a text.

The recall allows us to answer the following question:

What proportion of real positive text has been correctly identified?

$$Rappel = \frac{TP}{TP + FN} \tag{5}$$

Precision. The proportion of correct predictions among the text that were predicted to be positive. It is the ability of our model to predict a true class to a text.

Precision answers the following question:

What proportion of positive text was correct?

$$Precision = \frac{TP}{TP + FP} \tag{6}$$

F1-score. is the harmonic mean between precision and recall. The range of the F1-score is [0, 1]. It gives you an indication of the accuracy of your classifier (the number of instances it classifies correctly), as well as its stability.

$$F - Score = 2 \times \frac{Precision \times Rappel}{Precision + Rappel} = \frac{2 \times VP}{2VP + FP + FN} \tag{7}$$

6.3 Results and Analysis

To evaluate and test the contribution presented in this paper, an experimentation phase is required. The objective of this phase is to evaluate the performance of our implemented approaches. This will also allow us to identify the constraints and insufficiencies of our approaches.

Our contribution is based on a comparison between the two approaches of vectorization (BOW and TFIDF) and also the algorithms used (LR, SVM, ANN) on our dataset, in the first test we use the (train/test split) technique to split our dataset, and in the second test we use the (cross validation) technique, and we compare the accuracy, precision, recall and F-score to see the proportion of correct document predictions.

Results 1: Train/Test Split. For Train/Test split: we took 80% texts for training and 20% for testing phase (Table 2).

Table 2. Evaluation metrics by Train/Test Split.

	ACCURACY			PRECISION			RECALL			F1-Score		
	LR	SVM	ANN	LR	SVM	ANN	LR	SVM	ANN	LR	SVM	ANN
BOW	93.68	93.29	94.37	92.74	92.48	94.11	92.53	92.02	93.23	92.63	92.21	**93.64**
TF-IDF	93.58	94.47	**94.57**	93.39	**94.17**	93.68	91.78	93.11	**93.62**	92.45	93.57	93.63

For Train/Test Split, we found that TF-IDF gives better results in Accuracy, Precision, Recall, or F-Score compared to Bow. For all three algorithms.

Results 2: Cross Validation. For Cross validation: K = 10, at every iteration we took 90% texts for training and 10% for testing phase (Table 3).

Table 3. Evaluation metrics with Cross Validation.

	ACCURACY			PRECISION			RECALL			F1-Score		
	LR	SVM	ANN	LR	SVM	ANN	LR	SVM	ANN	LR	SVM	ANN
BOW	93.92	93.19	93.88	93.66	92.95	93.77	93.17	92.35	93.08	93.33	92.57	93.35
TF-IDF	93.51	**94.41**	94.14	93.75	**94.24**	94.11	92.13	93.47	**93.48**	92.79	93.79	**93.71**

For cross-validation, we also found that TF-IDF provides greater accuracy, precision, recall and F-score compared to BOW. For all three algorithms.

Analysis

- We observe that TF-IDF gives better results than BOW, for the two techniques (Train/Test Split and cross validation). This means the best method for representing our texts and the best vectorization approach for this dataset is TF-IDF.

- In the results 1 for Train/Test Split technic, we found the best accuracy value of 94.57% carried out by the ANN algorithm, we can conclude that ANN gives the best ratio for correct textual predictions. But in results 2 for Cross-validation technique the accuracy of SVM algorithm equal 94.41%, so the SVM is the better than LR and ANN.
- The precision realized by SVM for both techniques (Train/Test split 94.17 and Cross-validation 94.24) is still the best compared to LR and ANN. In this case we can see that SVM capable of predicting a true class for texts more accurately than LR and ANN.
- The best performance of our model to detect the class of a text is 93.62% for Train/Test split and 93.48 for cross-validation these results done by the ANN algorithm for both techniques.
- We can also see that the accuracy of the Artificial Neural Networks 94.57% is higher than the accuracy of SVM and LR for Train/Test Split, but for cross-validation the support vector machine 94.41% is better than LR and ANN.
- We can summarize that when we use the TF-IDF representation with the Test/Train Split technique and the artificial neural network algorithm, we can find the best classification with the accuracy of 94.57%, in addition, the support vector machine with TF-IDF and Cross-validation gives the accuracy of 94.41%, this also means that the SVM achieves a good classification.

Comparison

In this part we compare our work and results with the results of Mr. Bahassine's work [4]. He used in his paper several vectorization approaches such as (Chi-Square, Information gain "IG", ImpCHI and Mutual information "Mi") and two algorithms (Decision Tree "DT" and Support Vector Machine "SVM"), with different sizes of features (20, 40, 100, 300, 500, 700, 900, 1000 and 1400). In this case we have selected the best value for each Feature Selection and algorithm. And to select the best FS we will focus only on F1-Score value.

We have summarized his studies in the following Table 4:

Table 4. Evaluation metrics for the comparison part.

	PRECISION		RECALL		F1-Score	
	DT	SVM	DT	SVM	DT	SVM
Chi-Square	76.00	88.80	76.80	88.40	75.70	88.50
IG	79.00	89.70	79.50	89.70	77.80	89.50
ImpCHI	78.90	**90.80**	79.40	**90.50**	78.10	**90.50**
MI	77.80	84.50	74.20	84.40	73.00	83.50

following these results, we can note that the SVM gives the best results with ImpCHI for F1-score 90.50%.

To analyze the results correctly. We will choose our best results for ImpCHI and compare them with the results mentioned above (Table 5).

Table 5. Comparison of results

	PRECISION	RECALL	F1-Score
Our results	**94.24**	**93.62**	**93.71**
Compared results	90.80	90.50	90.50

As we can see in the table above, we have acquired best results compared to the ImpCHI and SVM approach. We might say now that the combination of TF-IDF and SVM or ANN gives better results compared to ImpCHI and SVM for all the evaluation metrics used (Accuracy, Precision, Recall and F1-score).

7 Conclusion

This paper focuses on two subjects, the first is the building of a classifier able to classify the Arabic text using machine learning algorithms and text representation. And the second part is a comparative part with another paper that works on the same dataset to evaluate the performance of our classifier. The results obtained confirm that the approaches we have developed gave a better classification compared to the results we evaluated against them. In addition, the accuracy results indicate the best classification performance, 94.57% with the artificial neural network algorithm, and an interesting result of 94.41 with the support vector machine.

In the future works, we will be interested to increase the obtained results by proposing other approaches for both phases: vectorization and classification.

References

1. Duwairi, R.M.: Arabic text categorization. Int. Arab J. Inf. Technol. 4(2), 125–132 (2007)
2. Mesleh, A.: Support vector machines based arabic language text classification system: feature selection comparative study. In: Sobh, T. (ed.) Advances in Computer and Information Sciences and Engineering, pp. 11–16. Springer, Dordrecht (2007). https://doi.org/10.1007/978-1-4020-8741-7_3
3. Hrala, M., Král, P.: Evaluation of the document classification approaches. In: Burduk, R., Jackowski, K., Kurzynski, M., Wozniak, M., Zolnierek, A. (eds.) Proceedings of the 8th International Conference on Computer Recognition Systems CORES 2013. Advances in Intelligent Systems and Computing, vol. 226, pp. 877–885. Springer, Heidelberg (2013). https://doi.org/10.1007/978-3-319-00969-8_86
4. Abu-Errub, A.: Arabic text classification algorithm using TFIDF and chi square measurements. Int. J. Comput. Appl. **93**, 40–45 (2014). https://doi.org/10.5120/16223-5674
5. Bahassine, S., Madani, A., Al-Sarem, M., Kissi, M.: Feature selection using an improved Chi-square for Arabic text classification. J. King Saud Univ. Comput. Inf. Sci. **32**, 225–231 (2020)

6. Boukil, S., Biniz, M., El Adnani, F., Cherrat, L., Moutaouakkil, Abd Elmajid El.: Arabic text classification using deep learning technics. Int. J. Grid Distrib. Comput. **11**(9), 103–114 (2018)
7. Simeone, O.: A Brief Introduction to Machine Learning for Engineers, 168083472X, pp. 6–7 (2017). ISBN 9781680834727
8. Hilbe, J.M.: Practical Guide to Logistic Regression, pp 3–4. Taylor & Francis, Abingdon (2016). ISBN 9781498709576, 1498709575
9. Caropreso, M., Sebastiani, F., Ricerche, C.: Statistical Phrases in Automated Text Categorization (2001)
10. Bisong, E.: Building Machine Learning and Deep Learning Models on Google Cloud Platform: A Comprehensive Guide for Beginners, pp. 247–248. Apress, Berkeley (2019). ISBN 978-1-4842-4469-2, 978-1-4842-4470-8
11. Tambade, S., Somvanshi, M., Chavan, P., Shinde, S.: SVM based diabetic classification and hospital recommendation. Int. J. Comput. Appl. **167**, 40–43 (2017)
12. Rimouche, N., Hadjira, H.: Amélioration du produit scalaire via les mesures de similarités sémantiques dans le cadre de la catégorisation des textes. Université abou Beker Belkaid Tlemcen (2016)
13. Mohammed, B., Brahim, B.: L'apprentissage profond (Deep Learning) pour la classification et la recherche d'images par le contenu, UNIVERSITE KASDI MERBAH OUARGLA Faculté des Nouvelles Technologies de l'Information et de la Communication (2017)
14. Sahin, Ö.: Text Classification (2021). https://doi.org/10.1007/978-1-4842-6421-8_3
15. Jalam, R.: Apprentissage automatique et catégorisation de textes multilingues, pp. 9–10. Université Lumière Lyon 2 (2003)

Analyzing Moroccan Tweets to Extract Sentiments Related to the Coronavirus Pandemic: A New Classification Approach

Youness Madani[(✉)] [iD], Mohammed Erritali, and Belaid Bouikhalene

Sultan Moulay Slimane University, Beni Mellal, Morocco
{y.madani,m.erritali,B.BOUIKHALENE}@usms.ma

Abstract. At the end of 2019, the world has known the covid-19 crisis that negatively affected the health, economic, social, and psychological status of people. Since the beginning of this crisis, users express their ideas, opinions, and sentiments about the coronavirus on all social networks such as Facebook, Twitter, Instagram, etc. For example, until May 8th, 2020, the number of tweets published on Twitter is equal to 628,809,016. In this paper, our proposed method analyzes and classifies covid-19 tweets published in morocco for extracting sentiments. Our approach uses the advantages of new proposed tweets features using a dictionary-based approach and a Python library for developing a new recommendation approach. As Experiments, Our proposed approach outperforms the well-known machine learning classifiers. We find also that based on the epidemiological situation in morocco, the sentiments of Moroccan users changed.

Keywords: Sentiment analysis · Covid-19 · Recommendation system · Collaborative filtering · Classification

1 Introduction

The COVID-19 epidemic has had a huge impact on the general lifestyles of people around the world. People express their views on COVID-19 more frequently on social media when cities are on lockdown. In Twitter, for example, millions of users share their opinions and sentiments about coronavirus every day and in all languages which produces a large amount of data.

In addition to analyzing and controling the physical health of people during this pandemic. The analysis of mental health has an important role to fight the coronavirus pandemic. The rapid number of exponential cases around the world has become the apprehension of panic, fear, and anxiety among people. As of November 2, 2020, 225070 people tested positive(infected with coronavirus) in Morocco.

Although the number of articles related to covid-19 is increased, few of them deal with the analysis of users' sentiments about covid-19 on Twitter. It's for

M. Fakir et al. (Eds.): CBI 2021, LNBIP 416, pp. 33–42, 2021.
https://doi.org/10.1007/978-3-030-76508-8_3

this reason that we decided to propose a new approach to detect and analyze sentiments of Moroccan users about covid-19 since the beginning of this epidemic in Morocco.

For analyzing the sentiment of Moroccan covid-19 tweets, we based on the sentiment analysis domain by proposing a new method. Sentiment Analysis(SA) is a subdomain of Natural language processing, which consists of extracting opinion, attitude, and sentiment from a written text, by classifying it such as: positive, negative, neutral; or by calculating a degree of polarity(a sentimental degree).

The proposed approach in this work is based on a new recommendation approach to classify each tweet into three classes: positive, negative, and neutral. Our method proposes new four tweets' features using a dictionary-based approach(SenticNet dictionary) and a python library, and also it takes advantage of natural language processing.

For the collection of the tweets to classify, we based on keywords such as "covid-19" and "coronavirus", we also using geolocation data to collect only tweets published in Morocco. The collected tweets are divided into periods, to analyze sentiments in relation to the health situation in Morocco.

The rest of this paper is organized as follows: in Sect. 2, we describe the literature review related to our domain. Section 3 presents our approach by describing all the steps to classify tweets. In Sect. 4, we implement the experimental results to show the performance of our model. And finally, it's the conclusion of our article.

2   Literature Review

In this section, we will present a literature review on the usage of sentiment analysis technics to analyze and extract the sentiments of social media users related to the covid-19 epidemic.

Authors in [1] used automated extraction of COVID-19–related discussions from social media and a natural language process (NLP) method based on topic modeling to uncover various issues related to COVID19 from public opinions. They investigate how to use LSTM recurrent neural network for sentiment classification of COVID-19 comments. For that, researchers present a systematic framework based on NLP that is capable of extracting meaningful topics from COVID-19– related comments on Reddit, and for the classification of the comments, they propose a deep learning model based on Long ShortTerm Memory (LSTM) for sentiment classification of COVID-19–related comments. Experiments demonstrated that the research model achieved an accuracy of 81.15% – a higher accuracy than that of several other well-known machine-learning algorithms for COVID-19–Sentiment Classification.

Aslam et al. [2] published a new article extract and classify sentiments and emotions from 141,208 headlines of global English news sources regarding the coronavirus disease (COVID-19). Authors take into account news with keyword coronavirus between the time frame 15 January 2020 to 3 June 2020 from top rated 25 English news sources. Each headline is classified into three classes: positive, negative, or neutral. For the classification of each headline, authors used the R package "sentiment" by relying on lists of words and phrases with positive and negative connotations.

In the article of [3], researchers identify public sentiment associated with the pandemic using Coronavirus specific Tweets and R statistical software, along with its sentiment analysis packages. The authors used a number of machine learning algorithms to extract sentiments from tweets (Linear Regression Model, Naïve Bayes Classifier, Logistic Regression) and also some textual methods. The aim of this article is to demonstrate insights into the progress of fear-sentiment over time as COVID-19 approached peak levels in the United States. Experimental results show a strong classification accuracy of 91% for short Tweets, with the Naïve Bayes method. We also observe that the logistic regression classification method provides a reasonable accuracy of 74% with shorter Tweets, and both methods showed relatively weaker performance for longer Tweets.

The study of [4] focuses on the sentiment analysis of tweets of the Twitter social media using Python programming language with Tweepy and TextBlob library. Authors collect tweets based on two specified hashtags keywords: $\#COVID-19$ and $\#coronavirus$ from the users who shared their location as 'Nepal' between 21st May 2020 and 31st May 2020. The result of the study concluded that while the majority of the people of Nepal are taking a positive and hopeful approach, there are instances of fear, sadness and disgust exhibited too.

In the work [5]; the authors analyze discussions on Twitter related to COVID-19. Researchers used tweets originating exclusively in the United States and written in English during the 1-month period from March 20 to April 19, 2020. For the classification, They applied machine learning methods to classify tweets into three classes(positive, negative, and neutral). For a dataset of 902,138 tweets, the proposed model classified 434,254 (48.2%) tweets as having a positive sentiment, 187,042 (20.7%) as neutral, and 280,842 (31.1%) as negative.

In [6], Muthusami et al. analyze and visualize the influence of coronavirus (COVID-19) in the world by executing such algorithms and methods of machine learning in sentiment analysis on the tweet dataset to understand very positive and very negative opinions of the public around the world. To analyse tweets, They used machine learning approaches, and results show that the LogitBoost algorithm performed better with accuracy of 74.

Authors of [7] use Twitter to analyse sentiments related to covid-19 epidemic. For the collection of the tweets to classify, researchers are based on two specified hashtag keywords, which are ("COVID-19, coronavirus") using the tweepy library. The date of searching data is seven days from 09-04-2020 to 15-04-2020. And by using TextBlob library in python, the sentiment analysis operation has been done.

The work of Chakraborty et al. [8] presents a new research on analysing sentiments in tweets during the period of covid-19 epidemic. The datasets used are obtained by searching using the keywords: $\#corona$, $\#covid19$, $\#coronavirus$,

coronavirus and $\#covid - 19$. For the classification, authors propose a model using deep learning classifiers with admissible accuracy up to 81%, and an implementation of a Gaussian membership function based fuzzy rule base to correctly identify sentiments from tweets. The accuracy for the said model yields up to a permissible rate of 79%.

In [9] authors analyse the sentiments and their evolution of people in the face of this public health crisis based on Chinese Weibo. For constructing the dataset of work, authors obtained the top 50 hot searched hashtags from January 10, 2020 to May 31, 2020, and collected 1,681,265 Weibo posts associated to the hashtags regarding COVID-19. For the classification, They use 7 classes(fear, anger, disgust, sadness, gratitude, surprise, and optimism) to annotate each Weibo post. To detect sentiments of users in Weibo, researchers use three methods, i.e., LSTM, BERT, and ERNIE, and experimental results show that ERNIE classifier has the highest accuracy and reaches 0.8837.

Lamiaa Mostafa in the paper [10], propose a Sentiment Analysis Model that will analyze the sentiments of students in the learning process within their pandemic using Word2vec technique and Machine Learning techniques. The proposed model use a method that starts with the processing process on the student's sentiment and selects the features through word embedding then uses three Machine Learning classifiers which are Naïve Bayes, SVM and Decision Tree. Results show that the Naïve Bayes classifier gives best results with an accuracy equal to 87% by using the DF word embedding method, and 91% using skip-gram method.

3   Research Methodology

As presented earlier the main objective of this work is to analyze the Moroccan tweets during the period of the covid-19 epidemic. The idea is to propose a new approach to improve the results of a sentiment analysis approach that will classify each tweet into three classes(positive, negative, and neutral).

We want to analyze the sentiments of Morrocan users from the beginning of the covid-19 crisis in morocco until the end of August 2020. For that, we collect tweets from March 2020 until October 2020 in the form of periods.

Our work needs old tweets from march to October 2020. For that, we developed a new python program that gives us the possibility to retrieve old tweets(because some libraries like Tweepy allow users to collect tweets for only one week later). Our program is based on GetOldTweets3(A Python 3 library and a corresponding command-line utility for accessing old tweets). The collected tweets are divided into periods, for example, the first period is from March,1st to march,15 2020, and the last period is from October,1st to October 15, 2020. For the collection, we based on keywords such as: "covid-19", "coronavirus". We based on the geolocate data to retrieve only Moroccan tweets.

Our work uses tweets expressed in four languages: Spanish, English, French, and Arabic(the most used languages by Moroccan users). That is, our approach of collection collects all tweets related to covid-19 expressed in one of these 4 languages and published in morocco.

Figure 1 shows the different steps for collecting the tweets to analyze.

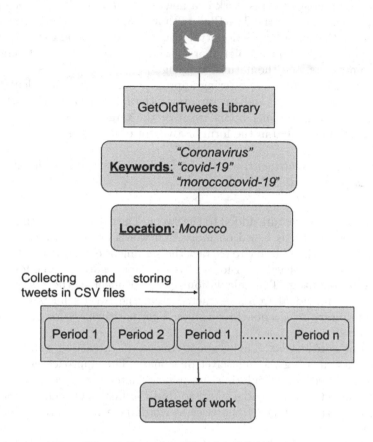

Fig. 1. Steps for collecting the tweets

After the construction of our dataset of work. Another important step in this article consists in choosing a labeled dataset that will help us to analyze the performance of our proposed model and also in the experimental results step. From all that in this article, we worked with the Twitter US Airline Sentiment TUAS dataset which consists in Analyzing how travelers in February 2015 expressed their feelings on Twitter. We use this dataset because its form is similar to what we need in our approach, each tweet is classified into three classes: negative, neutral, and positive. The classification in this dataset is based on tweets' text.

The proposed approach is based on the tweet text, an important step consists in preparing the tweet for the analysis. For that, we used a number of text preprocessing methods for cleaning the tweets [11] such as: removing the words with the "@" character, removing the "#" character, Tokenization, Removing stopwords, Stemming, etc.

3.1 Description of Our Method

The proposed approach in this work is a new collaborative filtering recommendation approach that uses the TextBlob python library and a dictionary based-approach using the SenticNet dictionary[1]. The proposed method uses advantages of recommendation systems, a dictionary-based approach to extract sentiment polarity from tweets, and the natural language processing methods.

Our proposed collaborative filtering approach begins with the calculation of ratings of the target tweet and the tweets of the labeled dataset(in which we will look for the k-nearest neighbors tweet of the target tweet). Each tweet in this work will be presented in the form of a vector of four elements as presented below:

$$Tweet\_vector = [sentiment, sentiment\_hashtag, sentiment\_sentic\_text,$$
$$sentiment\_sentic\_hashtag]$$

where:

- **Sentiment**: a new feature added to the tweet. The calculation of this element of the tweet's vector is based on individual words of the cleaned tweet and the TextBlob python library to extract the sentiment of the tweet's text. This element accepts three values: 1 for positive, -1 for negative, and 0 for neutral.
- **sentiment_hashtag**: The calculation of this element of the tweet's vector is based on individual words extracted from the tweet's hashtags and the TextBlob library, it accepts three values: 1 for positive, -1 for negative, and 0 for neutral.
- **sentiment_sentic_text**: This element consists in extracting sentiment from the tweet's text using the SenticNet dictionary. This approach goes through the tweet's words by calculating the polarity intensity of each word which has a value between -1 and 1. And to find the final sentiment of the tweet we calculate the average of the tweet's words polarity as presented in the following formulas:
- **sentiment_sentic_hashtag**: To calculate this element, we based on hashtags' words and the SenticNet dictionary to extract sentiments in the form of three values 0, 1, or -1.

The Definition of tweets' vectors(by calculating these 4 new features for the tweet to classify and the tweets of the labeled dataset) is the first step of our approach. The vectors' elements are like the product ratings in a normal CF system.

Calculating these four new features is equivalent to giving a rating to each one. In the dataset of training and test(Twitter US Airline Sentiment), for each tweet, we construct a vector with the new features and also with the label feature that have three values: 0 for neutral, -1 for negative, and 1 for positive, which means that every tweet's vector of the dataset of training and test will have the following format:

$$Tweet\_vector = [sentiment, sentiment\_hashtag, sentiment\_sentic\_text,$$
$$sentiment\_sentic\_hashtag, label]$$

---

[1] https://sentic.net/.

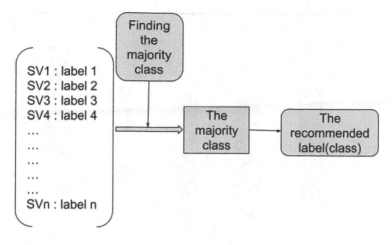

Fig. 2. Recommended label

After we have the necessary ratings(vectors) to start our recommendation approach, the next step in our new sentiment analysis collaborative filtering approach is the search of the k-nearest neighbor's tweets(from the dataset of training and test) of the target tweet(tweet to classify), that is to say, finding the similar vectors(from the vectors of the tweets of the Twitter US Airline Sentiment dataset) of the target tweet's vector.

By using a threshold, we keep as similar vectors to the target tweet's vector the ones that have a similarity value greater than or equal to 0.5. These retrieved similar vectors are the k-nearest tweets' vectors to the target tweet vector.

The final step of our CF approach is to recommend the relevant class(sentiment) to the target tweet. For that, we based on the labels of the similar vectors by searching the majority class (Fig. 2).

As presented in Fig. 2, after we find the k-nearest tweets' vectors we look for the majority class that will give us the relevant class of the target tweet.

4   Experimental Results

This section will present experimental results. As presented in the last section our method uses a recommendation approach based on Textblob and a dictionary-based method by proposing 4 new tweets' features to retrieve sentiments from covid-19 tweets in morocco.

To demonstrate the strengths of our proposed approach, we compare it with 4 machine learning algorithms (SVM support vector machine, NB naive Bayes, RF random forest, DT decision tree) by calculating the classification rate. Fig. 3 shows the results obtained.

According to Fig. 3, the proposed approach using a recommendation method with the four proposed new measures(new tweets' features) gives good results

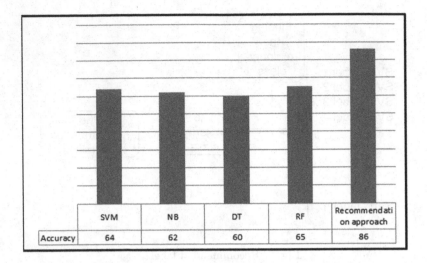

Fig. 3. The comparison result

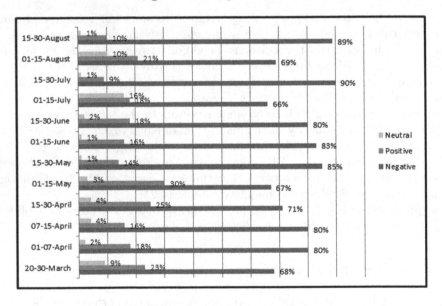

Fig. 4. Classification results

with an accuracy that reaches 86%. Our model outperforms the well-known machine learning algorithms, the best of them gives only 65% accuracy.

To extract the sentiments of people on the covid-19 epidemic, we apply our proposed approach to the tweets of each period, which is applying all the steps described earlier.

Figure 4 shows the results obtained after we classify the tweets of all the periods using our proposed approach.

From Fig. 4, we remark that In all periods the degree of negativity is greater than the value of positivity, which demonstrates that people have a negative feeling with a lot of fear of this new epidemic. Another remark shows that The percentage of negativity is changed from a period to another, and that is due to the change of the figures (number of cases contaminated by the virus in Morocco in a given period ... etc.).

5   Conclusion

In addition to the medical and health crisis due to the spread of the coronavirus, there is also a negative development of the feelings of people all over the world. Analyzing how the covid-19 crisis affects the sentiments of users on Twitter is becoming an important research axis from the beginning of this crisis until today, especially at the start of the second wave. In this article, we have proposed a new method based on a new collaborative filtering approach that uses four new tweets' features using the TextBlob python library and a dictionary-based approach with the SenticNet dictionary. Our approach outperforms the well-known classification methods with an accuracy equal to 86%. By applying our method on Moroccan covid-19 tweets, we find that the majority of published content on Twitter related to covid-19 is negative.

The future work consists of proposing a new approach that deals with the elongation of words in tweets and uses it as a new parameter to improve the quality of tweets' classification.

References

1. Jelodar, H., Wang, Y., Orji, R., Huang, S.: Deep sentiment classification and topic discovery on novel coronavirus or COVID-19 online discussions: NLP using LSTM recurrent neural network approach. IEEE J. Biomed. Health Inform. **24**(10), 2733–2742 (2020). https://doi.org/10.1109/JBHI.2020.3001216
2. Aslam, F., Awan, T.M., Syed, J.H., et al.: Sentiments and emotions evoked by news headlines of coronavirus disease (COVID-19) outbreak. Humanit. Soc. Sci. Commun. **7**, 23 (2020). https://doi.org/10.1057/s41599-020-0523-3
3. Samuel, J., Ali, G.G., Rahman, M., Esawi, E., Samuel, Y.: Covid-19 public sentiment insights and machine learning for tweets classification. Information **11**(6), 314 (2020)
4. Pokharel, B.P.: Twitter Sentiment Analysis During Covid-19 Outbreak in Nepal, 11 June 2020) Available at SSRN https://doi.org/10.2139/ssrn.3624719, https://ssrn.com/abstract=3624719
5. Hung, M., et al.: Social network analysis of COVID-19 sentiments: application of artificial intelligence. J. Med. Internet Res. **22**(8), e22590 (2020)
6. Muthusami, R., Bharathi, A., Saritha, K.: Covid-19 outbreak: tweet based analysis and visualization towards the influence of coronavirus in the world. Gedrag en Organisatie **33**(2) (2020). https://doi.org/10.37896/GOR33.02/062
7. Manguri, K.H., Ramadhan, R.N., Mohammed Amin, P.R.: Twitter sentiment analysis on worldwide COVID-19 outbreaks. Kurdistan J. Appl. Res. **5**(3), 54–65 (2020)

8. Chakraborty, K., Bhatia, S., Bhattacharyya, S., Platos, J., Bag, R., Hassanien, A.E.: Sentiment analysis of COVID-19 tweets by deep learning classifiers-a study to show how popularity is affecting accuracy in social media. Appl. Soft Comput. **97**, 106754 (2020). https://doi.org/10.1016/j.asoc.2020.106754

9. Lyu, X., Chen, Z., Wu, D., Wang, W.: Sentiment analysis on Chinese Weibo regarding COVID-19. In: Zhu, X., Zhang, M., Hong, Yu., He, R. (eds.) NLPCC 2020. LNCS (LNAI), vol. 12430, pp. 710–721. Springer, Cham (2020). https://doi.org/10.1007/978-3-030-60450-9_56

10. Mostafa, L.: Egyptian student sentiment analysis using Word2vec during the coronavirus (Covid-19) pandemic. In: Hassanien, A.E., Slowik, A., Snášel, V., El-Deeb, H., Tolba, F.M. (eds.) AISI 2020. AISC, vol. 1261, pp. 195–203. Springer, Cham (2021). https://doi.org/10.1007/978-3-030-58669-0_18

11. Madani, Y., Erritali, M., Bengourram, J., et al.: A multilingual fuzzy approach for classifying Twitter data using fuzzy logic and semantic similarity. Neural Comput. Appl. **32**, 8655–8673 (2020). https://doi.org/10.1007/s00521-019-04357-9

Towards a Support System for Brainstorming Based Content-Based Information Extraction and Machine Learning

Asmaa Cheddak[1]([⊠]) [iD], Tarek Ait Baha[2] [iD], Mohamed El Hajji[2] [iD],
and Youssef Es-Saady[2] [iD]

[1] TIAD Laboratory, Sultan Moulay Slimane University, Beni-Mellal, Morocco
[2] IRF-SIC Laboratory, Ibn Zohr University, Agadir, Morocco
{t.aitbaha,y.essaady}@uiz.ac.ma, m.elhajji@crmefsm.ac.ma

Abstract. Brainstorming is an effective technique for seeking out ideas on a specific issue that can be expressed shortly and powerfully and then determine the best solution. As a method, It is especially popular in areas that rely on creativity such as industry and advertising. Many solutions are created in the service of digital brainstorming to enable better management, however, literature still reports that these techniques offer only partial solutions in themselves. In this work, we present an architecture of a support system for brainstorming activities based on content-based information extraction and Natural Language Processing. First results show that it is possible to make decisions automatically or to effectively help the user to make the right decisions.

Keywords: Brainstorming · Support system · Knowledge · Natural Language Processing · Content-based information extraction

1 Introduction

Brainstorming is an effective technique for seeking out ideas and finding optimal solutions to a specific problem with the minimum of resources. It has become a very common tool used by large international corporations. However, its classic use has become difficult for a human facilitator faced by a large number of participants, their language differences as well as their time and the geographical areas. Hence the need to develop a decision support system comes from as a preferred solution preferred by most companies [1]. Brainstorming systems ensure remote communication and backup of brainstorming ideas and meetings. However, they don't allow automatic brainstorming [2].

Brainstorming systems are interactive decision support systems (IDSS). These kinds of systems aim to improve decision making. They help in the decision and do not replace the decision-maker. Before starting a Brainstorming session, the decision problem should be clear to all participants. Those participants share

M. Fakir et al. (Eds.): CBI 2021, LNBIP 416, pp. 43–55, 2021.
https://doi.org/10.1007/978-3-030-76508-8_4

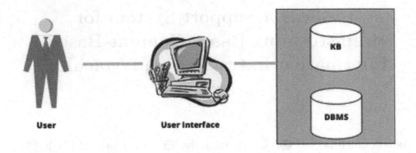

Fig. 1. The architecture of IDSSs according to Sprague et Carlson (1982)

their ideas with the system presently or remotely, at the same or different times. The system conceives the data to be analyzed from the ideas of the participants. Then, it uses methods, knowledge bases, and models to help the decision-maker make the best decision [4]. Decision support systems generally use the architecture consisting of a Human/Machine interface and a database management system consisting of a knowledge base and a model base [2], as shown in Fig. 1.

Artificial Intelligence (AI) was applied to all sectors of activity: transport, health, energy, industry, logistics, finance, and even commerce. Text analysis is also a subject of artificial intelligence. Whether oral or written, artificial intelligence is getting better at understanding and using language to automatically respond to various requests. Nowadays, it is used in different domains. As an example, managing customer relationships on the Internet or by telephone. Conversational Agents or ChatBots are intelligent systems that manage to maintain a conversation in natural language. Furthermore, AI can render a great service to Brainstorming systems by using its algorithms based on recent deep learning methods which require a large amount of data for the model to converge towards the desired results [3].

The proposed system in this paper addresses the recognition of handwritten texts from sticky notes generated during a brainstorming session. Then, it extracts domain-specific entities from notes and builds relationships among entities to extract contextual entities. The content-based information is stored in the knowledge base. Finally, the ideas are indexed for later use. The originality of our system comes from the fact that it ensures an automatic annotation and classification of the ideas generated, during a brainstorming session, to suggest the optimal solutions to a specified problem.

The paper is organised as follows: Sect. 2 outlines the related works in Brainstorming systems, handwriting text recognition, and information extraction, Sect. 3 presents the proposed Information Extraction system, Sect. 4 discusses some first results and finally Sect. 5 concludes the paper.

2   Related Works

The brainstorming technique consists of bringing together a group of collaborators so that they collectively propose as many new ideas as possible on a given topic. Conceived by Alex Faickney Osborn in the late 1950s, the technique of brainstorming has become adapted by many companies, in communication, improvement of production, administration, commerce, management, organization, etc. [1]. The objective is to get the maximum number of ideas and suggestions from a group of people to solve clearly defined problems. This creative technique gives people more freedom and spontaneity to be able to generate news and good ideas to solve a problem [1].

Several methods were adapted in different situations during the generation of ideas and were the result of research and studies on Brainstorming. Despite the obstacles posed by these methods, many studies in social psychology and group psychology have shown that groups generate better ideas with nominal brainstorming than with verbal brainstorming [5]. In this technique, a set of participants work individually to generate ideas in the presence of others but without interacting verbally. Besides, Faced with the blocking of the production of ideas, apprehension of evaluation or parasitism, while stimulating the production of good ideas, the electronic approach comes to offer the member of the Brainstorming group simultaneous work, anonymously and can also be at a distance [6]. Although it offers solutions such as remote work, the comparative studies of the results of electronic brainstorming and nominal brainstorming are identical for small groups [5]. Another type of method was proposed to improve Brainstorming such as Brainwriting and Brainswarming. Brainwriting [7] is a technique where the generation of ideas is simultaneous and written. It comes to remove the blockage of ideas, minimize social laziness and encourage the transformation of shared ideas. In a Brainswarming session, presented by Kevin Maney [8], every participant writes his idea on the sticky notes and adds it on a whiteboard (a physical or virtual one) where the goal of the session is placed at the top of a graph while the available resources are at the bottom of the graph. Brainwarming then consists of looking for relationships between resources and the goal to find a good solution.

Writing is a crucial common communication tool. Currently, within the digital age, physical manuscripts are progressively integrated into the technological environment, in which, through the method of handwriting recognition, machines have the ability to understand the text in scanned images and represent them within the digital form for later use. Opportunely, with the utilization of the Hidden Markov Model (HMM) for Text Recognition [9], Handwriting Text Recognition (HTR) systems have increasingly developed. Nowadays, it becomes possible to perform more, through Artificial Neural Networks (ANN), the recognition process at character, word, line, and paragraph levels of text segmentation. In this way, notable improvements in handwriting text recognition accuracy have been achieved especially through Convolutional and Recurrent Neural Network (CRNN). Besides, in order to improve the results of the optical model, a post-processing stage, using NLP techniques [10], is connected to the text decoding. In the last few years, fields of study in NLP, such as Grammatical Error Correction and Machine Translation, have produced promising results through neural network approaches. Encoder-Decoder models (Sequence to Sequence) [11], has developed within the area of NLP for tasks that need large linguistic knowledge. These models were, then, increased with the use of the Attention mechanism, reaching even better results. Therefore, a recent work [12] proposed a spell corrector module as a post-processing step of an HTR system to reach competitive results to the traditional methods which consist of decoding and remove the linguistic relationship between the decoding stage and the optical model.

On the other hand, Information Extraction (IE) techniques have been applied to the semantic annotation to different languages and in a variety of domains (e.g., news filtering in Turkish or mining of biomedical text) [13][14]. Currently, multiple works with diverse offerings in the field of the semantic web exist. Various general-purpose tools were developed to maintain the annotation, and many ontologies were introduced for different domains. Tables 1,2,3 show some of these tools.

Table 1. General-purpose tools

Tool/Approach	Main focus	Drawbacks
OpenCalais [15]	A service for annotation of named entities in documents	The DBpedia semantic information is not used in the disambiguation of instance annotating a given term Ignores the link, which includes important information about a term
DBpedia Spotlight [16]	A Semantic annotation tool, based on DBpedia, for data entities in a document	
Gate [17]	A tool for text organization used to assist the text annotation Offers basic text-processing functions (NER, Lemmatisation, POS, etc.)	
Ontea [18]	A tool for semantic data extraction from documents that uses regular expressions guides as a text analysis tool and identifies semantically equivalent elements through a predefined domain ontology	

Table 2. Specific domain tools

Tool/Approach	Main focus	Domain
MetaMap [19] Whatizi [20]	Are based on a strategy to search terms in the thesaurus Determines the concept sequences occurrences in text segments through the term strict coincidence	Biomedical

Table 3. Semantic annotation approaches based on information retrieval techniques

Tool/Approach	Main focus
KIM [21]	A platform for knowledge management, annotation, and indexed and semantic search The advantage of this tool is the identification of named entities based on ontologies
Castells et al. [22]	Provide an information search model through ontologies for the classification of annotations
Berlanga et al. [23]	Introduce a semantic annotation scheme for a corpus by the use of numerous knowledge bases
Fuentes-Lorenzo et al. [24]	A tool to enhance online search engine results, by performing a much better classification of query results

Information Extraction is a technique for the identification of entities from unstructured text. Extraction techniques mainly focus on rules, machine learning, and ontology based approaches to cater to unstructured text. *Rule-based* techniques identify hidden features within the text by using predefined rules. These techniques are applied in various domains, like RE through background knowledge, extraction of clinical data from medical texts, extraction of product reviews by using practical information and sentence dependency trees, the extraction of composed entities from biomedical fields using BioInferand GENIA corpora. And various others. These methods may process unstructured text repeatedly to generate important information.

Machine learning-based techniques support the extraction of existing and supplementary information from unorganized texts through Hidden Markov Models (HMM), and Conditional Random Fields (CRF). They need large datasets for training and evaluation goals. The above-mentioned techniques fail to associate information alongside context, which produce a loss of information.

Ontology-based techniques treat this weakness in information extraction. In fact, It employs domain-specific knowledge in the extraction of pertinent information from texts. In order to promote the entity extraction process, KIM [21], and TextPresso only use the information existing in the domain ontology. TextPresso is specified to the biomedical domain and uses Gene Ontology

(GO), which comprises about 80% of the vocabulary, during extraction. All new deduced information will be lost. This weakness has been fixed by mixing the new extracted information with existing domain knowledge, which results in a better one.

IE is often an early stage in the pipeline for various high-level NLP tasks such as QuestionAnswering Systems, Machine Translation, and so on. Various sub-tasks have been involved in IE, such as Named Entity Recognition (NER), Named Entity Linking (NEL), Coreference Resolution (CR), Event Extraction, Relation Extraction (RE), and Knowledge Base Construction.

Various low-level tasks in NLP such as Part-Of-Speech (POS) tagging, chunking, parsing, and NER are fundamental building blocks of complex NLP tasks such as Knowledge Base construction, text summarization, Question Answering systems, and so on. Hence, the effectiveness of these low-level tasks highly determines the performance of high-end tasks.

3 Proposed System

Many brainstorming systems offer solutions for remote working, freedom, and spontaneity when generating ideas in brainstorming sessions. However, none allows to automatically classify the generated ideas and especially does not allow to automatically propose the best solution according to the classification of the generated ideas. In this paper, we propose a system designed to automatically brainstorm as much as possible. In fact, this system first receives participants' ideas written on post-its, recognize the handwritten texts, extract the relevant information, then classify ideas according to their content.

In this section, we describe an overview of the proposed system. First, the optical and the Encoder-Decoder models are detailed according to the offline Handwritten Text Recognition and Spelling Corrector systems, respectively. Then, the Information Extraction module is detailed.

The following Fig. 2 represents the architecture of our solution.

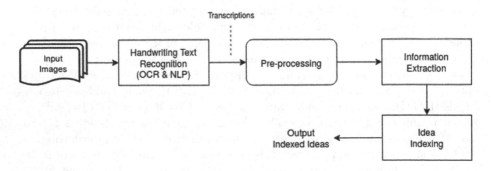

Fig. 2. System architecture

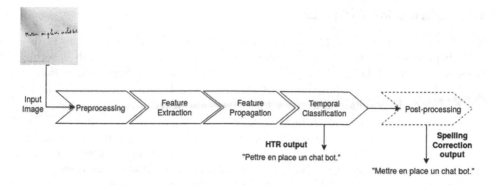

Fig. 3. HTR system workflow

3.1   Handwriting Text Recognition

For the Handwriting Text Recognition phase, two independent systems are set up: First, Handwriting Text Recognition (HTR), which gets a scanned image as input and renders the transcription, then, the Spelling Corrector receives the transcription as input and renders text with the necessary corrections. Figure 3 shows the HTR system workflow. The optical model is responsible for observing the scanned image, understanding the context, then decoding it in texts. First of all, the preprocessed scanned image serves the CNN layers to extract features. This process is done by the use of the CNN layers and the gated mechanism to extract more significant information. Secondly, the sequence of features extracted are used within RNN layers to map and create a matrix including character probabilities. To competently manage long text sequences, we used Bidirectional Gated Recurrent Unit (BGRU) layers. Finally, the probability matrix can process by calculating the loss function using the Connectionist Temporal Classification (CTC) loss algorithm to adjust the weights for learning the model, and text decoding, or by the use of the CTC decoding algorithm.

The Spelling Corrector which is responsible for correcting the output transcriptions of the optical model is represented as a post-processing step in the HTR system. It is composed of Encoder-Decoder architecture, particularly the Seq2seq approach, which uses GRU layers. The Encoder is a Bidirectional Gated Recurrent Unit (BGRU) layer, which encodes an input of variable-length token sequences into a fixed-length representation, and the Decoder is Gated Recurrent Unit (GRU) layer, which decodes this fixed-length representation into another variable-length symbol sequences. Then, the Attention mechanism is implemented in the output of the Encoder and Decoder, which lets the model create a context vector that represents in a better way the input sentence at each step of decoding.

3.2   Information Extraction

Based on the system SAJ [25] proposed by Awan et al., which presents an E-Recruitment system that extracts context-aware information from job descriptions using Linked Open Data, job description domain ontology, and domain-specific dictionaries. Our proposed IE system exploits Brainstorm domain ontology and domain-specific dictionaries for extraction of entities. IE module combines various operations to achieve extraction from participant notes in the Brainstorming session. As shown in Fig. 4, Transcription recognized from the note image is input. Then, the text is segmented into predefined classes exploiting a self-generated dictionary. NLP techniques and the dictionary are used in the identification of entities. These are rerouted to the context builder. The output of this process is, then, saved in the Knowledge Base. The following subsections present the Information Extraction module.

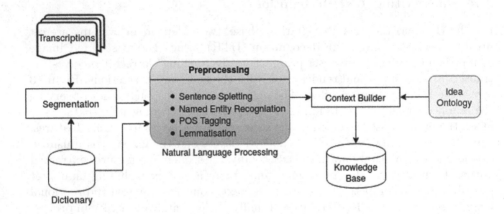

Fig. 4. Information Extraction module

Dictionary. The dictionary is used to support text segmentation and the entities extraction. It is a mixture of rules for identifying fragments and entities in participant notes (Idea).

Segmentation. The goal of segmentation, which is used to classify text in a participant note, is to reassure the correctness of the extracted entities and is within the correct text segment. A starting and ending index location are used in marking text segments in the notes. The text is placed in predefined classes, such as title, abstract, and description, with others. Now, a dictionary-based strategy is used for identifying text categorization, using a list of possible rules and heading values that may appear in a participant's notes.

Entity Extraction. After the segmentation of the participant notes, they are delivered to entity extraction. Through Natural Language Processing techniques, the first step is extracting all sentences from the text and then tagging each token with its corresponding Part of Speech (POS) tag (e.g. noun, verb, or adverb). Entity extraction is a method of extracting relevant information from unstructured text, like names, places, organizations, or dates. Besides extracting the basic entities, the domain dictionary, which includes patterns/action rules for the extraction of entities, is used during the extraction of domain-specific entities. These patterns/action rules are formed by Java Annotation Pattern Engine (JAPE) grammar. JAPE grammar uses features for building pattern/action rules. This feature collection contains aspects such as Part-Of-Speech tags, a dictionary of words, and simple action rules. Priorities for planning the execution order are also included in the JAPE rules. These rules consist of two parts. The left hand side (LHS), which identifies the patterns to be matched based on information generated previously, and the right hand side (RHS), which identifies the annotation set to be created for the text segment that matches the pattern on the LHS.

Context Builder. The extracted entities are transmitted to the context builder, which creates not only hierarchical but also associative relationships, such as skos: related and others, between extracted entities using the Idea ontology as shown in Fig. 5. An OWL-based Idea Ontology was designed by Christoph Riedl [26]. Its primary goal is to facilitate interoperability between the various tools necessary to support the full lifecycle of an idea in an open innovation environment. Indeed, the Idea Ontology schema assists structuring and forming the context of the extracted entities. The schema classes are CoreIdea, Title, Abstract, Description, as shown in Fig. 6. All three represent a textual description of the idea but vary in length and detail. Thus, this ontology can support very simple tools such as, in our case, an E-brainstorming, where an idea usually consists of no more than one sentence, up to more advanced tools that allow longer descriptions.

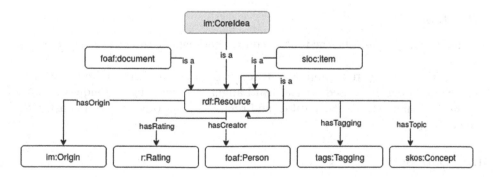

Fig. 5. Overview of the Idea Ontology

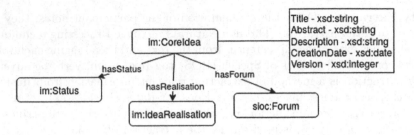

Fig. 6. Overview of the Core Idea elements

Knowledge Base. The goal of the knowledge base is to store the data delivered from the context builder after integration. Knowledge bases create an organized collection of data that is closer to how the human brain organizes information and add a semantic model to the data, which includes a formal classification with classes, subclasses, relationships, and instances (ontologies and dictionaries), on one hand, and rules for interpreting the data, on the other. Currently, GraphDBis being used as a knowledge base to store information. Graph databases implement W3C standards for describing the data and its semantics such as RDF (for representing graph data), SPARQL (the query language for distributed knowledge bases), SKOS (the knowledge organization system), and so on.

3.3 Idea Indexing

After storing the data in the knowledge base, the following step is to index them for the purpose of searching and helping to improve decision making. By going beyond just keywords, concepts, and tags, the knowledge base improves content indexing and advanced search. It provides the knowledge behind individual concepts and allows search engines and other content retrieval applications to interpret a text and match it to advanced queries.

4   First Results and Discussions

4.1 Tools

In this section, we describe the tools used in experiments to execute the proposed system.

GATE Tool: is a text-mining tool referred to as General Architecture for Text Engineering. It is used for Natural Language Processing techniques in our system to achieve extracting entities and relations from the texts and construct the ontology.

4.2 Data Sets

The experiment is performed on the Recognition and Indexing of Handwritten Documents and Faxes (Reconnaissance et Indexation de données Manuscrites et de fac similÉS, RIMES) dataset widely used for unrestricted french handwriting recognition. Concerning the HTR system, the partitioning of the dataset follows its partitioning methodology, determined in three subsets (training, validation, and testing). While the formation of new data, for the spelling corrector, is considered just for the training and validation subsets, while the test subset is employed to evaluate the final model. This database contains a collection of over 12,000 text lines written by diverse writers from 5600 handwritten mails in French. The difficulty in this set is to deal with the recognition of several accented characters since the images have a white background and more fair writing.

The partitioning for the HTR system is as follows: 10,192 lines of text for training, 1134 for validation, and 778 for testing. For the Spelling Corrector, 173,830 is used for training and 19,314 for validation.

4.3 Results

In this section, we present the first results obtained in the processed dataset, revealing the error rates of the optical model and Spelling Corrector.

The most popular evaluation metrics in HTR systems are used: First, Character Error Rate (CER), which is determined as the minimum number of editing operations, at the character level, a word must match its respective ground truth. Second and third, Word Error Rate (WER) and Sequence Error Rate (SER) are determined in the same way but at the word and sentence level, respectively.

The RIMES dataset has a particular scenario characterized by accentuation marks. Thus, the optical model reached about 96% of the correct characters, 84% of the correct words, and 50% of the correct sentences. Then, the Spelling Corrector reached about 99%, 98%, and 94% of the correct characters, words, and sentences, respectively, as shown in Table 4.

Table 4. Error rates results (CER) Character Error Rate; (WER) Word Error Rate ; (SER) Sequence Error Rate.

	CER	WER	SER
Optical Model	4,10%	16,01%	51,92%
Spelling Corrector	0,30%	1,093%	6,25%

5 Conclusion

In this paper, we propose an architecture of a support system for brainstorming sessions based on content-based information extraction supported by NLP techniques and exploits Idea ontology, and domain specific-dictionaries. In order to decrease the information loss in the extraction process, the system builds context among extracted entities. Our system recognizes handwriting texts from the ideas, written in post-its, generated during a brainstorming session, then segments the text into predefined classes through a self-generated dictionary. NLP techniques and a dictionary assist the identification of entities. Idea context is built using an idea domain ontology. Then, content-based information is stored in the knowledge base. Finally, the ideas are indexed for later use.

References

1. Le Brainstorming, une synthèse, 10 September 2016. https://medium.com/@jchnrd/le-brainstorming-une-synth
2. Zarate, P.: Des Systèmes Interactifs d'Aide á la Décision aux Systèmes d'Aide á la Décision: Contributions conceptuelles et fonctionnelles. Interface homme-machine [cs.HC]. Institut National Polytechnique de Toulouse - INPT, fftel-00274718f (2005)
3. CLASSIFICATION SUPERVISÉE DE TEXTES COURTS ET BRUITÉS: APPLICATION AU DOMAINE DES MÉDIAS SOCIAUX MÉMOIRE réalisé par BILLAL BELAININE AVRIL 2017 á l'UNIVERSITÉ DU QUÉBEC À MONTRÉAL Service des bibliothèques
4. Cours: Introduction aux Systèmes Interactifs d'Aide á la Décision (SIAD) 2009 par Bernard ESPINASSE. Professeur á l'Université d'Aix-Marseille
5. Barki, H., Pinsonneault, A.: Small group brainstorming and idea quality is electronic brainstorming the most effective approach? Small Group Res. **32**(2), 158–205 (2001)
6. Mukahi, T., Chimoto, J., Ui, T.: A study on the influence of personality and anonymity on electronic brainstorming. Proceedings of the 3rd Asia Pacific Computer Human Interaction, pp. 363–366 (1998)
7. Paulus, P.B., Yang, H.C.: Idea generation in groups: a basis for creativity in organizations. Organ. Behav. Hum. Decis. Process. **82**(1), 76–87 (2000)
8. Maney, K.: The new art of Brainswarming. [document PDF], IdeaPaint. Retrouvé le 23 avril 2014 (2013). http://www.ideapaint.com/landing-pages/brainswarming
9. Sánchez, J.A., Romero, V., Toselli, A.H., Villegas, M., Vidal, E.: A set of benchmarks for handwritten text recognition on historical documents. Pattern Recognit. **94**, 122–134 (2019)
10. Pirinen, T.A., Lindén, K.: State-of-the-art in weighted finite-state spell-checking. In: Gelbukh, A. (ed.) CICLing 2014. LNCS, vol. 8404, pp. 519–532. Springer, Heidelberg (2014). https://doi.org/10.1007/978-3-642-54903-8_43
11. Sutskever, I., Vinyals, O., Le, Q.: Sequence to sequence learning with neural networks. In: Advances in Neural Information Processing Systems, p. 4 (2014)
12. Neto, A.F.D.S., Bezerra, B.L.D., Toselli, A.H.: Towards the natural language processing as spelling correction for offline handwritten text recognition systems. Appl. Sci. **10**, 7711 (2020)
13. Küçük, D., Yazıcı, A.: Exploiting information extraction techniques for automatic semantic video indexing with an application to Turkish news videos. Knowl.-Based Syst. **24**(6), 844–857 (2011)
14. Friedman, C., Kra, P., Yu, H., Krauthammer, M., Rzhetsky, A.: GENIES: a natural-language processing system for the extraction of molecular pathways from journal articles. Bioinformatics **17**(Suppl 1), 74–82 (2001)
15. OpenCalais (2014). http://www.opencalais.com/

16. Mendes, P.N., Jakob, M., García-Silva, A., Bizer, C.: DBpedia spotlight: shedding light on the web of documents. In: Proceedings of the 7th International Conference on Semantic Systems (I-Semantics 2011), pp. 1–8. Association for Computing Machinery, New York (2011)
17. Cunningham, H., Maynard, D., Bontcheva, K., Tablan, V., Ursu, C., Dimitrov, M.: Developing Language Processing Components with GATE (a User Guide) (2003)
18. Laclavík, M., Seleng, M., Ciglan, M., Hluch, L.: Ontea: platform for pattern based automated semantic annotation. Comput. Inf. **28**(4), 555–579 (2009)
19. Caracciolo, C., Stellato, A., Morshed, A., et al.: The AGROVOC linked dataset. J. Web Semant. **4**(3), 341–348 (2013)
20. Rebholz-Schuhmann, D., Arregui, M., Gaudan, S., Kirsch, H., Jimeno, A.: Text processing through web services: calling Whatizit. Bioinformatics **24**(2), 296–298 (2008)
21. Popov, B., Kiryakov, A., Kirilov, A., Manov, D., Ognyanoff, D., Goranov, M.: KIM – semantic annotation platform. In: Fensel, D., Sycara, K., Mylopoulos, J. (eds.) ISWC 2003. LNCS, vol. 2870, pp. 834–849. Springer, Heidelberg (2003). https://doi.org/10.1007/978-3-540-39718-2_53
22. Castells, P., Fernandez, M., Vallet, D.: An adaptation of the vector-space model for ontology-based information retrieval. IEEE Trans. Knowl. Data Eng. **19**(2), 261–272 (2007)
23. Berlanga, R., Nebot, V., Pérez, M.: Tailored semantic annotation for semantic search. J. Web Semant. **30**, 69–81 (2015)
24. Fuentes-Lorenzo, D., Fernandez, N., Fisteus, J.A., Sanchez, L.: Improving large-scale search engines with semantic annotations. Expert Syst. Appl. **40**(6), 2287–2296 (2013)
25. Ahmed Awan, M.N., Khan, S., Latif, K., Khattak, A.M.: A new approach to information extraction in user-centric E-recruitment systems. Appl. Sci. **9**, 2852 (2019)
26. Riedl, C., May, N., Finzen, J., Stathel, S., Kaufman, V., Krcmar, H.: An idea ontology for innovation management Int. J. Semant. Web Inf. Syst. **5**(4), 1–18 (2009). Available at SSRN. https://ssrn.com/abstract=1648841

Classification of Documents Using Machine Learning and Genetic Algorithms

Chaima Ahle Touate$^{(\boxtimes)}$ (iD) and Hicham Zougagh

Information Processing and Decision Support Laboratory, Faculty of Sciences and Technics,
Sultan Moulay Slimane University, Beni Mellal, Morocco

Abstract. In the past few years, there has been rampant growth in the amount of complex documents that stand in need of a deeper understanding of machine learning methods to classify them in many applications. The success of these methods depends on their ability to understand complex patterns and nonlinear relationships in data. Yet, finding the right structures, architectures, and techniques for text classification is often a challenge for researchers. In this article, we present an automated document classification system based on two axes; the first regard the processing of natural language (NLP), along with the second that focuses on Machine Learning (ML) algorithms. In addition, a hybrid system that combines the best of classification models in a single strong system with a very high percentage of accuracy that we came to give rise to with the genetic algorithms (GA).

Keywords: Document classification · NLP · ML · Hybrid system · GA

1 Introduction

Textual information is becoming more and more important in the daily activity of researchers and companies, as well as the needs for intelligent access to huge textual databases and their manipulations which have greatly increased, on the one hand, On the other hand, the limits of a manual approach which is costly in terms of working time, not very generic, and relatively inefficient, have motivated research in this field. Therefore, the search for operational solutions and the implementation of effective tools to automate the classification of these documents becomes an absolute necessity. Many research works focus on this aspect thus giving a new impetus to research in the field which knows a real evolution since the last two decades thanks to the introduction of techniques inherited from machine learning which have significantly improved the rates of good classification. It is at this level that our problem of documents classification is positioned. Within this, our work aims to give two contributions. The first concerns the implementation of a set of vectorization and classification approaches to develop a well-structured document classification system. This contribution ends with a comparative study between the different approaches adopted to produce a successful system. To improve the classification accuracy rate, the second contribution uses the hybridization technique of the best classifiers developed, that make use of the genetic algorithms for the choice of parameterization that offers the top of classification accuracy rate. Our

© Springer Nature Switzerland AG 2021
M. Fakir et al. (Eds.): CBI 2021, LNBIP 416, pp. 56–72, 2021.
https://doi.org/10.1007/978-3-030-76508-8_5

paper will be organized as follows; Sect. 2, describes the related works; Sect. 3, details the architecture of the developed classification system. Section 4, set out the proposed architecture based on hybridization, followed finally by a conclusion and future scope in Sect. 5.

2 Related Works

In this section we first present the related works to our problem. Aas and Eikvil (1999) [1] have described the supervised text classification in ML steps, namely, pre-processing, vector space model creation, dimensionality reduction (feature selection and projection), training of a classification function, and performance measurement. In their work, text classification was used to present several schemas of feature weighting, e.g. Boolean, term frequency, inverse document frequency, and entropy. In addition, they summarized and elucidated six machine learning methods: Rocchio's algorithm, naive Bayes, k-nearest neighbor, decision tree, support vector machine (SVM), and ensemble learning, including bagging and boosting algorithms. Furthermore, they described the performance measures for binary, multi-class, and multilabel classification tasks, there work has introduced us to these six machine learning methods that have proven their efficiency by other papers work results like in Aggarwal & Zhai (2012) [2] that focused on specific changes which are applicable for the text classification. They used, as text classification algorithms, Decision Trees, Pattern (Rule)- based Classifiers, SVM Classifiers, Neural Network Classifiers, Bayesian (Generative) Classifiers, nearest neighbor classifiers. They discussed the methods for features selection in text classification and described these methods for text classification. In addition, Colas & Brazdil (2006) [3] sought about old classification algorithms in text categorization. They, also, found systematically the weaknesses and strength of SVM, Naïve Bayes and KNN algorithms in text categorization and examined how the number of attributes of the feature space effected on the performance. In a different setting, multi-class classification is tried by combining kernel density estimation with k-NN (Tang & Xu, 2016) [4]. It improves the weighting principle of k-NN, thereby increasing the accuracy of classification. It has also been proven efficient for complex classification problems.

Majority voting (MV) algorithm presented by [5] is based on the thought that k-experts opinion could be better than one, which means if their individual judgments are appropriately combined a trusted classification is provided as an example if two or three classifiers are agree on a category for a test document, the results of voting classifier is that class respecting the weights of vote for every class. In our work we got inspired by the thought of the voting method so we decided to improve it by providing a better architecture by using the famous genetic algorithms to produce hybrid optimal system with better précised voting weights choice.

3 General Architecture of a Classification System

Text classification (C.T) is the process of assigning one or more categories from a predefined list to a document. The goal of the process is to be able to automatically categorize a set of new texts. Document classification consists in learning, from examples characterizing thematic classes, a set of discriminating descriptors to allow a given document to be placed in the class (or classes) corresponding to its content [6, 7]. The automatic text classification process has two phases that can be distinguished as follows:

The first phase, therefore, consists of formalizing the texts so that they are understandable by the machine and usable by learning algorithms.

The categorization of documents is the second phase, this step is, of course, decisive because it is this which will or will not allow the learning techniques to produce a good generalization from the pairs (Document, Class). To improve the performance of the models, and evaluation of the quality of the classifiers and the comparison of the results provided by the different models is carried out at the end of the cycle.

Figure 1 illustrates the text categorization process with its steps that can be schematized as follows:

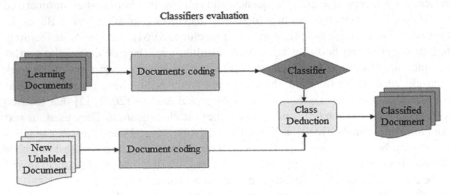

Fig. 1. Text classification process

3.1 Developed System Architecture

The classification system we are presenting is an automated document text classification system that operates in two main phases: Learning phase, Prediction phase.

In the learning phase, we start by Data acquisition, where labelled data is collected, prepared and transmitted for preprocessing. Data preprocessing is made to facilitate the work with new simple data. Next comes the vectorization which is the point of transformation of our work because the machine does not understand the text, only understands numbers, therefore, it builds a vector representation of the corpus. Later, we arrive at the main stage which is the classification, the corpus will be used by the classification algorithms to produce a well-trained classifier model which will play the main role in the phase of prediction of new unclassified documents. After the evaluation of the efficiency of the models, we select the best one to be used as a classifier of new data. The prediction phase is shorter we enter the new unlabeled document we preprocess and vectorize it; finally, we pass it to the model directly to predict its class.

The figure below illustrates the architecture of the developed classification system (Fig. 2):

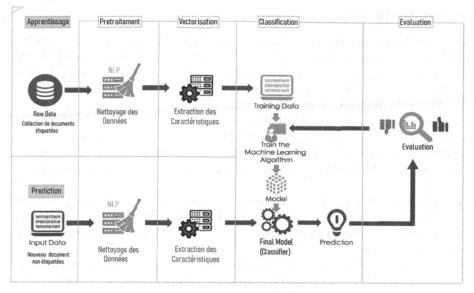

Fig. 2. Architecture of the developed system

3.2 Preprocessing Phase

Text preprocessing is a critical phase in the classification process, since an imprecise knowledge of the population can cause the operation to fail. This phase redacts as much as possible unnecessary information from the documents so that the knowledge retained is relevant. Pretreatment is typically performed in six sequential steps:

- Segmentation.
- Removing frequent words.
- Removing rare words.
- Morphological treatment.
- Syntactic processing.
- Semantic processing.

3.3 Vectorization Phase

Since documents contain mostly text and it's difficult to use the text itself to categorize a particular document, we go for feature vectors. A document classification system can be easily formed using vectors. A simple document is converted to an entity vector to indicate a document with a vector of numbers. Feature vectors can be created using several methods like Bag of Words (Bow), and TF-IDF. Bag of Words creates a set of vectors containing the count of word occurrences in the document [8]. While TF-IDF is a numerical statistic that is intended to reflect how important a word is to a document in a collection or corpus [9], to calculate this measure we did as [10].

3.4 Classification Phase

As for the classification part, after the pretreatment, the vectorization steps by BOW or TFIDF. We receive a characteristic matrix of our data. The next sub-step of the classification is to divide our corpus into a part for training and a part for testing, for this procedure we used Test Split method, where we fixed in our work the Train size on 80%, and 20% for the Test. On top of that, the Stratified K folds Cross Validation method, which splits the dataset into k parts (folds in English) roughly equal. In turn, each of the k parts is used as a test game. The rest (that is, the union of the other k-1 parts) is used for training, the choice of k value was chosen after several test that lead us to keep the default value K = 10. In our experiment, we used model-based classifiers, those are:

- RF (Random Forest): Random Forest, as the name suggests, consists of a large number of individual decision trees that function as a whole. Each individual tree in the random forest spits out a class prediction, and the class with the most votes becomes our model's prediction [11].
- MNB (Multinomial Naïve Bayes): Naive Bayes is based on Bayes' theorem and an attribute independence assumption [12]. used to calculate conditional probabilities. In the case of CT, the Naive bayes method is used as follows: we look for the classification that maximizes the probability of observing the words in the document.

$$P(c|d) = \frac{P(d|c)P(c)}{p(d)} \tag{1}$$

$$C_{NB} = argmax \ P(c_j) \prod_{i \in positions} P(x_j|C_j) \tag{2}$$

With d: the document c = {c1, c2..., cj}: fixed set of classes.

- SVM (Support Vector Machine): is a supervised machine learning algorithm the objective of the support vector machine algorithm is to find a hyperplane in an N dimensional space (N — the number of features) that distinctly classifies the data points. To separate the two classes of data points, there are many possible hyperplanes that could be chosen. Our objective is to find a plane that has the maximum margin, i.e. the maximum distance between data points of both classes. Maximizing the margin distance provides some reinforcement so that future data points can be classified with more confidence. [13].

We also used instance-based classifier as:

- KNN (K-Nearest Neighbors): is used to test the degree of similarity between documents and k training data and to store a certain amount of classification data, thereby determining the category of test documents. This method is an instant-based learning algorithm that categorized objects based on closest feature space in the training set [14].

4 Proposed Architect of the Document Classification System

4.1 Proposed Architect Based on Hybridization

The architecture of the new hybrid system is less complicated as it uses the same process as the last system developed with some changes on the classification phase. As we knew after vectorization, one classifier assumes to predict the class of the new document. But in our hybrid approach, we assemble 3 classifiers, so here's how it works, each classifier has a percent vote, the system prediction is based on a majority vote as long as each classifier has a weight to take into consideration. This weight which is presented by the parameters α, β, δ is unknown and can't be resolved manually and uneasy and time consuming to generate randomly by humans therefore we decided to use a well-known algorithm: the genetic algorithms. The fig below shows the general structure of the elaborated hybrid system (Fig. 3):

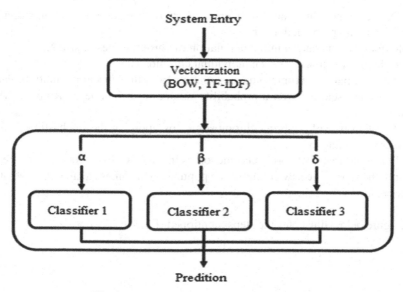

Fig. 3. Proposed Hybrid System Architecture

4.2 Genetic Algorithms

GAs are a heuristic solution-search or optimization technique, originally motivated by the Darwinian principle of evolution through (genetic) selection. A GA uses a highly abstract version of evolutionary processes to evolve solutions to given problems. Each GA operates on a population of artificial chromosomes. These are strings in a finite alphabet (usually binary). Each chromosome represents a solution to a problem and has a fitness, a real number which is a measure of how good a solution it is to the particular problem [19].

Starting with a randomly generated population of chromosomes, a GA carries out a process of fitness-based selection and recombination to produce a successor population, the next generation. During recombination, parent chromosomes are selected and their genetic material is recombined to produce child chromosomes. These then pass into the successor population. As this process is iterated, a sequence of successive generations evolves and the average fitness of the chromosomes tends to increase until some stopping criterion is reached. In this way, a GA "evolves" a best solution to a given problem.

In the genetic algorithms the process is as follows:

Step 1-

- Represent the problem variable domain as a chromosome of a fixed length.
- Initialize the size of a chromosome population N, the crossover probability Pc and the mutation probability Pm.

Step 2- Define a fitness function to measure the performance, or fitness, of an individual chromosome in the problem domain.

Step 3- Randomly generate an initial population of chromosomes of size N.

Step 4- Calculate the fitness of each individual chromosome.

Step 5- Select a pair of chromosomes for mating from the current population. Parent chromosomes are selected with a probability related to their fitness. (Roulette wheel selection).

Step 6- Create a pair of offspring chromosomes by applying the stochastic operators – crossover and mutation.

Step 7- Place the created offspring chromosomes in the new population. Step 8- Repeat Step 5 until the size of the new chromosome population becomes equal to the size of the initial population N.

The figure below illustrates the steps mentioned (Fig. 4):

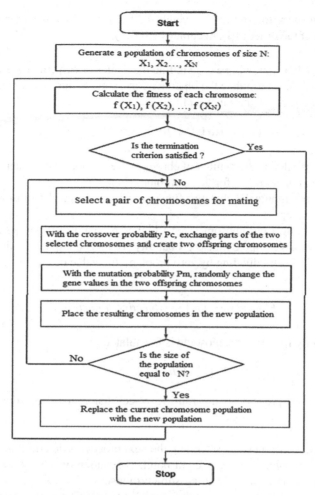

Fig. 4. Genetic algorithms process

4.3 Description of Classification Phase

The hybrid classification phase begins with the generation of the parameters α, β, γ by the genetic algorithms, in order to study these parameters according to our approach to be able to classify our new unlabeled documents. For this work, we use genetic algorithms to solve our problem which has the form of the following equation. Suppose there is equality:

$$\alpha C1 + \beta C2 + \gamma C3 = 1$$
$$c \in \mathbb{C}; \alpha, \beta, \gamma \in D; \alpha + \beta + \gamma = 1 \tag{7}$$

- $\mathbb{C} = C1, C2, C3$: The set of classifiers chosen after the model evaluation phase. These classifiers have the best Accuracy scores.

– α, β, γ: parameters to control the weight of each classifier in order to satisfy a hybridization of the order of 1 (Accuracy = 100%).

The Genetic Algorithm will be used to find the values of **α**, **β**, and **γ** that satisfy the equation. We must first formulate the objective function, for this problem the objective is to minimize the value of the function f (x) where

$$f(x) = ((\alpha C1 + \beta C2 + \gamma C3) - 1) \tag{8}$$

– Initialization: We define the number of chromosomes in population, then we generate random value of gene α, β, δ for those chromosomes.
– Evaluation: We compute the objective function value for each chromosome produced in initialization step.
– Selection: The fittest chromosomes have higher probability to be selected for the next generation. To compute fitness probability we must compute the fitness of each chromosome. To avoid dividing by zero problem, the value of the objective function is added by 1.

$$Fitness[i] = 1/(1 + Objective\_Function[i]) \tag{9}$$

The probability for each chromosome is formulated by:

$$P[i] = Fitness[i]/Total \tag{10}$$

For the selection process we use roulette wheel that choose according to random spins.

– Crossover: After chromosome selection, the next process is determining the position of the crossover point. We use one cut point, i.e. randomly select a position in the parent chromosome then exchanging sub-chromosome. Parent chromosome which will mate is randomly selected and the number of mate Chromosomes is controlled using crossover rate (ρc) parameters. The cut point is chosen by generating random numbers between 1 to (length of Chromosome − 1).
– Mutation: Number of chromosomes that have mutations in a population is determined by the mutation rate parameter. Mutation process is done by replacing the gen at random position with a new value. Mutation process is done by generating a random integer between 1 and total gen. If generated random number is smaller than mutation rate(ρm) variable then marked the position of gen in chromosomes. Finishing mutation process, we can now evaluate the objective function after one generation, if we reached the satisfying values, we stop else we redo all (Fig. 5).

4.4 Experimental Results

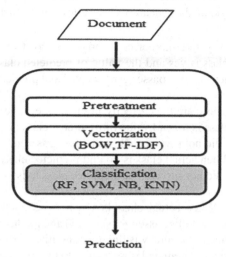

Fig. 5. Classification phase

Datasets Description

An accurate classifier is entirely dependent on getting the right training data, which means putting together examples that best represent the outcomes you want to predict. We used in our work four datasets of standard tests, details about the datasets are seen on the Table 1 below:

Table 1. Description of the datasets.

	Dataset name and source	Year of collection	Total number of distinct instances	Number of distinct classes	Distribution of the instance according to each class	
Bi- Class	SPAM [15]	2012	5572	2	Spam	747
					Ham	4825
	IMDB [16]	2011	50000	2	Positive	25000
					Negative	25000
Multi-Class	BBC NEWS [17]	2004–2005	2225	5	Politics	417
					Business	510
					Entertainment	386
					Tech	401
					Sport	511
	Multi- Lang [18]	2018	22000	22	22 different languages	1000 for each

Evaluation Methods
Classifiers are evaluated using well established performance metrics such as Pass Rate, Precision, Recall, and F-Measure are used to evaluate classifier performance. The measures mentioned are defined by means of the following characteristics:

True Positives (TP) - These are the correctly predicted positive values which means that the value of actual class is yes and the value of predicted class is also yes. E.g. if actual class value indicates that this passenger survived and predicted class tells you the same thing.
True Negatives (TN) - These are the correctly predicted negative values which means that the value of actual class is no and value of predicted class is also no. E.g. if actual class says this passenger did not survive and predicted class tells you the same thing.
False Positives (FP) – When actual class is no and predicted class is yes. E.g. if actual class says this passenger did not survive but predicted class tells you that this passenger will survive.
False Negatives (FN) – When actual class is yes but predicted class in no. E.g. if actual class value indicates that this passenger survived and predicted class tells you that passenger will die. Accuracy - Accuracy is the most intuitive performance measure and it is simply a ratio of correctly predicted observation to the total observations.
Accuracy - Accuracy is the most intuitive performance measure and it is simply a ratio of correctly predicted observation to the total observations.

$$Accuracay = \frac{TP + TN}{TP + TN + FP + FN} \tag{3}$$

Precision - Precision is the ratio of correctly predicted positive observations to the total predicted positive observations.

$$Precision = \frac{TP}{(TP + FP)} \tag{4}$$

Recall (Sensitivity) - Recall is the ratio of correctly predicted positive observations to the all observations in actual class.

$$Recall = \frac{TP}{(TP + FN)} \tag{5}$$

F1 score - F1 Score is the weighted average of Precision and Recall. Therefore, this score takes both false positives and false negatives into account.

$$F1\_Score = \frac{2 * Recall * Precision}{(Recall + Precision)} \tag{6}$$

Results and Discussion

Our experiments were implemented under the following hardware and software platforms: Hardware: PC with 8 Go de RAM, Intel(R) Core (TM) i7-7500U CPU,2.70 GHz. Software: Windows 10 Professional, Python 3.7, Jupiter Notebook.

The algorithm's parameters have been chosen after several tests on the accuracy improvement to suit the case we study. The following parameters are the used parameters according to each database:

SPAM, BBC NEWS, MULTI-LANG I Random Forest: number of estimators = 1000, K Nearest Neighbors: K = 1.

IMDB I Random Forest: number of estimators = 2000, K Nearest Neighbors: K = 17.

Result 1- Variation of the Accuracy rate according to BOW and TFI-IDF on the datasets under the Test-Split training technique (Fig. 6).

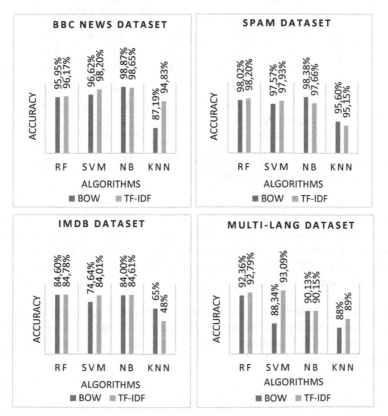

Fig. 6. Variation of the Accuracy rate according to BOW and TFI-IDF on the datasets under the Test-Split training technique.

Result 2- Variation of the Accuracy rate according to BOW and TFI-IDF on the datasets under the Stratified K-Folds Cross Validation training technique (Fig. 7).

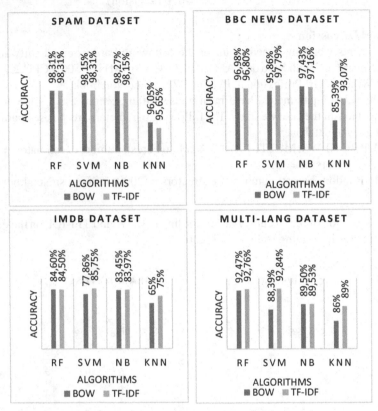

Fig. 7. Variation of the Accuracy rate according to BOW and TFI-IDF on the datasets under the Stratified K-Folds Cross Validation training technique.

We can see through the results (Result 1, Result 2) in the figures that the Accuracy for the different datasets and the type of classification wither it is a binary classification or a multiclass classification testifies to a superior performance of the algorithms with the second approach (TF-IDF) on most datasets. As for the BOW approach it is also considerable also with its Accuracy.

Result 3- Discusses the best Accuracy retained for each classifier according to the different vectorization approaches (BOW and TF-IDF) and specifically presents how training techniques (Test-Split and Stratified K-Folds Cross Validation) affect performance of each algorithm (Fig. 8).

Fig. 8. Variation of the Accuracy according to the algorithms and the two training techniques

After analyzing (Result 3) we noticed the following:

Each algorithm reacts to training techniques differently depending on the nature of the database being processed and the nature of the multi-class or binary classification.

For the Multi-Class Classification, the Test-Split technique gives good performance according to the previous results in datasets BBC NEWS and MULTI-LANG, as for the binary classifications the case is controversial since the two techniques are admissible on dataset IMDB, while the Stratified technique Cross Validation is more famous on dataset SPAM.

Result 4- Comparison between the algorithms on the four datasets (Fig. 9).

We have reached the point of making the decision on the choice of the best candidate classifiers that will be able to achieve the most precise, efficient prediction model for classifying new unlabeled documents. From the figures above the suitable choice adopted for the developed system is as follows:

Naive Bayes with a 98.38% Accuracy, with the Bow Approach and the Test Split training technique for a spam detection classifier, and a 98.87% Accuracy, for a topic detection classifier.

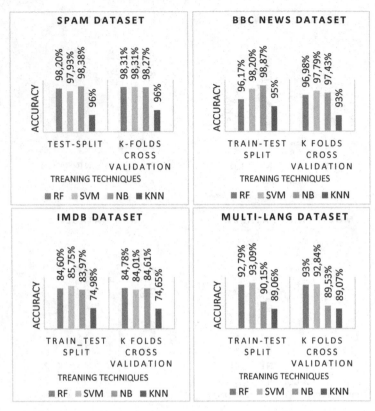

Fig. 9. Comparison between the algorithms on the four datasets.

SVM with an Accuracy of 85.75%, with the TF-IDF approach and Test-Split training technique for an emotion detection classifier, and an Accuracy of 93.09%, for a language detection classifier.

After choosing the best classifiers we decided in order to increase the accuracy rate to a maximum percentage that will allows us to achieve a perfect classification system, the idea was to adapt a hybrid approach on our new improved proposed system. Therefore, we began our first step to reach our hybrid idea by finding the vote weights that are the principal keys to the success of the approach by generating them properly with the use of genetic algorithms.

Before we begin the process, genetic algorithms take into concern parameters that have been chosen according to several tests to reach our purpose because must of the parameters doesn't have restricted values interval like: the Population size, Chromosome length, Maximum of generation, in the other hand the mutation probability is given according to real probability values interval in nature, typically in the range between [0.001,0.01] since its quite small in the nature, as for the cross over probability it is chosen in the range between [0,1].

The chosen GA Parameters:

– Population size = 14
– Chromosome length = 3
– Maximum of generation = 10
– Crossover probability = 0.25
– Mutation probability = 0.01

After executing the GA code written under the python programming language. The table follows illustrate the best results we achieved after the execution, for the different types of detection: Spam, emotion, topic and language with the use of a well-chosen classifiers from previous results (Table 2).

Table 2. Genetic Algorithm execution results

Classifiers For	Chosen Classifiers (Accuracy)	α	β	γ	Accuracy
SPAM Prediction	RF (98.31%)	0.0391913	0.14844778	0.83096192	1.0000
	SVM(98.31%)				
	NB (98.15%)				
Sentiments Prediction	RF (96.98%)	0.3667331	0.04237745	0.77153062	1.0000
	SVM(98.20%)				
	NB (98.87%)				
Topics Prediction	RF (96.98%)	0.1345197	0.61585623	0.267763	0.9999
	SVM(98.2%)				
	NB (98.87%)				
Languages Prediction	RF (84.78%)	0.4071405	0.66844072	0	1.0000
	SVM(85.7%)				
	NB (84.61%)				

The results proves that an accuracy of 100% can be reached by the assemblies of classifiers votes according to specific weight though some values seem to over fit but these are approximate values.

5 Conclusion

In this paper, we studied the performance comparison of the classifiers consistent with BOW and TF-IDF feature vectors, Test Split and Stratified Cross Validation training techniques we can say that the models built with the TF-IDF showed the best performances and that each algorithm reacts to training techniques differently counting on the character of the datasets processed and also the nature of the classification (binary or multiple). We furthermore recommended a hybrid approach to realize a classification system with a 100% Accuracy that we attained with the genetic algorithms.

References

1. Aas, K., Eikvil, L.: Text categorisation: A survey Technical report, Norwegian Computing Center (1999)
2. Aggarwal, C.C., Zhai, C.: A survey of text classification algorithms. In: Aggarwal, C., Zhai, C. (eds.) mining Text Data, pp. 163–222. Springer, Boston (2012). https://doi.org/10.1007/978-1-4614-3223-4_6
3. Colas, F., Brazdil, P.: Comparison of SVM and some older classification algorithms in text classification tasks. In: Bramer, M. (ed.) Artificial Intelligence in Theory and Practice. IFIP International Federation for Information Processing, vol. 217, pp. 169–178. Springer, Boston (2006). https://doi.org/10.1007/978-0-387-34747-9_18
4. An, Y., Tang, X., Xie, B.: Sentiment analysis for short Chinese text based on character-level methods. In: Proceedings of the 9th International Conference on Knowledge and Smart Technology (KST). IEEE. Chonburi, Thailand (2016)
5. Li, Y.H., Jain, A.K.: Classification of text documents. Comput. J. **41**, 537–546 (1998)
6. Brown, E.W., Chong, H.A.: The GURU system in TREC-6. In: The Sixth Text Retrieval Conference (TREC-6), pp. 535–540. National Institute of Standards and Technology (NIST) (1998)
7. Sebastiani, F.: A tutorial on automated text categorization. In: Proceedings of ASAI-99, 1st Argentinian Symposium on Artificial Intelligence. Buenos Aires, AR (1999)
8. HuilGol, P.: Quick Introduction to Bag-of-Words (BoW) and TF-IDF for Creating Features from Text, 28 February 2020
9. Rajaraman, A., Ullman, J.D.: Mining of Massive Datasets, pp. 1–17 (2011)
10. Hakim, A.A., Erwin, A., Eng, K.I., Galinium, M., Muliady, W.: Automated document classification for news article in Bahasa Indonesia based on term frequency inverse document frequency (TF-IDF) approach (2015)
11. Yiu, T.: Understanding random forest. Towardsdatascience.com, 12 June 2019
12. Duda, R.O., Hart, P.E.: Pattern Classification and Scene Analysis. Wiley, New York (1973)
13. Gandhi, R.: Support vector Machine-Introduction to Machine learning Algorithms, 7 June 2018
14. Tam, V., Santoso, A., Setiono, R.: A comparative study of centroid-based, neighborhood-based and statistical approaches for effective document categorization. In: Object Recognition Supported by User Interaction for Service Robots, vol. 4. IEEE (2002)
15. https://archive.ics.uci.edu/ml/datasets/SMS+Spam+Collection
16. https://ai.stanford.edu/~amaas/data/sentiment/
17. https://mlg.ucd.ie/datasets/bbc.html
18. https://www.kaggle.com/zarajamshaid/language-identification-datasst
19. McCall, J.: Genetic algorithms for modelling and optimisation. J. Comput. Appl. Math. **184**(1), 205–222 (2005)

Toward Student Classification in Educational Video Courses Using Knowledge Tracing

Houssam El Aouifi[1]([✉]) [ID], Youssef Es-Saady[1] [ID], Mohamed El Hajji[1,2] [ID],
Mohamed Mimis[1,2], and Hassan Douzi[1]

[1] IRF-SIC Laboratory, Ibn Zohr University, Agadir, Morocco
`houssam.elaouifi@edu.uiz.ac.ma`,
`{y.essaady,m.elhajji,h.douzi}@uiz.ac.ma`
[2] CRMEF-SM, Agadir, Morocco

Abstract. Using videos as a learning resource has received a lot of attention as an effective learning tool. Knowledge Tracing is intended to track students' knowledge acquisition when they answer a serie of problems. In this paper, we describe an experiment to model students' knowledge acquisition in educational video courses. For this purpose, Deep Knowledge tracing is used to classify and predict learners' performance as they interact with an educational video course in the subject of "C programming language". Learners' responses in previous quizzes were analyzed in order to forecast their next responses. The implementation of DKT in our dataset, led to an AUC (Area Under the Receiver Operating Characteristic Curve) of 0.73 which is a notable performance.

Keywords: Deep knowledge tracing · Video course · Performance prediction · Video quizzes

1 Introduction

Thanks to the accelerated evolution of technology, online learning is now part of the academic programs of many institutions around the world. As the current global crisis caused by the Covid 19 pandemic and the application of quarantine by countries, the majority of schools and educational institutes have adopted distance learning which has greatly contributed to its widespread use.

The use of video courses in education as a learning resource can be highly efficient. It allows students to learn, encourage diverse learning styles, and facilitates creative methods of teaching. At this point, tracking learning mastery process and course completion by learners remains very significant.

On one hand, educational platforms such as learning management systems (LMS), Intelligent Tutoring Systems (ITS) and MOOC (Massive Open Online Courses), provide a large scale student trace data, that can be used in learning process benefits. On the other hand, millions of learners have appreciated video

© Springer Nature Switzerland AG 2021
M. Fakir et al. (Eds.): CBI 2021, LNBIP 416, pp. 73–82, 2021.
https://doi.org/10.1007/978-3-030-76508-8_6

streaming from various platforms (e.g., YouTube) on a different product of terminals (e.g., TV, desktop, smart phone, tablets); which create traces of billions of interactions [1]. Based on the large volume and the variety of educational data, predicting learners' performance becomes an emerging field of research. Numerous investigations have been carried in this field based on various criterias and aspects [2–7].

Models such as knowledge Tracing is provided to track student's acquisition and to assess learning process. It helps to estimate the students' knowledge and to further forecast their future results [8]. knowledge Tracing (KT) is the process of modeling student knowledge over time in order to reliably predict how learners will perform in future interactions and activities [8].

Predicting students' performance based on knowledge Tracing has been widely explored in various tasks with different factors and data, such as, contextual estimation of student guessing and slipping were used in predicting student performance outside of intelligent tutoring systems [9], predicting student's inquiry skill acquisition [10], predicting future trainee's performance [11], students' performance and progress estimation [12,13], predicting learners' needs [14].

Hence, in this paper, Deep Knowledge Tracing is used to classify and predict learners' performance as they interact with educational video courses. Learners' responses in previous quizzes were analyzed in order to forecast their next responses.

The rest of this paper is organized as follows; The related works are discussed in Sect. 2. Methodology is described in Sect. 3. Experiment and first results are discussed in Sect. 4. Sect. 5 concludes the paper.

2   Related Works

Interactive Educational Environments (IEE) have allowed researchers to trace student's knowledge across different skills and provide suggestions for a better learning path. Presumptions on what a student understands and doesn't understand enable a tutoring system to adjust its input and guidance dynamically to maximize learning depth and effectiveness [15].

In this section, we first describe Knowledge Tracing task. Then, we present the most important models to tackle and deal with the Knowledge Tracing (KT) task.

The Knowledge Tracing (KT) task evaluates a student's state of knowledge based simply on the correctness or incorrectness of the responses of a student in the exercise solving process [16]. Knowledge Tracing models help tracking students' mastred skills and to concentrate practice on the skills that they have not yet learned. Commonly, the Knowledge Tracing task might be formulated like the following; student's interactions $X_t = (x_1, x_2, x_3..., x_n)$ on a specific learning work in order to predict his next interaction x_{t+1} and attempts to model learners' mastery of knowledge being tutored [18]. Over the past years, different methods were developed starting from probabilistic models to deep neural networks for instance: Deep Knowledge Tracing (DKT), Learning factors analysis (LFA), and Bayesian Knowledge Tracing (BKT).

2.1   Bayesian Knowledge Tracing (BKT)

Bayesian Knowledge Tracing (BKT) model was suggested by [16], considered as the most famous approach for constructing temporal models of student's learning. Bayesian Knowledge Tracing (BKT) model the latent knowledge state of a learner as a serie of binary values, any of them indicates grasping or not a particular concept [16]. Bayesian Knowledge Tracing (BKT) presents the knowledge state of knowledge components (KCs) based on Hidden Markov Model (HMM). Four types of model parameters are used in Bayesian Knowledge Tracing:

- $p(L_0)$ (or p-init): Initial knowledge state.
- $p(T)$ (p-transit): the probability of learning the knowledge component.
- $p(S)$ (p-slip): the probability of answering incorrectly a skill.
- $p(G)$ (p-guess): the probability of guessing correctly a unlearned skill.

In BKT, the formula used to update student's knowledge of skills are as follows: First of all, the system sets the first probability of the student knowing the skill a priori in Eq. 1. The conditional probability is then measured using either Eq. 2 or 3 based on whether the student correctly answered the question. As in Eq. 4, this conditional probability is then used to update the probability of skill mastery.

$$P(L_1) = P(L_0) \tag{1}$$

$$PL(L_n|Correct) = \frac{P(L_n - 1)(1 - P(S))}{P(L_n - 1)(1 - P(S)) + (1 - P(L_n - 1))P(G))} \tag{2}$$

$$PL(L_n|Incorrect) = \frac{P(L_n - 1)(1 - P(S))}{P(L_n - 1)(1 - P(S)) + (1 - P(L_n - 1))P(G))} \tag{3}$$

$$P(L_n) = P(L_{n-1} - Outcome) + (1 - P(L_{n-1} - Outcome))P(T) \tag{4}$$

2.2   Learning Factors Analysis (LFA)

Learning Factors Analysis (LFA) was presented by [17], a method for assessing and improving cognitive models based on how exactly a student's performance corresponds to expected learning skill. Learning Factors Analysis (LFA) combines a statistical model, human expertise and a combinatorial search. LFA models multiple students and multiple skills. The Learning Factors Analysis standard format is specified by the following equation:

$$\ln(\frac{P}{1 - P}) = \sum \alpha_i X_i + \sum \beta_j Y_j + \sum y_j Y_j T_j \tag{5}$$

Where:

- P = the probability to get an item right.

- X = the covariates for students.
- Y = the covariates for skills.
- T = the covariates for the number of opportunities practiced on the skills.
- α = the coefficient for each student.
- β = the coefficient for each rule.
- γ = the coefficient for the interaction between a production and its opportunities.

2.3 Deep Knowledge Tracing (DKT)

Deep Knowledge Tracing (DKT) was presented by [8], it employs recurrent neural network (RNN) as its backbone model. It achieves impressive performance and outperforms traditional methods for the Knowledge Tracing task.

The Deep Knowledge Tracing model represents a student's knowledge as latent variables state using the hidden variable of a RNN. The model transforms the problem of knowledge tracing by implying that each question can be related to a skill_ID, with a total of N skills. The skill mastery state of a student at a certain time stamp is specified by the following equations:

$$h_t = \tanh(W_{hx}x_{t-1} + W_{hh}x_{t-1} + b_h) \tag{6}$$

$$p(s_t)\epsilon y_t = \sigma(W_{yh}h_t + b_y) \tag{7}$$

Where both sigmoid $\sigma(.)$ and tanh functions are used. The model is parameterized with an recurrent weight $matrix_{hh}$, input weight $matrix W_{hx}$, readout weight $matrix_{yh}$ and initial state h_0 [8].

As presented in Fig. 1, The input to RNN is a sequence of fixed length input vectors x_t of students' interactions. The output y_t is a vector of length equal to the number of tasks, each input depicts the predicted probability of the student correctly answering to that particular problem [8]. Various DKT enhancements have been proposed [18–20].

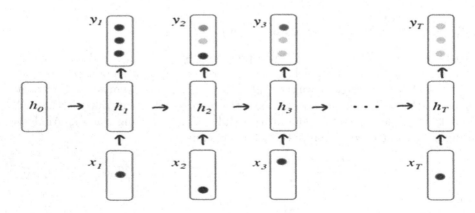

Fig. 1. The relation among parameters in a simple RNN. Adapted from [8].

In this work, we used the enhanced extension proposed in [18], called DKT+; which address two issues in the original DKT: The first problem is that the model is unable to rebuild the observed input. Second, the predicted performance for knowledges is not consistent through time steps. Introducing regularization to loss function in the original Deep Knowledge Tracing model helps improving model prediction consistency.

3   Methodology

In this section, we present the methodology followed for course presentation and data acquisition process.

3.1   Course Presentation and Learning Assessment Process

To capture students' data, a Learning Management System (LMS) based on Moodle platform [21] is used. We track the path followed by learners interacting with multiple video courses. A quiz test was proposed to assess learners' knowledge acquisition on each video course formed by multiple choice questions with multiple answer options. Figure 2 illustrates an abstract view of the course presentation and learning assessment process.

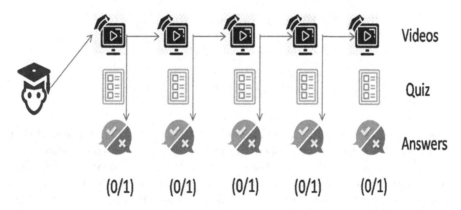

Fig. 2. Abstract view of the Course presentation and Learning assessment process.

Learner's answers are converted into binary format; 0 for failed quiz and 1 for passed quiz.

3.2   Data Acquisition and Preprocessing

For the data acquisition process, we accompanied the students during the entire semester in a module named "Introduction to C language Programming", from the virtual learning environment[1] at the polydisciplinary faculty of Taroudant of Ibn Zohr University.

The number of learners examined for this study was ($N = 64$). Five video courses on C programming language are presented to learners with different durations, each video course introduces a new concept of C language Programming to students. The details about video courses are given in Table 1.

Table 1. General information about video course.

Video courses	Duration
Loops in C	18 min50
Loop control statements	13 min
Arrays	19 min58
2D array	18 min04
Functions in C	27 min57

For each video lesson, a quiz was proposed, consisting of multiple-choice questions to test the learners' acquisition and obtain their scores. Students' scores are categorized with one of two states: '1' for scores above or equal to 10, and '0' for scores less than 10 (see Table 2).

Table 2. Class labels according to students grades.

Grade	Label
Grade $\geq$ 10	1
Grade $<$ 10	0

The learners' results differ from one video course to another. We stated that the number of students who passed the quizzes outnumber those who failed them. The data sample selected in the preceding step was preprocessed to transform raw data into a suitable format in order to proceed with DKT analysis. Indeed, a specific query statement is applied to combine different Logs. Only three attributes from three tables required were selected which are Student_Id, Quiz_ID, and Quiz_Grade. Then the data samples are converted to sequences grouped by user id.

The first line corresponds to user-ID, second line represents quizzes-ID, and the last line shows student's responses (see Table 3).

4   Experiment and First Results

In this section, we describe our experiment setup and our report experimental results.

---

[1] http://ecours-fpt.uiz.ac.ma/.

Table 3. Example of sequences grouped by user id.

User ID
QUIZ_1, QUIZ_2, QUIZ_3, QUIZ_4, QUIZ_5,....., QUIZ_n
1, 0, 1, 1, 1,...

4.1 Experiment Settings and Evaluation Measures

In this paper, prediction model were generated using the implementation of an enhanced version of the Deep Knowledge Tracing called DKT+ [18] offered by the authors to our dataset. The parameter settings are initially set as follows:

- A single layer RNN size = 200.
- The norm clipping threshold = 3.0.
- The dropout rate = 0.5.
- The learning rate = 0.0.

80% of the data is reported as a training set and the remaining 20% is reported for the validation set. In addition, 5-fold cross validation is performed on the training set to select the hyperparameter setting. To evaluate our model, we used Area Under the Receiver Operating Characteristic Curve (AUC), which offers a powerful binary prediction evaluation criterion. An AUC score of 0.5 shows that the consistency of the model is just as good as a random guess [22].

4.2 First Results

The implementation of Deep Knowledge Tracing+ in our dataset led to an AUC of 0.73 which is a notable performance. Figure 3 shows the training set AUC score for epoch of 1 to 19 during training phase. Figure 4 illustrates the evolution of AUC score and loss during the validation phase.

The model resulting AUC is about 0,72%. After the 5-th epoch the training model starts to converge as well as the validation model. The best testing result

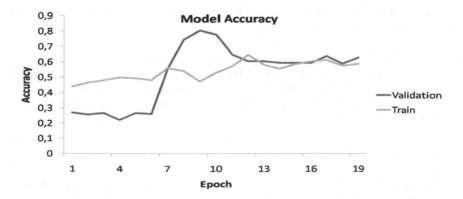

Fig. 3. Evolution of the AUC and loss during training

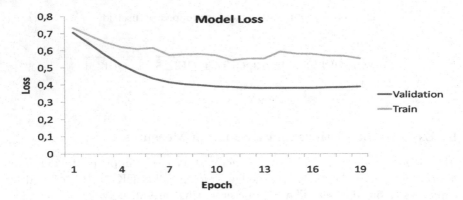

Fig. 4. Evolution of AUC and loss and during validation.

occurred at: 8-th epoch, with testing AUC of 0,80. To prevent over-fitting and optimize our model, we used regularization parameters ($\lambda_{Waviness1}$, $\lambda_{Waviness2}$) proposed by authors of DKT+.

The results of the prediction offer a significant illustration of the capabilities of DKT+ in predicting the probabilities of a student answering correctly a future problem in educational video courses with high credibility.

5   Conclusion and Future Work

In this paper, we used deep knowledge tracing, to predict students' performance enrolled in a video course on "C programming language"; by gathering information about students' previous responses on various quizzes in order to forecast their next responses. The model prediction result demonstrates the strength of Knowledge Tracing (KT) task in video learning. Improving Knowledge Tracing (KT) task implies that educational videos can be recommended and customized to learners depending on their particular demands, also it is possible to skip or delay video course content that is predicted to be too easy or too hard. For future work, we seek to enhance the performance of knowledge tracing task by adding students' viewing behavior interactions with video courses as an important input feature of the model.

Acknowledgement. This work was supported by Al-Khawarizmi Program for research Support in the Field of Artificial Intelligence and its Applications (Morocco).

The authors wish to thank the participants of the study who kindly spent their time and effort. We also want to thank **Pr. Fouzia Boukbir** (English teacher) for her help and devotion.

References

1. Giannakos, M., et al.: Video-based learning and open online courses. Int. J. Emerg. Technol. Learn. (iJET) **9**, 4 (2014)

2. Bayer, J., Bydzovska, H., Geryk, J., Obsivac, T., Popelinsky, L.: Predicting dropout from social behaviour of students (2012)
3. El Aouifi, H., et al.: Predicting learner's performance through video viewing behavior analysis using graph convolutional networks 1–6 (2020). https://doi.org/10.1109/icds50568.2020.9268730
4. Mimis, M., El Hajji, M., Es-Saady, Y., Oueld Guejdi, A., Douzi, H., Mammass, D.: A framework for smart academic guidance using educational data mining. Educ. Inf. Technol. **24**(2), 1379–1393 (2018). https://doi.org/10.1007/s10639-018-9838-8
5. El Aouifi, H., El Hajji, M., Es-Saady, Y. et al.: Predicting learner's performance through video sequences viewing behavior analysis using educational data-mining. Education and Information Technologies (2021). https://doi.org/10.1007/s10639-021-10512-4
6. Pereira, F.D., et al.: Early performance prediction for CS1 course students using a combination of machine learning and an evolutionary algorithm. In: 2019 IEEE 19th International Conference on Advanced Learning Technologies (ICALT). IEEE (2019). https://doi.org/10.1109/ICALT.2019.00066
7. Yu, C.-H., Wu, J., Liu, A.-C.: Predicting learning outcomes with MOOC clickstreams. Educ. Sci. **9**, 104 (2019)
8. Piech, C., et al.: Deep knowledge tracing. Adv. Neural Inf. Process. Syst. **28**, 505–513 (2015)
9. Baker, R.S.J., et al.: Contextual slip and prediction of student performance after use of an intelligent tutor. In: De Bra, P., Kobsa, A., Chin, D. (eds.) UMAP 2010. LNCS, vol. 6075, pp. 52–63. Springer, Heidelberg (2010). https://doi.org/10.1007/978-3-642-13470-8_7
10. Sao Pedro, M., Baker, R., Gobert, J.: Incorporating scaffolding and tutor context into Bayesian knowledge tracing to predict inquiry skill acquisition. In: Educational Data Mining 2013 (2013)
11. Jastrzembski, T.S., Gluck, K.A., Gunzelmann, G.: Knowledge tracing and prediction of future trainee performance, Florida State University, Tallahassee, Department of Psychology (2006)
12. Yang, F., Li, F.W.: Study on student performance estimation, student progress analysis, and student potential prediction based on data mining. Comput. Educ. **123**, 97–108 (2018). https://doi.org/10.1016/j.compedu.2018.04.006
13. Cui, Y., Chu, M.-W., Chen, F.: Analyzing student process data in game-based assessments with Bayesian knowledge tracing and dynamic Bayesian networks. J. Educ. Data Min. **11**, 80–100 (2019). https://doi.org/10.5281/zenodo.3554751
14. Chanaa, A., El Faddouli, N.-E.: Predicting learners need for recommendation using dynamic graph-based knowledge tracing. In: Bittencourt, I.I., Cukurova, M., Muldner, K., Luckin, R., Millán, E. (eds.) AIED 2020. LNCS (LNAI), vol. 12164, pp. 49–53. Springer, Cham (2020). https://doi.org/10.1007/978-3-030-52240-7_9
15. Khajah, M., Lindsey, R.V., Mozer, M.C.: How deep is knowledge tracing? arXiv preprint arXiv:1604.02416 (2016)
16. Corbett, A.T., Anderson, J.R.: Knowledge tracing: modeling the acquisition of procedural knowledge. User Model. User-Adapt. Interact. **4**, 253–278 (1994). https://doi.org/10.1007/BF01099821
17. Cen, H., Koedinger, K., Junker, B.: Learning factors analysis – a general method for cognitive model evaluation and improvement. In: Ikeda, M., Ashley, K.D., Chan, T.-W. (eds.) ITS 2006. LNCS, vol. 4053, pp. 164–175. Springer, Heidelberg (2006). https://doi.org/10.1007/11774303_17

18. Yeung, C.-K., Yeung, D.-Y.: Addressing two problems in deep knowledge tracing via prediction-consistent regularization. In: Proceedings of the Fifth Annual ACM Conference on Learning at Scale (2018). https://doi.org/10.1145/3231644.3231647
19. Wang, Z., Feng, X., Tang, J., Huang, G.Y., Liu, Z.: Deep knowledge tracing with side information. In: Isotani, S., Millán, E., Ogan, A., Hastings, P., McLaren, B., Luckin, R. (eds.) AIED 2019. LNCS (LNAI), vol. 11626, pp. 303–308. Springer, Cham (2019). https://doi.org/10.1007/978-3-030-23207-8_56
20. Xiong, X., et al.: Going deeper with deep knowledge tracing, International Educational Data Mining Society (2016)
21. Rice, W.: Moodle 1.9 E-Learning Course Development. Packt Publishing Ltd. (2008)
22. Myerson, J., Green, L., Warusawitharana, M.: Area under the curve as a measure of discounting. J. Exp. Anal. Behav. **76**, 235–243 (2001). https://doi.org/10.1901/jeab.2001.76-235

Assessment of Lifestyle and Mental Health: Case Study of the FST Beni Mellal

Juienze Melchiore Magolou-Magolou(✉) and Abdellatif Haïr(✉)

Department of Computer Science, Faculty of Science and Technology, Sultan Moulay Slimane University, Beni Mellal, Morocco

Abstract. Lifestyle habits are defined as behaviors of a sustainable nature which are based on a set of elements incorporating cultural heritage, social relations, geographic and socio-economic circumstances as well as personality. Mental health encompasses the promotion of well-being, the prevention of mental disorders, and the treatment and rehabilitation of people with these disorders. In order to address this issue, we propose a solution which consists of the development of an extended autonomous computer model for large textual data. This model will make it possible to give a psychological, emotional or even a lifestyle character from tweets or a web forum. So we turned to the notions of sentiment analysis and Text Mining using Deep Learning. This work (which will be limited to a Moroccan context) concerns the development of a computer model that allows to determine the habits of life and the Health of the students of the Faculty of Sciences and Technologies at the Sultan Moulay Slimane university in Beni Mellal. We started by developing a script to retrieve posts made by students from a Facebook group. The choice of Facebook and not Twitter is due to the fact that the twitter community among the students is relatively small. Afterwards, we built our deep learning model and we tested it with data from twitter comprising of thirteen (13) classes (anger, joy, sadness, disgust etc.). We also submitted these textual data to automatic learning algorithms (naive Bayesian, K nearest neighbors).

Keywords: Web scraping · Sentiment analysis · Text classification · Machine learning · Deep learning

1 Introduction

The web is filled with a large amount of data, and the exploration as well as manipulation of this data, prove to be the new challenges of our century and the future. With the arrival of blogs, social networks and websites, several disciplines and methods of data science have emerged, such as sentiment analysis, text mining and web scraping. All these methods and disciplines share a common objective which is the extraction and manipulation of the data in order to retrieve information from it.

Sentiment analysis [1] is a growing field which covers many disciplines. It can be achieved by several approaches: machine learning, deep learning and another approach which uses automatic natural language processing (NLP). All these approaches can be

© Springer Nature Switzerland AG 2021
M. Fakir et al. (Eds.): CBI 2021, LNBIP 416, pp. 83–93, 2021.
https://doi.org/10.1007/978-3-030-76508-8_7

combined depending on the case and the objectives to be achieved. Text Mining [2] is an industrial approach which simply means it's a statistical approach for text classification [3] using automated learning or deep learning. Web scraping [4] is a technique for extracting content from websites, via scripts or programs, and transforming it and making it useful in other contexts.

In this article, we aim to use state-of-the-art of text classification [5] configured specifically for the moroccan language context. This is due to the fact that students of the Faculty of Sciences and Technologies of Beni Mellal use a rather complex way of communicating on social media platforms such as Facebook. This complexity in the analysis of the text is due to the fact that a Student can write and tweet using Arabic and French letters to talk in the Arabic dialect language locally known as "Darija". In this case, we used web scraping [4] by programming a script to retrieve and collect textual posts from any Facebook group related to the Faculty. After retrieving the posts, we stored them in a database and then explored the data to discover the contents and we also structured it so that it can later be utilized by Machine Learning and Deep Learning algorithms. However, this data manipulation and exploration was only partially done because we retrieved 3504 posts from the public group "fstbenimellal" which were posted between May 2008 and August 2020. Moreover, because of the lack of a large enough team, we were only able to browse a hundredth of this database. Afterwards, we built our text classification models based on the deep learning approach together with Recurrent Neural Networks (RNN) and Convolutional Neural Networks (CNN). We considered a semantic model with word-embedding weighting [5], one-hot-encoding [5] while using the word dictionary concept. Moreover, we also used the transfer of learning. In order to have an overview of our prediction model, we make a comparison of the results of other machine learning algorithms (SVM, KNN, Naives Bayesian). The output results of the models are expressed by the precision, recall and classification rate metrics.

And finally, we will test the implementation of our models (3 models) applied on a database already classified using Tensorflow-keras. Then, we conclude with a perspective to improve this work.

2 Related Work

Text classification is a field of research that is expanding rapidly, and the automation of this operation has become a challenge for the scientific community. Work has progressed considerably over the past twenty years and several models have emerged such as supervised binary classification, supervised multi-class classification and ordered classification (classification of texts in order of relevance for each category) [6]. There are mainly two areas of research that deal with intelligent word processing and each has its own methods [7]: Approaches resulting from data analysis and statistics and "black box" type approaches.

The approaches derived from data analysis and statistics are based on linguistics and statistics. They employ methods such as principal correspondence analysis, which is subject to be interpreted by humans depending on the obtained results. There are many software developed for this purpose, for example IRaMuteQ (which is under an open-source license and has an API to interact with the R language), Lexico3, Hyperbase.

The "black box" approach processes documents automatically without human intervention. They integrate low-level functions: lexical analysis, surface syntax analysis, search for information by key words. Popularized search engines work with this approach.

3 Proposed Method

In order to successfully build our three Deep Learning models, we first submitted our textual data to classical Machine Learning algorithms, in particular SVM, KNN and Naïve Bayes. And for these algorithms in particular, we used the textual representation of TF-IDF (term frequency -inverse document frequency) [8]. In order to increase the prediction capacity of these algorithms, we used the Grid search approach to adjust the hyper parameters of Machine Learning algorithms. Grid search is a tool from the SKLEARN machine learning python library.

Scikit-learn (Sklearn) is the most used and most robust python machine learning library. It provides a wide-range selection of efficient tools for machine learning and statistical modeling including classification, regression, clustering and dimension reduction through a consistency interface in python. This library (which is largely written in python) is built to interact with other libraries such as NumPy, SciPy et Matplotlib. It also offers a variety of algorithms such as SVM, Naïve bayes et KNN.

After using the Machine Learning algorithms, we obtained the following results (Table 1):

Table 1. Comparative results of the classification rate before and after a parameter adjustment.

Algorithms	Accuracy before grid search	Accuracy after grid search
SVM	50.7%	72.9%
KNN	66.4%	65.8%
NB	72.6%	No adjustment parameters

Let's explain the adjustment parameters on SKLEARN (version 2.3) by Grid Search used for our case study:

For the SVM we adjusted the "regularization parameter" noted C which we varied according to the values 1, 10, 100, 1000. Here the correct adjustment value of C is 10. The kernel here as a good value is the "RBF" (Radial basis function kernel) which has been compared to a "linear" kernel. We have also adjusted the coefficient of the nucleus called "gamma", the correct value is gamma $= 0.1$ compared with the following values 0.1, 0.2, 0.3, 0.4, 0.5, 0.6, 0.7, 0.8 and 0.9.

For the KNN we adjusted the "metric" that returned the Euclidean distance comparison with the Manathan distance. We also added the k of the neighbors and as a good value retained, k $= 27$, compared with values ranging from 3 to 99.

So, our next goal was to build a Deep Learning model with a classification rate of over 72.9%. In order to achieve this, we used Recurrent Neural Network (RNN), Transfer Learning and Convolutional Neural Networks (CNN). These models (that we built in place of using the TF-IDF approach [8]) have options for the representation of our texts, deep Learning models on the Word-Embedding representation [5] and one-hot-encoding [5] using a dictionary approach of words.

For the rest of the Deep Learning models, let's first present our work environment and the tools used: We used Anaconda, Jupiter and TensorFlow (version 2.1).

Anaconda is a free and open-source distribution of Python and R, applied to the development of applications dedicated to data science and machine learning, which aims to simplify package management and deployment.

Jupyter is a web application used to program in more than 40 programming languages, including Python, Julia, Ruby, R, or even Scala. Jupyter makes notebooks and these notebooks are used in data science to explore and analyze data.

TensorFlow is an open-source machine learning tool developed by Google and released under the Apache license and has an interface for Python, Julia and R. TensorFlow is one of the most widely used AI tools in machine learning and its current version comes with keras which is one of the most used frameworks for deep learning.

3.1 First Model

This model is based on recurrent neural networks using LSTM (Long short-term memory) as memory cells [9]. We used word-embedding as the vector representation of the words. Figure 1 illustrates our model:

This model contains 4 layers:

An embedding layer that makes it possible to create a link between the words which are read inside, in the form of numbers and each number represents a term. Here, the input is of type (none, 100), 100 is the size of the vector of tweets (or text from a forum) that we have set and the "none" is the size of the batch. The output is of type (none, 100,100), the "none" is batch size, the first 100 represents the number of time steps which is equal to the size of the tweet vector. The second 100 is the size of word-embedding. So, the dimension of the word-embedding matrix is the number of words in our vocabulary divided by the number of the depth of word-embedding.

Two LSTM layers: the first layer takes as input the output of the embedding layer and as output data corresponding respectively to batch size, time steps and the input dimension (number of neurons in a layer). The last LSTM layer takes its input from the output of the other above. and as an output, it gives, respectively, the batch size and input dimension for the next layer.

The Dense layer is the output layer, it gives us the values (probability) of our thirteen classes to predict.

The "Input layer" is not a layer but it is just a representation of the shape of our data when at the start of our first layer (input layer).

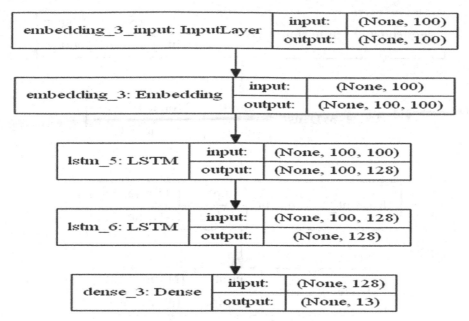

Fig. 1. First model of Deep Learning.

3.2 Second Model

This model is based on neural networks using simple dense layers. It uses Transfer Learning [10] in order to learn from a stable learning model that classifies articles in the journal "Reuters" with a classification rate of 80%. For the vector representation of words, it was combined with the one-hot-encoding representation. The Fig. 2 represents our inherited model.

In this model, we are going to take all these already trained layers but we will not take its output layer because the "Reuters" journal categorizes these documents into 46 classes while in our case we have 13 classes. This model has an 80% classification rate for 8,500 documents. This is, primarily, the reason we thought of collecting these already trained layers. In order to adapt it to our situation, we will just change its output and see the benefit that we can derive from it.

This model contains 3 dense layers:

The first is the input layer which takes as input, the values (none,10,000) of the size of the batch and the dimension of our tweets (encoded on 10,000 columns). The second dense layer takes as input the output of the first and at the output, it gives the size of the batch and the number of neurons in this layer which is 512. The last layer used but its output layer will replace by 13 neurons (corresponding to our problem).

The Dropout layer is not a layer but is a technique in neural networks which consists of solving the problem known as "gradient explosion", i.e., the gradient can no longer converge. The trick behind this approach is to switch off certain neurons as they go through the learning phase to prevent them from further seeking to get specialized.

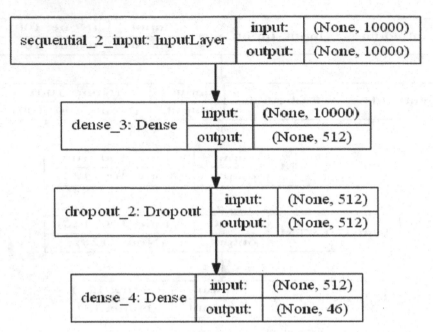

Fig. 2. Inherited model from Transfer Learning.

3.3 Third Model

This is based on convolutional neural networks (CNN) coupled with the vector representation of word-embedding words and the use of pre-trained words [11] from the "Glove" file. GloVe is an unsupervised learning algorithm for obtaining vector representations of words developed by MIT. The Fig. 3 represents our model.

This model contains 5 layers:

The first layer (embedding) allows us to create a link between the words which are read inside in the form of the numbers and of which each number represents a term. Here the input is of type (none, 100), 100 is the size of the vector of tweets that we have set and the "none" is the size of the batch. Its output is of type (none, 100, 50) the "none" is batch size, the 100 is the number of the temporal steps which is equal to the size of the tweet vector (dimension of the input data). The 50 is the plunge size that we have established. Thus, the dimension of the embedding matrix is the number of words in our vocabulary divided by the number of the embedding depth.

The second layer, conv1D is a convolution layer. It takes as input the output of the embedding layer of the form batch size (none), temporal steps (100), data dimension (50 which is the number of embedding columns of matrix embedding). At its output it is of the form batch size (none), new time step (96 which is defined and depending on the size of the kernel. we have taken 5 in this case), number of filters (we used 128 in this case).

The third layer, GloabMaxPooling layer, takes as input the outputs of the conv1D layer and as output the batch size (none) and the characteristics (here equal to 128, the number of filters).

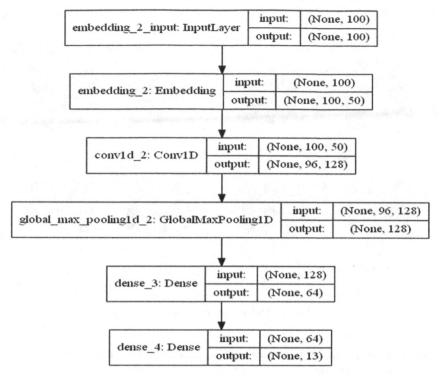

Fig. 3. Third model of Deep Learning.

The fourth layer (the Dense layer) takes input from the output of GlobalMaxPooling. Its output is in the form of batch size (none) and dimension of the input data (here equal to 128, the number of neurons in this layer).

The last layer gives as output our expected prediction results on our thirteen classes corresponding to the outputs of the thirteen neurons of this class.

4 Framework

To establish the above presented model and in order to obtain a Deep Learning model with a classification rate greater than 72.9%, we established this structure (see Fig. 4) using state-of-the-art text classification.

In this working structure, we introduce the data of the forums, we do the preprocessing of the text and at the text cleaning level we only apply simple filters, that is to say we only have to remove the simple punctuation marks (comma, period...). This simple punctuation is justified by the fact that in the Moroccan dialect language "Darija", grammatical and spelling rules are organized at the writing level.

Then we vectorize our data by word-embedding, TF-IDF (only for Machine Learning algorithms) and one-hot-encoding representations. Once our textual data is vectorized, we enter it into our models (RNN, ANN, CNN) and we compare the metrics, particularly,

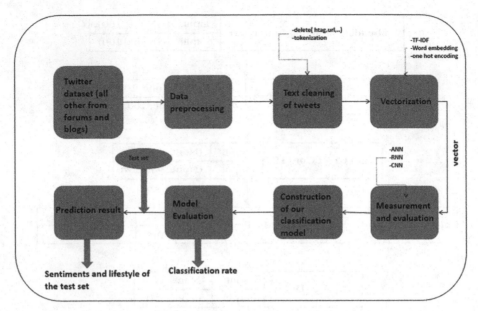

Fig. 4. Architecture of the system.

the classification rate on the evaluation and training data. We then make a prediction on the models with the test data.

These models are built gradually so that the latest model is always more efficient than those which precede it. We also have to make the adjustments of the parameters for each respective model in order to increase its scores.

5   Results

After training our models, we obtained the following results:

Table 2. Scores of deep learning.

Models	Precision	Recall	F-Measure	Accuracy
First model	24%	49%	32%	49%
Second model	70%	72%	70%	72%
Third model	72%	74%	71%	74%

The number of layers and the number of neurons per layer which define our models and which we have chosen with great care (see Fig. 1, Fig. 2 and Fig. 3), had an influence on the results of their performance (see Table 2). Here we focus on the other hidden parameters which are not shown in the Figures (see Fig. 1, Fig. 2 and Fig. 3) but which also influenced our models.

For the first model, the following parameters were adjusted and their values returned: The optimizer is equal to "Adam", this is the chosen gradient descent algorithm that we compared with the other "SGD" and "RMSprop" gradient descent algorithms. The epoch is equal to 10, this is the number of epochs on which this model turned because after that time this model could no longer learn. The "Dropout" is equal to 0.1, this is a parameter that we have adjusted and added to avoid gradient explosion, this value was compared with the following values 0.2, 0.3, 0.4 and 0.5.

For the second model, the adjusted parameters are as follows: The optimizer is the "Adam", it is the chosen gradient descent algorithm that we compared with the other "SGD" and "RMSprop" gradient algorithms. The epoch is equal to 16, this is the number of epochs on which this model ran because after that time this model could no longer learn.

For the third model, the adjusted parameters are as follows: The optimizer is the "Adam", it is the chosen gradient descent algorithm that we compared with the other "SGD" and "RMSprop" gradient algorithms. The epoch is equal to 13, this is the epoch number on which this model ran because after that time this model could no longer learn. In the second layer, the selected activation function is "tanh" compared with the "reread" function. In the fourth layer, the selected activation function is "reread" compared with the "tanh" function.

These adjusted parameters are at the origin of the performance of our models given in summary in our table (see Table 2). The classification rate of the third model is 74% and the F-Measure is 71%; which means that our model manages to class our posts among our thirteen classes with an overall success rate of 74% and a balance between false positives and true negatives of 71%. These two scores are higher than the other two models which makes the third model the best among the three.

In this table we are able to visualize the performance of our latest model. All the scores of the latest model are higher than predecessors. These scores are given by the metrics Precision, Recall, Accuracy and F-Measure.

Precision is the fraction of relevant instances among the retrieved instances, while recall is the fraction of retrieved relevant instances among all relevant instances. Both precision and recall are therefore based on an understanding and measure of relevance. F-measure is calculated from the precision and recall of the test. Accuracy is the classification rate.

We have pre-processed our data sets from twitter using our specific approaches. We used several approaches to represent tweets, all based on a vocabulary approach from all of our tweets depending on the specificities of our models. We used Word embedding and one-hot-encoding representations for the first, third and second, respectively. We tested our Deep Learning models against each other and compared the Classification Rate, Precision, Recall, and F-Measure for our various models offered. The best performance was provided by our third model integrating convolutional neural networks and the use of "pre-trained words".

6 Conclusion

Lifestyle and mental health are two relatively dense and complex concepts to determine in an individual because they sometimes show subjectivity and contextualization linked to geographic, social and economic environment. However, there are common things that are shared by all, such as feelings, emotions and some basic needs of human life. These needs also punctuate the lifestyle of a human being. So, to provide a solution to this problem which consists of determining these through tweets (or any other written on forums) we considered using approaches of classification of texts and analysis of feelings but in a multi-class approach using Deep Learning.

We worked to apply our Deep Learning model constructs to a dataset of tweets thereby allocating each tweet to a very specific sentiment class.

We preprocessed our data slightly according to the context of our work (removal of hashtag, URL, comma, points…) but we did not do word filtering or word level processing (lemmatization, stemming). Then we vectorized our tweets using TF-IDF, word-embedding and hot-one-encoding techniques. To build our models, we relied on SVM, KNN and Naïve Bayesian Machine Learning models. We obtained the following classification rates 72.9%, 66% and 72.6% respectively. Then we built our Deep Learning models in order to improve the classification rate of our Machine Learning algorithms. The three Deep Learning models we proposed obtained respective scores of 49%, 72% and 74%.

And finally, we propose, in hope of continuity of this work, to envisage the design of a dictionary or an NLP system for "Darija" (Moroccan dialect) based on learning by reinforcement to allow processing at the level of the sentence which can also be turned to neural networks called "autoencoders".

References

1. Cambria, E., Poria, S., Gelbukh, A., Thelwall, M.: Sentiment analysis is a big suitcase. IEE Intell. Syst. **32**, 74–80 (2017)
2. Mooney, R.: Text mining. In: Zaverucha, G., da Costa, A.L. (eds.) SBIA 2008. LNCS (LNAI), vol. 5249, p. 6. Springer, Heidelberg (2008). https://doi.org/10.1007/978-3-540-88190-2_5
3. Stein, A., Jacques, P., Valiatib, J.: An analysis of hierarchical text classification using word embeddings. Inf. Sci. **471**, 216–232 (2019)
4. Karthikeyan, T., Karthik, S., Ranjith, D., Vinoth, K., Balajee, J.: Personalized content extraction and text classification using effective web scraping techniques. Int. J. Web Portals (IJWP) **11**, 12 (2019)
5. Ge, L., Moh, T.: Improving text classification with word embedding. In: 2017 IEEE International Conference on Big Data (Big Data), Boston, MA, pp. 1796–1805 (2017). arXiv:1708. 02657v2. https://arxiv.org/pdf/1708.02657.pdf. Accessed 17 Aug 2017
6. Mondher, B., Ohtsuki, T.: A pattern-based approach for multi-class sentiment analysis in Twitter. IEEE Access 20617–20639 (2017)
7. HAL Id: lirmm-00321401. https://hal-lirmm.ccsd.cnrs.fr/lirmm-00321401. Accessed 4 Sept 2019
8. Zhang, Y.-T., Gong, L., Wang, Y.-C.: An improved TF-IDF approach for text classification. J. Zheijang Univ.-Sci. A **6**, 49–55 (2005). https://doi.org/10.1007/BF02842477

9. Greff, K., Kumar, R., Koutník, J., Steunebrink, B., Schmidhuber, J.: LSTM: a search space odyssey. IEEE Trans. Neural Netw. Learn. Syst. **28**(10), 2222–2232 (2015)
10. Sinno, J., Qiang, Y.: A survey on transfer learning. IEEE Trans. Knowl. Data Eng. **10**, 1345–1359 (2019)
11. Luo, Y., Tang, J., Yan, J., Xu, C., Chen, Z.: Pre-trained multi-view word embedding using two-side neural network. In: Twenty-Eighth AAAI Conference on Artificial Intelligence, Quebec, vol. 28, no. 1 (2014). https://ojs.aaai.org/index.php/AAAI/article/view/8956

The Search for Digital Information by Evaluating Four Models

Mouhcine El Hassani[✉] iD and Noureddine Falih

LIMATI Laboratory, Polydisciplinary Faculty, University of Sultan Moulay Slimane, Mghila, BP 592 Beni Mellal, Morocco

Abstract. As information becomes more and more abundant and accessible on the web, researchers do not have to dig through books and libraries. Web pages are rich in textual information, the web search engines provide Internet users with various files corresponding to the searched keywords. This large number of digital data makes manual sorting difficult to do, so it is necessary to automate collection of useful information using techniques based on artificial intelligence. In today's digital age, great importance is given to information retrieval techniques via Internet. Therefore, it appears essential to preconize a credible and performing system dealing with all textual information, in order to deduce structured and useful knowledge. This work focuses on four models used in the field of information retrieval, and highlights their limits of use, with a view to developing new techniques that can fill the gaps detected. At the end, the evaluation parameters will be discussed to enhance human intervention in decision making.

Keywords: Extraction of knowledge · Boolean model · Text mining

1 Introduction

Textual information responding to a search request is ranked according to its relevance scores. However, the obtained data is generally not structured; the search does not use SQL queries on databases. Web pages differ from text documents; they contain hyperlinks and anchor texts. Hyperlinks are very useful for research and play an important role in ranking algorithms. Similarly, the anchor texts associated with hyperlinks are necessary, because an anchor text is frequently a more accurate description of the page to which the hyperlink is pointing. In HTML, the content of a web page is structured in blocks of fields (title, metadata, body, etc.).

Some fields are very important compared to others, they are useful for indexing and referencing web pages by search engine robots, however spamming or redirection to advertising web pages remains a problem, and can be harmful to the quality of the results of a search query on the web.

In addition, textual information written in different languages, and in the same database poses a major classification problem.

Through four models we will highlight the importance of human intervention for a good classification of textual data.

© Springer Nature Switzerland AG 2021
M. Fakir et al. (Eds.): CBI 2021, LNBIP 416, pp. 94–106, 2021.
https://doi.org/10.1007/978-3-030-76508-8_8

2   The Proposed Process of Finding Information from Documents

Information is only useful if it provides answers to the user's question. For this to happen, an information retrieval system [1] should first retrieve all files and documents responding to client requests, and then index and organize them before storing and publishing them. The figure below represents the general architecture of the information retrieval process [2] (Fig. 1):

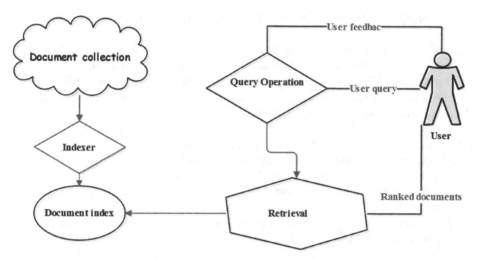

Fig. 1. A general architecture of the information retrieval process [2]

A user sends a request to the recovery system; the latter consults the indexed documents beforehand, and classifies them according to the relevance score meeting the needs of the client and then returns a set of results.

To fully understand the principle of finding information, the user expresses his needs in different forms, either a query composed by independent keywords or linked by logical operators such as AND, OR, NO…, some research applications of information uses them to view indexes of databases containing web pages or Collected files. Search engines have their own syntax for handling such requests. In most cases, the user starts his search by writing simple sentences, and depending on the language and writing style, however, the word order of the entered text influences the quality of the obtained results and as a results, the recovered document may not contain all terms of the query. In this case, another type of queries called proximity queries is required. The system calculates the proximity (distance) of the component terms, and classifies the founded pages and documents, taking into account this proximity factor and the order of the terms; the user can set this order in advance. Other searched expressions can be put between two quotation marks thus forcing the search engine to find the complete document and return the results or URL links of similar pages.

Asking questions about natural language is another way to meet user's needs. However, this technique is difficult to implement in practice, the principle is to prepare a set of questions with their answers, then structure them according to a model information system. After preparing the query of the question and format it, taking into account the prepositions of the used natural language, the user then run the query and obtained his results.

In the following sections focuses on four different models define factors of relevance [3] and proximity exploited in the queries, in order to extract documents and web pages that meet the needs of users.

3   Research Method of Different Information Retrieval Models

Four information search models can be distinguished: the Boolean model, the vector space model, the probabilistic model and the connectionist and linguistic.

Documents "D" or collected web pages can be considered as a set of words or terms, their positions are ignored in sentences. Each term "ti" is linked to a weight "wij", all of these distinct terms represent a collection V:

- $V = \{t_1, t_2, t_3, \ldots, t_N\}$ with N = the size of the document.
- For a document $d_j \in D$, we associate a weight w_{ij} with the term t_i
- If t_i exists in d_j then $w_{ij} = 1$ otherwise $w_{ij} = 0$
- Each document dj represents a vector of the terms: $d_j = \left(w_{1j}, w_{2j}, w_{3j}, \ldots, w_{Nj}\right)$.

Therefore, we can represent the collection of documents "D" by a matrix "table" vectors di "or objects" whose attributes are wij.

3.1   The Boolean Model for Information Retrieval

This model was developed by Edward Alan Fox [4] in 2006. A search query is represented by an expression of terms linked by And, Or, and Not logical operators.

To better explain the basic principle, we assume three documents d_1, d_2 et d_3 of the collection D, and the terms of the query t1 and t2. Each document dj is represented by a vector of the terms linked to the weights wij with $i \in \{1, 2\}$ and $j \in \{1, 2, 3, 4\}$ where:

- $d_1 = (w_{11}, w_{21})$
- $d_2 = (w_{12}, w_{22})$
- $d_3 = (w_{13}, w_{23})$
- $d_4 = (w_{14}, w_{24})$

The table below represents the Boolean information retrieval model (Table 1):

When searching for the terms t1 and t2 in the documents di, and according to the Boolean expression used in the query, we choose the documents that have the largest number of 1 in the table. The implementation of this model in practice is easy, but the results are not satisfactory, because the frequency and the proximity of the terms in a document are not taken into account composed queries by complex logical expressions of the terms, are used by search engines to extract certain relevant documents and exclude others. Hence, the appearance of the vector space model takes into account the frequency of repetition of terms in the search for information.

Table 1. Boolean information retrieval model

	Terms of the request		Request similarity	
	t1	*t2*	*t1 OR t2*	*t1 AND t2*
d1	1	1	1	1
d2	1	0	1	0
d3	0	1	1	0
d4	0	0	0	0

3.2 The Vector Space Model for Information Retrieval

A text is a grouping of terms, it is likened to a vector in a vector space [5], we measure the degree of importance of each term in the document by a real number, and a frequency representing its repetition in the document. A document responding to a query is the one whose vector is close to that of a query, so the measure of relevance is the calculation of the cosine of the two vectors.

For a vector space V: $V = \{t_1, t_2, t_3, \ldots, t_n\}$ representing n terms ti, we consider a document D such:
D = (a1, a2, a3,..., an), with ai represents the weight of ti in D.
Let the query Q: Q = (b1, b2, b3,..., bn), with bi the weight of ti in Q.
We define the score or the reminder "R", also noted as similarity, by a function such:
R (D, Q) = Sim (D, Q).

The figure below schematizes the vector space model by a representative matrix.

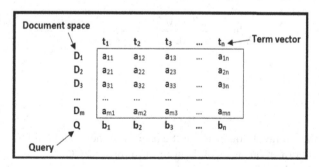

Fig. 2. Matrix representation of the elements of the vector space model

Some similarity or distance factors can be used, such as the scalar product of two vectors or the cosine, as well as the Jaccard index or the Sørensen-Dice index 1945, also called coefficient of Dice:

- The scalar product of document D and query Q is defined as:

$$Sim(D, Q) = D.Q = \sum_i (a_i * b_i) \tag{1}$$

- Similarly, we define Cosinus

$$Sim(D, Q) = \frac{\sum_i (a_i * b_i)}{\sqrt{\sum_i a_i^2 * \sum_i b_i^2}} \tag{2}$$

- For the Sørensen-Dice index, we have:

$$Sim(D, Q) = \frac{2 * \sum_i (a_i * b_i)}{\sum_i a_i^2 + \sum_i b_i^2} \text{Sim}(D, Q) = \frac{2 * \sum_i (a_i * b_i)}{\sum_i a_i^2 + \sum_i b_i^2} \text{Sim}(D, Q) = \frac{2 * \sum_i (a_i * b_i)}{\sum_i a_i^2 + \sum_i b_i^2} \tag{3}$$

- And for the Jaccard index we have:

$$Sim(D, Q) = \frac{\sum_i (a_i * b_i)}{\sum_i a_i^2 + \sum_i b_i^2 - \sum_i (a_i * b_i)} \tag{4}$$

In the matrix of Fig. 2, we compare the similarity (the cosine of the nail between two vectors) between two documents to classify them according to their terms ti, or else calculate the score of correspondence of the terms between the documents Di and the query Q.

Another way to evaluate the degree of relevance is to calculate a relevance score for each document Di relating to the query Q. The Okapi method proposed in 1976 by Robertson [6] and its variants are widely used weighting techniques in the search for information.

The calculation of the Okapi relevance score gives more precision than the cosine for short queries.

In this context, we define this score as follows:

$$okapi(Dj, Q) = \sum_{t_i \in Q, D_j} \ln \frac{N - df_i + 0.5}{df_i + 0.5} * \frac{(k_1 + 1)f_{ij}}{k_1(1 - b + b\frac{dl_j}{avdl} + f_{ij})} * \frac{(k_2 + 1)f_{iq}}{k_2 + f_{iq}} \tag{5}$$

With:

- t_i a term.
- f_{ij} is the number of raw frequencies of the term ti in the document Dj
- f_{iq} is the number of frequencies of the term ti in the query Q
- N is the total number of documents in D.
- df_i is the number of documents containing ti
- dl_j is the byte language of the Dj document
- avdl is the average length of documents in the D collection

With the parameters: k1 varies between 1 and 2, k2 between 1 and 1000 and b often equal to 0.75.

In this vector model, considered frequent words and empty like "of, and,... ", depending on the language of the dictionary used, must be removed from the terms to have relevant results that come close to the query.

Another disadvantage is that it may ignore the infrequent terms in the document, which have a considerable weight in the search as "terrorism...", however the model allows a good indexation of documents that respond in whole or in part to the request.

This model is relatively simple to implement, its complexity is linear like the Boolean model. But because of the size of the documents on the net and the size of the request in terms searched, it becomes expensive for the search engines. One possible solution is to process the documents in advance and to index them according to the terms in directories. Robots or autonomous processes do these operations automatically.

3.3 The Probabilistic Model of Finding Relevant Information

In this model, a document D can be useful to respond to a query Q of the user, the relevance of D with respect to a request is independent of the other documents, and the probability of relevance is defined as a ranking factor in the search for information.

Robertson (1977) in his article [7] presented the principle of probabilistic ranking (PRP), in which the relevance score of a document D depends on a query Q such as:

$$RSV(D, Q) = \frac{P(R|D)}{P(NR|D)} \tag{6}$$

With R: the set of documents relevant to the query Q and NR represent the irrelevant one.

P (R | D): probability that D is in R, and P (NR | D): probability that D is in NR.

To estimate the calculation of P (R | D), we use the Bayes theorem; it gives the probability of relevance before and after observing the document D.

As defined at the beginning, each document is described by the presence or absence of index terms, it is represented by a binary vector $= (x_1, x_2, x_3, \ldots, x_N)$, with:

$$x_i = \begin{cases} 0 & \textit{if the index term is missing} \\ 1 & \textit{if the index term is present} \end{cases} \tag{7}$$

So for this document, we can associate two possible situations W1 and W2 such that:

- $w_1 =$ the document is relevant
- $w_2 =$ the document is not relevant.

We try to calculate the probability of x by observing it randomly:

$$P(x) = P(x/w_1)P(w_1) + P(x/w_2)P(w_2) \qquad (8)$$

With:

- P $(x/w1)$ and P (x/w_2) represent the probability of relevance of x probability of irrelevance of x, respectively.
- P $(w1)$ and P $(w2)$ represent the probability of relevant documents and irrelevant documents, respectively.

The value P (x) allows us to quantitatively measure the density of x with respect to relevant documents or not.

So, for a good decision making, we use the following rule:
If P $(x/w1)$ > P $(x/w2)$) then x is relevant else x is not relevant end if.
Taking into account this rule, the relevant documents x are randomly extracted, and sorted in descending order of P (x).

The retrieval of documents is related to the minimum probability threshold imposed by the user and the cost of filing operations of the used application. It is often essential to know the size to better manage the path traveled during research based on heuristics. Hence, using trees, also called dendrograms, for the classification of data or the introduction of Clustering algorithms by mixing probability densities such as Dempster's EM (Expectation and Maximization) algorithm [8] becomes important.

In addition, there is an algorithm that uses the partitioning of the grid data space, like the STING algorithm [9] "Statistical Information Grid Approach to Spatial Data Mining", which builds a hierarchy of grid: a cluster is composed of dense and connected cells. For each cell of the current level, STING calculates the probability interval for which sub-cells are dense, and distances the others; likewise, he repeats the same treatment until he reaches the lowest level.

To begin this algorithm, we fix a database and we should answer a request of type selection based on the parameters of a cell "point" (Fig. 3).

This algorithm is flexible; it requires a single pass on the data and at each level of our hierarchy, and user can easily insert a new object in the appropriate cells.

The complexity of this algorithm depends on the number of cells k, it can be of the order of O (k), and number k is smaller than the number of data since the algorithm discards the absurd data.

In summary, the strong point of the probabilistic model is that it presents a solid theoretical basis, the optimal ranking of the research results is justified in the theory, but in practice, it is difficult to accurately calculate the probability, and the frequency of terms is ignored. As a result, the results cannot be convincing and require human intervention.

STING algorithm;
    Input: a database containing "cell" object parameters, and a SQL type query.
    Output: A set of cells representing a layer or grid.
Begin
    1. Determine a layer with which one begins.
    2. For each cell in this layer, we calculate the confidence interval (or margin estimate) of the probability that this cell will have the answer to the database query asked at the beginning.
    3. Depending on the value of the interval calculated above, we mark the cell as appropriate or inappropriate.
    4. If this layer is the bottom layer, go to step 6; otherwise, go to step 5.
    5. Descend the hierarchy structure by one level. Step 2 is followed for those cells that form the appropriate cells of the higher level layer.
    6. If the specifications of the query are verified, go to step 8; otherwise, go to step
    7. Look for the fall of these data in the appropriate cells and make a further transformation. Return the result that meets the requirement of the query. Go to step 9.
    8. Find the appropriate cell regions. Return those regions that meet the requirement of the query. Go to step 9.
    9. Stop.
End.

Fig. 3. Algorithm STING

3.4 The Statistical Language Model for Information Retrieval

This type of model [10] is based on probability and statistics to model the capture, position and probability of frequency of words in a language.

For a corpus, we try to classify the documents taking into account the criteria like the author, the date of publication, the themes etc…

The probability of the search query depends on the language model of each document, the processing of the spoken language and the frequency of the words as well as the logical rules of the language [11]. Statistics allow the design, configuration and evaluation of the information search model in the corpus. The relevance of the query results is related to statistical processing during indexing, spelling and grammar correction, as well as short automatic recognition of writing, speech and data translation. These push search engines to use more text volume in directories powered by robot processes and robust applications for machine learning and speech recognition languages.

4   Results and Discussion of Research Model Evaluations

Given the number of information retrieval models and constraints related to existing data types on the web, it is necessary to evaluate the performance, quality and reliability of the results retrieved by the model. The "recall" parameter [12] is defined by:

$$recall = \frac{number\ of\ relevant\ document\ returned}{number\ of\ relevant\ database\ documents} \tag{9}$$

This parameter is a percentage number indicating the model's ability to retrieve relevant documents that satisfy the user's request. This search is conducted in a database already available.

Another parameter called "precision", measures the ability of the search system to eliminate documents irrelevant to the query, it is defined as follows:

$$precision = \frac{number\ of\ relevant\ documents\ returned}{number\ of\ document\ returned} \tag{10}$$

Good results recovered correspond to a precision equal to 1, these are far from being verified in practice, so a recall equal to 1 means that the system is effective when ranking the relevant documents of the database.

Two other parameters "Silence" and "Noise" measure the opposite case:

$$Silence = 1 - recall \tag{11}$$

$$Noise = 1 - precision \tag{12}$$

The Silence parameter gives the percentage of the relevant document not returned and the Noise parameter indicates the percentage of irrelevant items returned by the search appliance.

For a Web search engine, often we only look at the first 20 pages, then we can calculate the details for the 5, 10, 15, 20, 25 and 30 pages returned. On the other hand, the "recall" parameter is not very interesting since we do not know the number of relevant documents in the search engine's search base. Another problem encountered during the evaluation is that, the relevance of a document is often determined by the user, which renders the evaluation unspecific.

The experimental evaluation on the data below is represented by the curve giving the precision and the recall of the documents (Fig. 4) (Table 2).

Note that from d5, the precision decreases as opposed to the recall, and it is better for the first returned documents that have small values of the callback parameter, this accuracy is bad for large callback values. Thus, the ideal is to choose the first three documents retrieved and to redo a human analysis on the relevance of each.

4.1   Our System and Comparative Analysis of Models

Using an application, made with the C-Sharp language, in a Windows platform, we tried to implement the four models and to do tests on databases from the server "https://archive.

Table 2. Recall and Precision values for a web search

Document	Score	Relevant	Precision	Recall
d1	9.92	Relevant	1.00	0.2
d2	9.77	Non relevant	0.50	0.2
d3	9.76	Relevant	0.67	0.40
d4	9.59	Relevant	0.75	0.60
d5	8.72	Non relevant	0.60	0.60
d6	6.85	Relevant	0.67	0.80
d7	6.51	Relevant	0.57	0.80
d8	4.32	Non relevant	0.63	1
d9	4.16	Non relevant	0.56	1
d10	3.47	Non relevant	0.50	1
d11	2.69	Non relevant	0.45	1
d12	2.04	Non relevant	0.42	1
d13	1.84	Non relevant	0.38	1
d14	1.67	Non relevant	0.36	1
d15	0.07	Non relevant	0.33	1

ics.uci.edu/ml/datasets/Bank+Marketing". The application is a knowledge extraction system from data; it allows processing, classifying and visualizing data. The example below shows the overall visualization of the two classes of credit applicants according to their martial states "single man, married…" according to the set of other attributes (Fig. 5):

A comparative evaluation known as benchmarking has been developed to identify the strengths and weaknesses of each model, see Table 3.

Using Business Intelligence Systems [13], these different information retrieval models are exploited in the search for digital marketing skills aimed at industry, commerce, healthcare and many others.

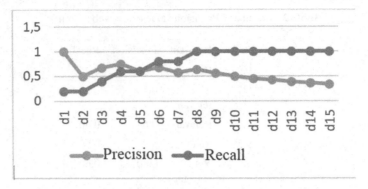

Fig. 4. Representation of precision and recall of documents.

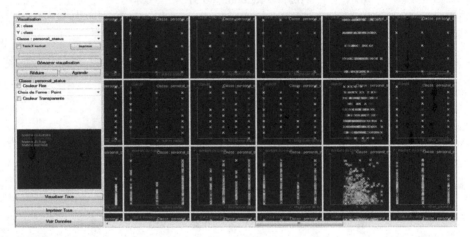

Fig. 5. Global visualization of two classes of credit applicants (single and married)

Rather than being based on intuition and human experience, the strengths of each model are used to find reliable and accurate reference parameters. The benchmark leverages data categorized through these models to enable companies to evaluate and develop their businesses. Manual Data recovery is not easy. While providing their services to users, web applications like Google Analytics are there to crosscheck and classify relevant data on the behavior and needs of users, afterwards these applications use the algorithms of continuous and automatic learning to extract the knowledge and reinforce the results of the models.

Table 3. Comparison of the four information search models.

Model	Strengths	Weaknesses
Boolean Model	• Simple to implement in practice • Linear complexity	• Unsatisfactory results • Ignore frequency and proximity of terms
Vector Space Model	• Simple to implement in practice • Measures the frequency of terms and proximity	• High cost, slow processing • Requires advance indexing of documents
Probabilistic Model	• Strong theoretical basis • Exists in several Clustering algorithms	• Difficulty calculating probability in practice • Frequency of terms ignored • Requires human intervention
Statistical Language Model	• Uses statistics for indexing	• Dependence of languages • Too many parameters to consider

5 Conclusion

This document describes the general process of information retrieval; it examines the main models of search for digital information that are the Boolean model, that of vector space, the probabilistic and statistical language.

By introducing the notion of score for the classification of documents, the constraints and the limits of each model were identified. For the assessment of degree of relevance, the calculation of the Okapi relevance score was introduced; it takes into consideration the number of frequencies of the terms of the request in the returned documents. Like the Boolean model, the vector is simple to realize but the size of the request in term searched for and the number of data to be treated, decrease the performances of the system.

The solution is to pre-prepare the databases by automating the indexing and clustering techniques. Subsequently, the probabilistic model exploits probability as a unit of measure of relevance without forgetting the use of Bayes' theorem for computation.

The STING algorithm can be very interesting in the classification, since it divides the data space into a grid and a layer and then calculates the probability of the cells responding to the user request. Thus, it presents a hierarchical structure of the relevant documents to improve the performance of the search system. In addition, the statistical language model has several factors that make it difficult to implement in practice. In this perspective, some search engines have already begun to exploit artificial intelligence for the recognition of voice and image.

Finally, the evaluation parameters of an information retrieval system have been discussed, but so far it has emerged that human intervention is necessary for decision-making.

References

1. Rijsbergen, B.: Information retrieval, pp. 3–6. Department of Computing Science, University of Glasgow. Published by the Press Syndicate of the University of Cambridge (2004)
2. Liu, B.: Web Data Mining: Exploring Hyperlinks, Contents, and Usage Data, Data-Centric Systems and Applications, pp. 213–214. Springer, Heidelberg (2011). https://doi.org/10.1007/978-3-642-19460-3
3. Ali, M., Zainal, A., Christiano, S., Suryo, I.: Indonesian news classification using Naïve Bayes and two-phase feature selection model. Indones. J. Electr. Eng. Comput. Sci. (IJEECS) 401–408 (2016)
4. Garcia, E.: The Extended Boolean Model (2016). Published at www.minerazzi.com
5. Manwar, B.: A vector space model for automatic indexing. Indian J. Comput. Sci. Eng. (IJCSE) 223–228 (2012)
6. Stephen, E., Robertson, K.S.: Relevance weighting of search terms. J. Am. Soc. Inf. Sci. **27**(3), 129–146 (1976)
7. Stephen, E.: The probability ranking principle in IR. J. Document. **33**(4), 294–303 (1977)
8. Dellaert, F.: The expectation maximization algorithm. College of Computing, Georgia Institute of Technology. Report number: GIT-GVU-02-20, pp. 1–7 (2002)
9. Saurav Kumar, S.: Statistical Information Grid. Department of Computer Science & Engineering Dual degree 4th year, pp. 17–18 (1997)
10. Alain, P., Boëffard, O.: Evaluation des Modèles de Langage n-gramme et n=m-multigramme. Dourdan J. TALN 1–4 (2005)
11. Iftakher, M.D., et al.: An investigative design of optimum stochastic language model for Bangla autocomplete. Indones. J. Electr. Eng. Computer Sci. (IJEECS) **13**(2), 671–676 (2019)
12. Nakache, D., Metais, E.: Evaluation: nouvelle approche avec juges. In: Conference: Actes du XXIIIème Congrès INFORSID, Grenoble, pp. 2–4 (2005)
13. Julyeta, P., Irene, R.: Vertical information system: a case study of civil servant teachers data in Manado city. Indones. J. Electr. Eng. Comput. Sci. **6**(1), 42–49 (2017)

Overview of the Main Recommendation Approaches for the Scientific Articles

Driss El Alaoui[1]([✉]), Jamal Riffi[1], Badraddine Aghoutane[2], Abdelouahed Sabri[1], Ali Yahyaouy[1], and Hamid Tairi[1]

[1] LISAC Laboratory, Department of Informatics, Faculty of Sciences Dhar El Mahraz, Sidi Mohamed Ben Abdellah University, 1796 Fez-Atlas, Fez, Morocco
[2] Team of Processing and Transformation of Information, Polydisciplinary Faculty of Errachidia, Moulay Ismaïl University, 11201 Zitoune Meknes, Morocco
http://fsdmfes.ac.ma/, http://www.fpe.umi.ac.ma

Abstract. With the explosive growth of the data that being produced and published on the Web every day by the scientific community, it is becoming difficult for researchers to find the most appropriate scientific articles for their needs. For alleviate such information overload, the recommender systems plays a key role in allowing users to access what interests them as quickly as possible.

This is why we are going to focus on finding the best approach that can be supported in scientific articles recommendation systems, to be able to guide the researchers in finding articles in an effective way.

This paper presents a comparison between the main Recommender Systems techniques that aims to recommend to users the relevant articles, according to preferences and habits. Preference and relevance are subjective and are generally derived from items previously consumed by users. We chose here three most used techniques; first, collaborative filtering, then content-based systems and finally a Hybrid recommendation. To evaluate the recommendation we have used classical measures in search of information: precision at top k, recall at top k, NDCG@k and novelty of the recommended items.

Keywords: Recommender systems · Scientific articles · Collaborative filtering · Content-based · Hybrid methods

1 Introduction

Recommender systems are first coined by Tapestry [1], the system is a subcategory of the filtering system, 'it is a way to understand the interests and behavior patterns of users by extrapolating their preferences and using them to predict which products follow those patterns, then present them to them in the form of recommendations, so that the individual feels that the system understands them, which builds the client's confidence in the service providers. Many areas rely heavily on recommendation systems to find the perfect link between the product/service and the ideal customer [2–4]. Table 1 provides examples of services or products recommended by recommendation systems:

© Springer Nature Switzerland AG 2021
M. Fakir et al. (Eds.): CBI 2021, LNBIP 416, pp. 107–118, 2021.
https://doi.org/10.1007/978-3-030-76508-8_9

Table 1. Examples of companies using the recommendation system

Systems	Products
Google Search	Advertisements
Netflix, YouTube	Streaming Video
GroupLens; Google News	News
MovieLens; IMDb	Watching movies
last.fm	Listening to music
Amazon.com	Distance selling
Facebook	Friends, Advertisements
Tripadvisor	Travel products

These systems are based on two types of information:

- Features: distinctive attributes of the articles (keywords, categories, etc.) and users (preferences, profiles, etc.).
- User-item interactions: such as ratings, likes, comments, number of purchases, etc.

From this factor, we can distinguish several classifications of the recommendation system, but we will rely on the classic classification (see Fig. 1):

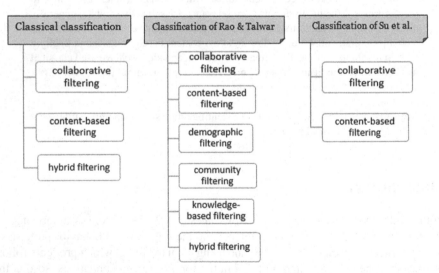

Fig. 1. Main classification of recommendation systems.

This document is organised as follows: Sect. 2 presents a general overview of recommendation systems, their categorization, as well as their advantages and disadvantages of each approach; Sect. 3 describes the recommendation system workflow, explains the evaluation metrics applied to the models, and why they are chosen. The construction of the recommendation systems and the different parameters that we tested. The results of these experiments, their evaluation and analysis are the subject of the Sect. 4. While the last section focuses on conclusions and future work.

2 Recommender Systems Approaches

2.1 Collaborative Filtering

The term Collaborative Filtering (CF) [5] was first coined by David Goldberg et al. [6] in 1992 in 1992 to describe an email filtering system called "Tapestry". Which was an electronic messaging system that allowed users to rate "good" or "bad" messages or to attach text annotations to those messages.

This approach was the first technology used in recommender systems, and it is still simpler and more efficient. The collaborative filtering process is divided into three stages: first, collecting user information, then using it to create a matrix for calculating user associations, and finally to provide high reliability recommendations. Some of the larger companies in their field rely heavily on systems of this kind, including YouTube, Netflix, and Spotify. Collaborative systems are broadly classified into two subcategories: Memory-based and Model-based.

Memory-Based. Memory-based algorithms use the entire database of user ratings to make predictions: these algorithms use user-to-user and item-to-item correlations or similarity. Memory-Based CF mechanism is used in many commercial systems, as it is easy to implement and is effective [7, 8]. To approximate users or items we can use statistical techniques as Pearson Correlation, Cosine Similarity, Euclidean Distance, Jaccard measure (Table 2).

Table 2. The similarity measures

Similarity measure	Formula	
Pearson correlation	$\text{sim}(A, B) = \dfrac{\sum_j \left(v_{A,j} - v_{\overline{A},j}\right)\left(v_{B,j} - v_{\overline{B},j}\right)}{\sqrt{\sum_j \left(v_{A,j} - v_{\overline{A},j}\right)^2 \left(v_{B,j} - v_{\overline{B},j}\right)^2}}$	(1)
Cosine similarity	$\text{sim}(A, B) = \sum\limits_{j=1}^{n} \dfrac{v_{A,j}}{\sqrt{\sum_{j=1}^{n} v_{A,j}^2}} \times \dfrac{v_{B,j}}{\sqrt{\sum_{j=1}^{n} v_{B,j}^2}}$	(2)
Euclidean Distance	$\text{sim}(A, B) = \sqrt{\sum_{j=1}^{n} (v_{A,j} - v_{B,j})^2}$	(3)
Jaccard measure	$\text{sim}(A,B) = \dfrac{|I_A \cap I_B|}{|I_A \cup I_B|}$	(4)

Model-Based. Model-based algorithms build statistical models of user/item rating models to provide automatic rating predictions.

Algorithms in this category take a probabilistic approach and assume the collaborative filtering process as calculating the expected value of a user's forecast, given their assessments of other elements. The process of model building is carried out by different machine learning algorithms such as Bayesian, Neural Networks, Clustering Models, Networks and Matrix Factorization based algorithm.

It should be noted that we have used this second approach to set up a collaborative filtering recommendation system (Fig. 2).

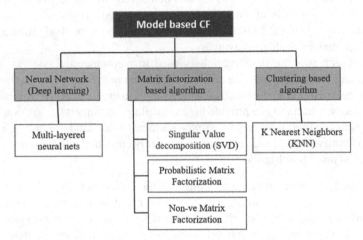

Fig. 2. Types of model based collaborative filtering approaches

2.2 Content-Based Filtering

This second approach makes use of the comparison between items and the past preferences of a particular user [9]. That is, the items, which have similar properties to the ones that the user liked or checked previously, are likely to be recommended.

Here we are using a technique of analysis and information retrieval called TF-IDF [10, 11]. This technique converts unstructured text into a vector structure, where each word is represented by a position in the vector, and the value measures the relevance of a given word to an article. We calculate the similarity between the articles, because all the elements will be represented in the same vector space model. For a term t in a document d, the weight $\mathbf{W_{t,d}}$ of term t in document d is given by:

$$W_{t,d} = TF_{t,d}\log\frac{N}{DF_t} \tag{5}$$

It is a formula in which the two values TF (Term Frequency) and IDF (Inverse Document Frequency, i.e. the number of documents containing the term) are multiplied between them. The result is the relative frequency of terms (or "term weight") in a document compared to all other documents that also contain the keyword in question during analysis. Before you can perform the TF-IDF analysis, the two mentioned factors must first be determined.

2.3 Hybrid Recommendation Systems

The both recommendation techniques already discussed in detail have strengths and weaknesses summarized in this Table 3:

Table 3. Strengths & weaknesses of each model

	Strengths	Weaknesses
Collaborative filtering (CF)	– Recommend a content without the need to understand the meaning or semantics of the content itself	– **Scalability:** millions of users, products and content – **Cold Start:** The launch of a service suffer at the beginning of the lack of users and information about them – **Sparsity (Rarity):** The number of products is huge, and the most active users will have noticed a very small subset of the entire database
Content-based (CB)	– This approach doesn't require a large number of users to make recommendations	– Requires descriptive content for the acquisition of subjective and qualitative characteristics

This led to the emergence of a so-called hybrid approach, which combines two or more several recommendation approaches. These combinations make it possible to benefit from the advantages of the approaches used, to overcome their disadvantages and to propose recommendations that are more relevant. For example, item-level cold start of collaborative approaches can benefit from the benefits of content-based approaches. Indeed, in a content-based approach, only the description of the item is used in the recommendation, so there is no need for the item to be noted by a number of users before it can be recommended. As is the case in collaborative approaches.

Robin D. Burke [12] has identified the different ways of hybridizing recommendation approaches, described in Table 4 below:

Table 4. Hybrid methods

Hybrid method	Description
Weighted	The results of multiple recommendation approaches are combined to produce a single recommendation [13]
Switching	The system switches between recommendation approaches depending on the current situation [14]
Mixed	Recommendations from several recommenders are mixed at the same time [15]
Feature combination	Features of different recommendation data sources are thrown together into a single recommendation algorithm [16]
Cascade	One recommender refines the results made by another [17]
Feature augmentation	Output from one approach is used as an input feature to another [18]
Meta-level	The model learned by one recommender is used as input to another [19]

3   Recommendation System Workflow and Evaluation

The diagram below (Fig. 3) represents the workflow of the recommender systems. First, the recommendation algorithms are provided by a training set, to produce a model, which is used to create a list of recommendations, containing items rated and ordered according to their importance to the user. For model evaluation, a test set is furnish to the learned model where scores are generated for each user-article pair. This ordered list of recommended articles with ratings are then used as an input to the evaluation algorithm to produce evaluation results.

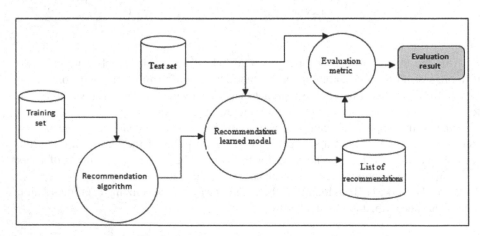

Fig. 3. Recommendation system evaluation

In this study, the recommenders are based on implicit data (implicit feedback), because we make an estimate of the preferences from some metrics (views, clicks, likes, comments…), what drives us to look into rank metrics, such as those used in information retrieval settings. Precision and recall are two metrics for evaluating and judging the relevance of each of the recommendation system models used in this work. It should be noted that these metrics used to evaluate models with a binary output (relevant and irrelevant elements). In our context, we are most interested in recommending top-N articles to the user. Therefore, it is more judicious to compute precision and recall metrics in the first N articles instead of all the articles.

It must be remembered that the Precision at k [20] is the proportion of recommended items in the top-k set that are relevant and the Recall at k is the proportion of relevant items found in the top-k recommendations. Mathematically Precision@k and Recall@k are defined as follows:

$$Precision@k = \frac{(Recommended\ items\ @\ k\ that\ are\ relevant)}{(Recommended\ items\ @\ k)} \tag{6}$$

$$Recall@k = \frac{(Recommended\ items\ @\ k\ that\ are\ relevant)}{(Total\ number\ of\ relevant\ items)} \tag{7}$$

Our recommendation system returns a result as an ordered list of articles; for this, it is desirable to take into account the order. Normalized Cumulative Actualized Gain (NDCG) [21] is a popular technique for measuring the quality of search results in the form of an ordered list. NDCG at position n is defined as follows:

$$NDCG@n = \frac{1}{IDCG} \times \sum_{i=1}^{n} \frac{2^{r_i} - 1}{\log_2(i+1)} \tag{8}$$

Where r_i is the relevance rating of document at position i. IDCG is set so that the perfect ranking has a NDCG value of 1. In our problem, r_i is 1 if the document is recommended correctly.

Accuracy is important, however in most cases, focusing only on accuracy generates less satisfactory recommendations. This leads us to use the non-accuracy metrics as novelty of the recommended items, which determines how unknown recommended articles are to a user. The novelty for a given user u defined the ratio of unknown articles in the list of top–N recommended articles [22, 23]:

$$Novelty(u) = \frac{\sum_{t \in R}(1 - knows(u, t))}{|R|} \tag{9}$$

Where R is the set of top–N recommended articles, and knows(u, i) is a binary function that returns 1 if user u already knows article i, and 0 otherwise. It should be noted that: Higher novelty values represents that less popular articles are being recommended, thus less well-known articles are likely being surfaced for users.

4 Experiments

4.1 Dataset

In this work, we used two datasets:

The first file contains information about articles shared between the Users. Each article has its sharing date (timestamp), the original url, title, content in plain text, and information about the user who shared the article (author).

The second file contains logs of user interactions on shared articles. User-article interactions are represented by the eventType column, which values like:

- VIEW: The user has opened the article.
- LIKE: The user has liked the article.
- COMMENT CREATED: The user created a comment in the article.
- FOLLOW: The user chose to be notified on any new comment in the article.
- BOOKMARK: The user has bookmarked the article for easy return in the future.

4.2 Comparing the Methods

Throughout these experiments, we treated 1139 users and 3047 articles.

To measure the performance of each models, we examine the top 5 and the top 10 recommended articles returned. We first find as Recall the results shown in Fig. 4:

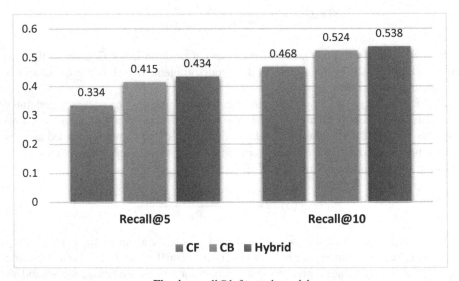

Fig. 4. recall@k for each model

Through this recall graph: we see that the hybrid model has a better performance in the both cases: (recall@5 = 43% and recall@10 = 54%), followed by CB than CF. and we can see that there is some increase in the recall value, for all approaches when the value of k goes from 5 to 10.

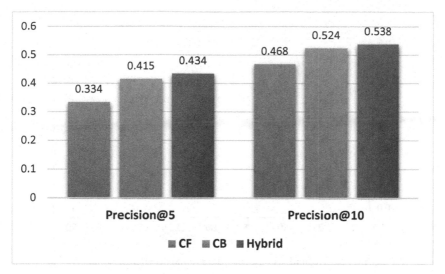

Fig. 5. precision@k for each model

Figure 5 show that the hybrid approach gives a more positive result in terms of precision compared to other approaches (CB and CF). Moreover, we notice that this precision decreases when we increase the k value for all the approaches studied.

In general, a perfect recommendation system will provide answers whose precision and recall are equal to 1. In reality, the recommendation algorithms are more or less precise and more or less relevant. It is possible to obtain a very precise system, but poorly performing. Similarly, an algorithm with a strong reminder, but low precision will provide many erroneous the relevant ones: it will therefore be difficult to exploit. Thus, in borderline cases, a recommender system that returns all of the documents in its base will have a reminder of 1 but poor precision; while a recommendation system that returns only the user's request will have a precision of 1 for a very weak reminder. The value of a recommender system is not reduced to a good score in precision or recall.

The Fig. 6, represents a curve that bring out the difference between the three approaches (content-based: CB, collaborative filtering: CF and hybrid). Each point on the NDCG@k curves corresponds to the rank position, ranging from top-1 (left) to top-200 (right). It can be seen that if k < 60, the collaborative filtering system is better than the other systems (content-based and hybrid, respectively). contrariwise if (k ≥ 60) the curves intersect and the hybrid system starts to give better results than the other systems depending on the results obtained, we can deduce that Hybrid system is the most accurate system compared to other systems, on the other hand if we take into account the order of the relevant items returned, we find that the CF system is the most effective When determining a small value of k; which is the number of ranks to be taken into account, but when k = 60: Hybrid (resp CB) approach Outperforms CF System with more ranks. However, since the users care more about recommendations at low ranks, we can say that the CF model gives better results.

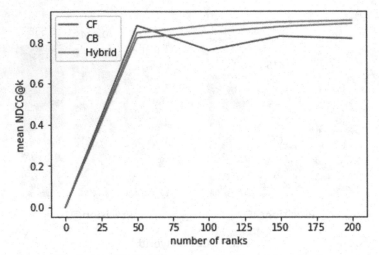

Fig. 6. Mean nDCG@k according to the number of ranks

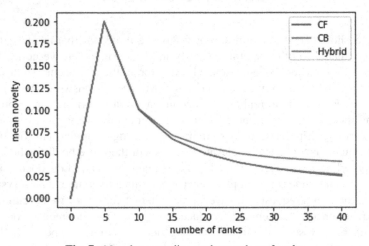

Fig. 7. Novelty according to the number of ranks

From the Fig. 7, we observe that the three curves arrive at the top (novelty = 0.200) when k = 5, and that the curves go down if we increase the value of k. To see things more clearly we take the Table. We find that the two approaches (Cf and CB) give the same results and that the hybrid approach has the highest values for all the points k. this means that the hybrid approach allows to recommend to users the least popular items, and that unknown items for a user are ranked at the top of the recommendation list (Table 5).

Table 5. Novelty according to the number of ranks for the three recommendation approaches

	Novelty@5	Novelty@10	Novelty@15	Novelty@20
Collaborative-filtering (CF)	0.200	0.100	0.067	0.050
Content based (CB)	0.200	0.100	0.067	0.050
Hybrid	0.200	0.101	0.071	0.058

5 Conclusion

In this paper, we wanted to make a comparative study between the main recommendation approaches used to suggest scientific articles. Through the experiments, we have found that: despite the distinction of the hybrid model in terms of recall and precision, the relevance of a recommendation system cannot be limited to good precision or a good reminder, which has pushed us to measure the nDCG@k to take into account the order in information search and we found in this case that the CF model gives better results.

Since accuracy are not enough to evaluate the performance of a recommender system, we used other recommendation measure technic; such as Novelty, which determines how unknown recommended articles are to a user, and we find that the hybrid approach gives slightly better results than other approaches, this means that the hybrid approach allows to recommend to users the new items. Finally, it should be mentioned that it is impossible to conduct an exhaustive study of recommendation algorithms because this field is constantly evolving. The work presented here is only an overview of some techniques that exist in this area of recommendation. However this doesn't preclude making further improvements and testing other algorithms of articles recommendation, to be able to make a more complete and reliable study possible, in order to combine the algorithms giving the best results and to find a more efficient hybrid solution.

References

1. Wu, C.Y., Ahmed, A., Beutel, A., Smola, A.J., Jing, H.: Recurrent recommender networks. In: Proceedings of the Tenth ACM International Conference on Web Search and Data Mining (2017)
2. Shakirova, E.: Collaborative filtering for music recommender system. In: IEEE Conference of Russian Young Researchers in Electrical and Electronic Engineering (2017)
3. Prateek, S., Yash, S., Pranit, A.: Movie Recommender System. Search Engine Architecture, NYU Courant (2017)
4. Bancu, C., Dagadita, M., Dascalu, M., Dobre, C., Trausan-Matu, S., Florea, A.M.: ARSYS - article recommender system. In: 14th International Symposium on Symbolic and Numeric Algorithms for Scientific Computing, SYNASC (2012)
5. Konstan, J.A., Miller, B.N., Maltz, D., Herlocker, J.L., Gordon, L.R., Riedl, J.: Applying collaborative filtering to Usenet news. Commun. ACM **40**(3), 77–87 (1997)
6. Goldberg, D., Nichols, D., Oki, B.M., Terry, D.: Using collaborative filtering to weave an information tapestry. Commun. ACM **35**(12), 61–70 (1992)
7. Ricci, F., Rokach, L., Shapira, B. (eds.): Recommender Systems Handbook. Springer, Boston (2015). https://doi.org/10.1007/978-1-4899-7637-6

8. Badrul, S., George, K., Joseph, K., John, R.: Item-based collaborative filtering recommendation. Published in the Proceedings of the 10th International Conference on World Wide Web, 15 May 2001 (2001)
9. Pazzani, M.J., Daniel, B.: Content-Based Recommendation Systems (2007)
10. Jones, K.S.: A statistical interpretation of term specificity and its application in retrieval. J. Document. **28**(1), 11–21 (1972)
11. Salton, G., Fox, E., Wu, W.H.: Extended Boolean information retrieval. Commun. ACM **26**(11), 1022–1036 (1983)
12. Burke, R.: Hybrid web recommender systems. In: Brusilovsky, P., Kobsa, A., Nejdl, W. (eds.) The Adaptive Web. LNCS, vol. 4321, pp. 377–408. Springer, Heidelberg (2007). https://doi.org/10.1007/978-3-540-72079-9_12
13. Mobasher, B., Jin, X., Zhou, Y.: Semantically enhanced collaborative filtering on the web. In: Berendt, B., Hotho, A., Mladenič, D., van Someren, M., Spiliopoulou, M., Stumme, G. (eds.) EWMF 2003. LNCS (LNAI), vol. 3209, pp. 57–76. Springer, Heidelberg (2004). https://doi.org/10.1007/978-3-540-30123-3_4
14. Billsus, D., Pazzani, M.J.: User modeling for adaptive news access. User Model. User-Adapt. Interact. **10**(2–3), 147–180 (2000)
15. Smyth, B., Cotter, P.: A personalised TV listings service for the digital TV age. Knowl.-Based Syst. **13**(2–3), 53–59 (2000)
16. Bas, C., Hirsh, H., Cohen, W.: Recommendation as classification: using social and content-based information in recommendation. In: Proceedings of the Fifteenth National Conference on Artificial Intelligence, pp. 714–720. AAAI Press (1998)
17. Burke, R.: Hybrid recommender systems: survey and experiments. User Model. User-Adapt. Interact. **12**(4), 331–370 (2002)
18. Melville, P., Mooney, R.J., Nagarajan, R.: Content-boosted collaborative filtering for improved recommendations. In: Eighteenth National Conference on Artificial Intelligence, pp. 187–192 (2002)
19. Pazzani, M.J.: A framework for collaborative, content-based and demographic filtering. Artif. Intell. Rev. **13**(5–6), 393–408 (1999)
20. Schütze, H., Manning, C.D., Raghavan, P.: Introduction to Information Retrieval (2008)
21. Kalervo, J., Jaana, K.: Cumulated gain-based evaluation of IR techniques. ACM Trans. Inf. Syst. **20**(4), 422–446 (2002)
22. Konstan, J., McNee, S., Ziegler, C., Torres, R., Kapoor, N.: Lessons on applying automated recommender systems to information-seeking tasks. In: AAAI (2006)
23. Jones, N., Pu, P.: User technology adoption issues in recommender systems. In: Networking and Electronic Conference (2007)

Online Students' Classification Based on the Formal Concepts Analysis and Multiple Choice Questions

Moulay Amzil(✉) and Ahmed El Ghazi

TIAD Laboratory, Department of Computer Sciences, Faculty of Sciences and Techniques,
Sultan Moulay Slimane University, Beni Mellal, Morocco
moulaymasterid@gmail.com, hmadgm@yahoo.fr

Abstract. Students orientation in the university is an important research area. Actually, students fill out their choice manually except some specialties in which there is a preselection. These procedures are very classics, they based on the student's marks to calculate a general average that puts the student above or below a selection bar. Several problems arise from these procedures, the major problem is that a majority of students are misguided. As the solution of this problem, we purpose, in this paper, a new approach for automatic student's orientation witch incorporate the Formal Concept Analysis (FCA) techniques. On one hand, we have students as objects and the online proposed questions are the attributes, on the other hand we have the specialties as objects and the questions as attributes. These objects and attributes help us to build a Concept Lattice Student (CLS) and Concept Lattice Specialty (CLSp). Then after we introduce two algorithms that explore the two concepts lattices and extract the pair (student, specialty) which is our classification result.

Keywords: Students' classification · FCA · Concept lattice

1 Introduction

In the recent years, the internet has replaced the human being in many areas such as e-learning, marketing and economics. In this way, many decisions had attributed to algorithms. Very fast, error rate recognized a priori and can make very complex decisions, these algorithms are growing more and more these ten last years [1]. The student's classification by specialties can be applied in several areas including online recruiting, recommender systems, relational learning, visualization and emotional analysis [2, 3].

In this direction, the choice of the suitable specialty is a fundamental decision for students. Afterward, it determines their future and it helps universities to better distribute and benefits from competences. In summary, our approach is based on the online multiple choice questions. The student is invited to answer these questions, then a binary matrix (1 if the answer is correct and 0 if not) is constructed as following: in lines we put the code of students and in column we put the questions. Whereas, a second binary matrix is constructed between the specialties and questions, these two matrices allow to construct

© Springer Nature Switzerland AG 2021
M. Fakir et al. (Eds.): CBI 2021, LNBIP 416, pp. 119–129, 2021.
https://doi.org/10.1007/978-3-030-76508-8_10

two Galois lattice [1, 4]. Thus, our algorithm explores the main Lattice (ML which is the students lattice) obtained from the first matrix to find the classes of students (relationship of {student, questions}). Then, a second part of this algorithm has to combine between classes of ML and classes of Second Lattice (SL) to obtain the final result as {students, attributed specialty}.

The remainder of the paper is outlined as follows: In Sect. 2, we describe the related works in students' classification area. Then, in Sect. 3 we give the main principles of Formal Concept analysis (FCA). Section 4 explains the proposed work of students' classification and the proposed algorithms. The descriptions of these data sets, the experimental results and their analyses are given in Sect. 4. Finally, the conclusions and future works are given in Sect. 5.

2   Related Works

FCA is a topic that has been treated in several recent works. F. Wang presents it in his paper [5] as a method for measuring the existing similarity between the FCA relations in text context. In his research paper [6], J. Muangprathub presents an important work in which an algorithm for detecting document plagiarism similarity is used. FCA can be applied also in technology innovation. In [7] Sung Eun Kwon presents a new approach to identify the mobile application.

There are a several researchers who have focused on students' classification. One of them is Nazlia in [8], this work presents an automatic analysis of exam questions according to bloom's taxonomy. In resume, this paper uses a data mining task to classify the proposed questions into two categories: knowledge and analysis. It uses a NLTK tagger for segmenting the questions text. Then, an algorithm is implemented for attributing the weights for each of categories described below based on text Mining task. This proposed approach has a limit that is when it detects two categories in the same question. In [8], it's noted that the category is attributed based on the matter experts which is not always evident.

Al-Radaideh try a classification model for predicting the suitable study of students in [9]. This paper proposed the use of recommendation system based on decision three to help students to make their study track choice. This system is based on the averages obtained in the last years of studies. Then, a table of rules is make to determine the interval of each of future specialties of the student. This system is too adequate for some schools that uses the preselection like the Sciences and Technics Faculties even in Morocco but not in faculties that uses an exam to direct their students.

Mashael has proposed in his paper [10] a Predicting Students Final GPA Using Decision Trees. This approach is based on in students' grade in previous courses. This classification helps college department to focus more on the particular course to help students to pass it with the best grade. In the end, they obtain a summary of the most important scours in the all years of study. This is very useful work to detect where the students have a problem and try to correct it in the future. But it can be better if they can regroup students and try to correct those who have a bad index in these courses.

Al-Shalabi presents Autocorrecting system in [11] that is capable of auto marking once students submit their answer online. Using stemming techniques and Levenshtein edit operations [11], the student's answer is evaluated with semantic meaning. A pre-processing phase is placed to process text and prepare it to the classification phase. The obtained result in this phase is the canonical form that is compared with the correct answer to obtain the level results. This system is enable to analyze and to cover the different meaning overlaps.

These works present several powerful classification and decision making systems. However, some of this technics is not very adapted for the question auto marking. As powerful as it is, the auto marking system is enable to learn the different way with which we can write an answer. In this sense, the multiple choice questions are adequates to obtain an extract auto marking and true students' classification system.

3   Theoretical Phase of FCA

Formal Concept Analysis (*FCA*): the *FCA* is a mathematical concept that links a set of objects to attributes by a binary relation [1, 5, 11]. A formal context is a triplet $K = (O, A, I)$, where O represents a set of objects, A is a set of items (or attributes) and I is a binary (incidence) relation; (*i.e.,* $I \subseteq O \times A$). Each couple $(o, a) \in I$ expresses that the object $o \in O$ contains the item $a \in A$. Here O is called one-valued context. It is worth linking between the power-sets P(A) and P(O) associated respectively to the set of items A and the set of objects O. This leads us to the definition of a formal concept [12, 13].

Definition 1. (Formal concept) *A pair* $c = (O, A) \in O \times A$ *of mutually corresponding subsets; i.e.,* $O = \psi(A)$ *and* $A = \varphi(O)$, *is called a formal concept, where O is called extent of c and A is called its intent.*

Definition 2. (Partial order formal concepts) *A partial order on formal concepts is defined as:* $\forall$ $c1 = (O1, A1)$ *and* $c2 = (O2, A2)$ *are two formal concepts,* $c1 \leq c2$ *if* $O2 \subseteq O1$, *or equivalently* $A1 \subseteq A2$ [12].

The concepts of a *FCA* lattice are arranged in a hierarchical order, often referred to as a partial order, based on a $\leq$ relation between concepts. A concept $(A1, B1)$ is a sub-concept of a concept $(A2, B2)$ if $A1 \subseteq A2$ or $B2 \subseteq B1$. Correspondingly, $(A2, B2)$ is a super-concept of $(A1, B1$, hence $(A1, B1) \leq (A2, B2)$.

The FCA permits to obtain a sort of hierarchical graph named Galois lattice [1]. The Fig. 1 presents an example of Galois Lattice between a numbers from 1 to 10 (objects) and attributes as peer(e), odd (o), divisible (c), first (p) and perfect square (s) [14]. As we see, we can deduce the content of each lattice by using union and intersection operators on the lattice of intent and extent of parents and child in the trellis [15, 16].

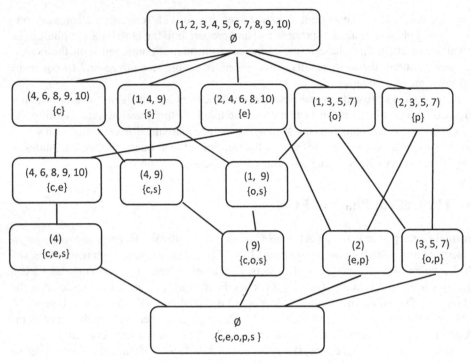

Fig. 1. Example of Galois trellis.

4   The Proposed Work

4.1   Students Classification

Three concepts are involved in our model: the object context, the relation context, and the *CLF*. Suppose that we have a set of Users witch are in this case the students S = {S$_1$, S$_2$, ..., S$_n$}, who answer to Q = {Q$_1$, Q$_2$, ..., Q$_n$} questions in online exam. A question can be part of specialty Sp = {SP$_1$, SP$_2$, ..., SP$_n$} and have a relationship between each other. To generally describe such collaboration data, we define an object context as a set of objects or entities of the same type, *e.g.,* a user context is a set of users and define a relation context as the interactions among objects contexts, *e.g.,* (Students, Questions) relation and (Questions, Specialties) relation.

We use online interface proposed for students to answer to many multiple choice questions, this step allows to obtain a first object context (Students, Questions) called $S_{tudents}$ (Table 1). The second object context is obtained by the administrator; it presents the questions correctly answered, this allows to obtain a second object context called $S_{pecialty}$ (Table 2). Then, we define a relation between the two objects contexts as r_{has_sp}, this make a connection between every student, has passed the exam, and the corresponding specialty.

Our proposed system is given by the Fig. 2 below (Figs. 3 and 4):

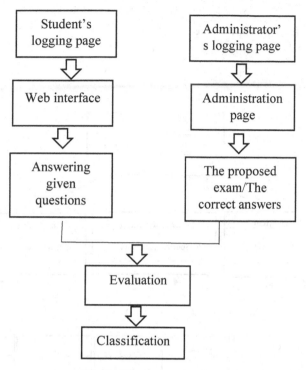

Fig. 2. The proposed system

Table 1. First relation context extracted from online exam

S*tudents*	Q1	Q2	Q3	Q4	Q5	Q6
S1 (user_1)	x	x				
S2 (user_2)	×	×				
S3 (user_3)	x	x				
S4 (user_4)			x	x		
S5 (user_5)			×	×		
S6 (user_6)			x	x		
S7 (user_7)					x	x
S8 (user_8)					×	×
S9 (user_9)					x	x

Our proposed algorithm takes as the input: Students Lattice and Question lattice and as output: List of students with the attributed specialty (AS). It works in the following steps:

Table 2. Second relationship context extracted from online exam.

$S_{pecialty}$	Q1	Q2	Q3	Q4	Q5	Q6
Sp1	x	x				
Sp2			x			
Sp3					x	x

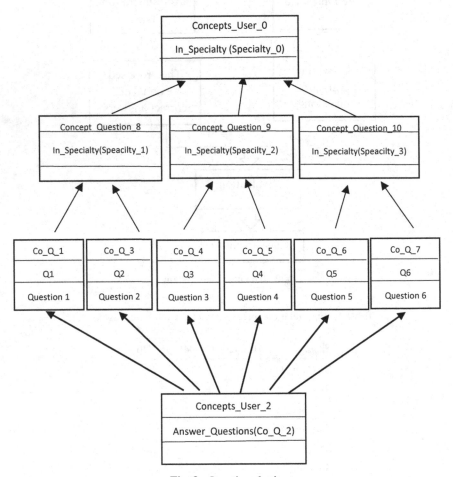

Fig. 3. Questions lattice

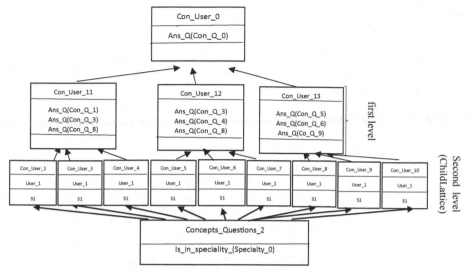

Fig. 4. Specialty lattice

Step 1: Find the number of lattice in first level of main lattice N.

Step 2: Extract the concept Ci of each intent of i in N and the intent of childLattice

Step 3: Go to 'Questions Lattice' and find the intent of Ci;

Step 4: Find the intent of Ci : Spi;

Step 5: Find the child of Ci and their Intent int(ChildOfCi);

Step 6: Use Function FindLabel to get the intents from the questions lattice.

Step 7: Group Students={IntentChildLatticeOf(i), intentChildOfCi, Spi}

These steps are summarized in the following pseudo-code:

Algorithm 1: GroupStudents

```
Input: Strudents_lattice(SL), Specialty_Lattice(SpL);
Output: Groupe {Students, Questions,Specialty);
Begin
Root→ SL
NS→Number of lattice in first lvel of SL;
Foreach i in NS do
                Groupi=  getExtent(i) ;  //  students
                            groups
                            (User_11,User_12,Usre_
                            13)
                Inti=getIntent(i) ;
//(intent of each group)
                For each  cpt in Groupi do
                            NLC=getChildLattice(Cpt);
                            //find     the     child
                            lattice     of     each
                            //Groupi
                        Group_sti= getIntent  (NLC);
                            // Extract  the  intens
                            of each child //lattice
                        FindLabel(Inti,
Questions_Lattice);
End.
```

Algorithm 2: FindLbael(Int, Questions_Lattice)

```
Begin
Foreach i in groupi
    Foreach j in Int do
            Group_sti=Group_sti U getIntent(j);
End.
```

---

5 Experiments Results

5.1 Datasets

We have tested our algorithm in three different datasets based on three specialties as following:

As we see in Table 3: the two first specialties are not linearly separable because there is a Q4 witch is common between two specialties. Inversely, the economics is independent from two other specialties.

Table 3. Distribution of questions through datasets

Biology and chemistry and geology (BCG)	Mathematics, Informatics and physics (MIP)	Economics
Q1	Q4	Q8
Q2	Q5	Q9
Q3	Q6	Q10
Q4	Q7	Q11

Table 4. Number of students by group of questions

{Q1, Q2, Q4}	{Q5, Q6, Q7}	{Q3, Q6, Q7}	{Q9, Q10, Q11}	{Q1, Q2, Q8}
25	25	20	16	14

We take a sample of 100 students and the answers of the eleven questions are distributed as it is given in Table 4:

The figure below presents the visualization of this results (Fig. 5):

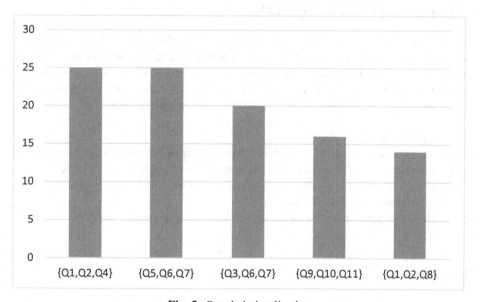

Fig. 5. Results' visualization

As classification obtained by our algorithm we have obtained the groups in Table 5:

The Table 6 illustrates the summary of the obtained results:

The work presented above is an automatic students classification by specialties based on the online questions. This project has been implemented in java language, this presents

Table 5. Classification results

{Q1, Q2, Q4}	{Q5, Q6, Q7}	{Q3, Q6, Q7}	{Q9, Q10, Q11}	{Q1, Q2, Q8}
25	25	20	16	14
BCG	MIP	MIP	Economics	BCG

Table 6. Summary of classification results

Group	{Q1, Q2, Q4, Q8}	{Q3, Q5, Q6, Q7}	{Q9, Q10, Q11}
Number of students	39	45	16
Specialty	BCG	MIP	Economics
percentage	39%	45%	16%

the advantage that it can be integrated easily in web page then in web site of universities. As interesting observation is that the obtained results are satisfying and prove that we can use this approach in real university life.

6   Conclusion and Future Works

In this work, we presented a new approach for students' classification based on Formal Concept Analysis and a new algorithm to combine different lattices and give a result. This technic can help our universities for decision making and to better distribute students between specialties. The proposed algorithm simplifies the exploration of lattices in a record time, this can be first steps to the automatic orientation for students.

- The application of this technic in:
- documents classification based on the semantic information retrieval to obtain a table of relation context. Then, apply the FCA techniques to obtain a Galois lattice.
- online disease diagnosis: take online disease symptoms from sick peoples and form a relation context and Galois lattice.
- Speech files indexation based on the speech recognition system and FCA.

References

1. Menguy, T.: Utilisation d'analyse de concepts formels pour la gestion de variabilité d'un logiciel configuré dynamiquement (2014)
2. Guesmi, S., Trabelsi, C., Latiri, C.: Community detection in multi-relational social networks based on relational concept analysis. Proc. Comput. Sci. **159**, 291–300 (2019). https://doi.org/10.1016/j.procs.2019.09.184
3. Keller, B.J., Eichinger, F., Kretzler, M.: Formal concept analysis of disease similarity. AMIA Jt Summits Transl. Sci. Proc. **2012**, 42–51 (2012)

4. Dau, F., Sertkaya, B.: Formal concept analysis for qualitative data analysis over triple stores. In: De Troyer, O., Bauzer Medeiros, C., Billen, R., Hallot, P., Simitsis, A., Van Mingroot, H. (eds.) Advances in Conceptual Modeling. Recent Developments and New Directions, pp. 45–54. Springer, Heidelberg (2011). https://doi.org/10.1007/978-3-642-24574-9_8
5. Wang, F., Wang, N., Cai, S., Zhang, W.: A similarity measure in formal concept analysis containing general semantic information and domain information. IEEE Access **8**, 75303–75312 (2020). https://doi.org/10.1109/ACCESS.2020.2988689
6. Muangprathub, J., Kajornkasirat, S., Wanichsombat, A.: Document plagiarism detection using a new concept similarity in formal concept analysis. J. Appl. Math. **2021**, 1 (2021). https://doi.org/10.1155/2021/6662984
7. Kwon, S.E., Kim, Y.T., Suh, H., Lee, H.: Identifying the mobile application repertoire based on weighted formal concept analysis. Expert Syst. Appl. **173**, 114678 (2021). https://doi.org/10.1016/j.eswa.2021.114678
8. Omar, N., et al.: Automated analysis of exam questions according to bloom's taxonomy. Proc. – Soc. Behav. Sci. **59**, 297–303 (2012). https://doi.org/10.1016/j.sbspro.2012.09.278
9. Al-Radaideh, Q.A., et al.: A Classification model for predicting the suitable study track for school students. Int. J. Res. Rev. Appl. Sci. **8**(2), 247–252 (2011)
10. Al-Barrak, M.A., Al-Razgan, M.: Predicting students' performance through classification: a case study. J. Theor. Appl. Inf. Technol. **75**, 167–175 (2015)
11. Al-Shalabi, E.F.: An automated system for essay scoring of online exams in arabic based on stemming techniques and levenshtein edit operations. arXiv:161102815 [cs] (2016)
12. Škopljanac-Mačina, F., Blašković, B.: Formal concept analysis – overview and applications. Proc. Eng. **69**, 1258–1267 (2014). https://doi.org/10.1016/j.proeng.2014.03.117
13. Belohlavek, R., Trnecka, M.: Basic level in formal concept analysis: interesting concepts and psychological ramifications. In: Twenty-Third International Joint Conference on Artificial Intelligence, pp. 1233–1239, Beijing, China (2013)
14. Ignatov, D.: Introduction to formal concept analysis and its applications in information retrieval and related fields. In: Braslavski, P., Karpov, N., Worring, M., Volkovich, Y., Ignatov, D.I. (eds.) RuSSIR. CCIS, vol. 505, pp. 42–141. Springer, Cham (2015). https://doi.org/10.1007/978-3-319-25485-2_3
15. Rocco, C.M., et al.: Introduction to formal concept analysis and its applications in reliability engineering. Reliab. Eng. Syst. Saf. **202**, C (2020)
16. Kuznetsov, S.: Machine learning and formal concept analysis. In: Eklund, P. (ed.) ICFCA. LNCS, vol. 2961, pp. 287–312. Springer, Heidelberg (2004). https://doi.org/10.1007/978-3-540-24651-0_25

How BERT's Dropout Fine-Tuning Affects Text Classification?

Salma El Anigri[1]([envelope]) [iD], Mohammed Majid Himmi[1] [iD],
and Abdelhak Mahmoudi[2] [iD]

[1] LIMIARF Laboratory, Faculty of Sciences, Mohammed V University,
Rabat, Morocco
{salma_elanigri,mohammed-majid.himmi}@um5.ac.ma
[2] LIMIARF Laboratory, Ecole Normale Supérieure, Mohammed V University,
Rabat, Morocco
abdelhak.mahmoudi@um5.ac.ma

Abstract. Language models pretraining facilitated fitting models on new and small datasets by keeping the previous pretraining knowledge. The task-agnostic models are to be fine-tuned on all NLP tasks. In this paper, we study the fine-tuning effect of BERT on small amount of data for news classification and sentiment analysis. Our experiments highlight the impact of tweaking the dropout hyper-parameters on the classification performance. We conclude that combining the hidden layers and the attention dropouts probabilities reduce overfitting.

Keywords: BERT · Fine-tuning · Text classification · Dropout

1 Introduction

Natural Language Processing (NLP) aims to create intelligent systems able to learn and understand the meanings embedded in characters, words, sentences, and documents. Among the main NLP tasks such as document summarization, and language translation, text classification remains a fundamental task in real-world applications such as sentiment analysis, spam detection, topic categorization, disease detection, etc. The performance of text classification systems requires valid preprocessing pipelines and appropriate text representations. Deep artificial neural network architectures achieved advances in text representation compared to classical probabilistic and statistical approaches. These architectures are accurate when text representations like Word2Vec, GloVe and FastText were used [2,13,14]. These representations map the intrinsic features and meanings of words (or sub-words) to a vector space, but they lack context information and do not respect the polysemy of words. Embeddings from Language Models (ELMo) addressed this issue using a two-layer bi-LSTM to capture the hidden features [15].

Using appropriate text representation, text classification has been investigated through deep neural architectures. Kim et al. [8] conducted a series of

© Springer Nature Switzerland AG 2021
M. Fakir et al. (Eds.): CBI 2021, LNBIP 416, pp. 130–139, 2021.
https://doi.org/10.1007/978-3-030-76508-8_11

experiments using convolution-based architecture built on top of Word2Vec embeddings for sentence classification. Zhang et al. [22] proposed a char-level text classification method where a sequence of encoded characters was given to a one-dimensional Convolutional Neural Networks (CNNs) then compared with word-level methods. Dai et al. [3] present a semi-supervised approach for sequence learning using LSTM recurrent networks for NLP tasks. This approach uses both a sequence autoencoder as an encoder and a recurrent language model as a decoder that stabilizes the model's learning and generalization. Huang et al. [7] proposed a graph-based text classification model that builds text-level graphs instead of affecting a single graph to the whole corpus.

Transformers [19] came to solve the problems faced by the seq2seq models that use convolutions or recurrence mechanisms. Transformers provided the self-attention mechanism to focus on the important information while processing multiple sequences in parallel. This architecture contributed to the creation of task-agnostic language models such as GPT [16], BERT [4], XLNet [20], RoBERTa [11], ALBERT [9], etc.

The pretrained language model representations could be applied to downstream tasks in two ways: the first one, as feature extractors like ELMo, the pretrained representations are passed to another task-specific architecture, and the second one, fine-tuning the pretrained representations on all NLP tasks with minimal changes in the architecture like adding a linear layer on top of the model for a classification task or fine-tuning the pretraining parameters. Howard et al. [6] proposed ULMFIT, a new transfer learning method for fine-tuning language models on all NLP tasks. Radford et al. [16] presented a transformer-based generative language model that can learn text representation in a unidirectional context (left-to-right) by pretraining the transformer decoder in an unsupervised manner. Then, Devlin et al. [4] proposed the Bidirectional Encoder Representations from Transformers (BERT), a language representation model that learns representation from unsupervised data in bidirectional context left-to-right and right-to-left. Zhilin et al. [20] proposed XLNet, a transformer-based model with an autoregressive pretraining method outperforming BERT on 20 NLP state of the art tasks such as document classification, natural language inference, and question answering.

While BERT is suited for sentence-level and token-level tasks [4], several limitations appeared when processing documents with long sequences (more than 512 tokens). Sun et al. [18] conveyed a series of exhaustive experiments on BERT to examine the BERT fine-tuning and pretraining methods on document classification. Adhikari et al. [1] proposed fine-tuning BERT and distilling the learned knowledge from BERT large to small bidirectional LSTM layer for improving document classification baselines. Dodge et al. [5] conducted a set of experiments for fine-tuning BERT on a set of downstream NLP tasks, and demonstrated that the choice of one hyper-parameter influences the performance of pretrained models. Furthermore, Lee et al. [10] proposed the Mixout regularization technique for fine-tuning the pretrained models by mixing the parameters between a pretrained language model and a target language model. The Mixout stabilizes the fine-tuning of BERT and improves the accuracy score on different NLP tasks.

In this paper, we discuss the effect of the dropout hyper-parameter tuning on BERT architecture. This consists of randomly select the input neurons to cancel from the network and reduce its dependence on specific weights. This regularization can be applied on both the BERT's hidden and attention layers. We will discuss the effect of dropout probabilities fine-tuning on the model's performance when applied to English text classification in two real-world problems: news classification and sentiment analysis.

2   Methodology

In this paper, we studied the impact of fine-tuning BERT using the dropout regularization on the text classification performance. Subsets of two datasets were used for this purpose: the AG's News dataset [22] for the task of news classification and the Yelp reviews - polarity dataset [22] for sentiment analysis. The two subsets were balanced and the accuracy was chosen as the evaluation metric.

2.1   Datasets

In many applications, understanding text documents such as news articles and customer reviews is crucial. On one hand, news classification consists in labeling each news article by its topic. News are a source for rich information and a lot of language models were pretrained on general data domains like Wikipedia. Real-world applications require processing of news resources for many tasks like fake news detection, news recommendation, news topic classification, headline generation, etc. We run our experiments on the AG's News dataset, which is a subset of the large AG's corpora[1] that contains more than 1 million news articles gathered from more than 2000 news sources. This dataset contains four news topics: World, Sports, Business, Sci/Tech. On the other hand, sentiment analysis aims to extract judgments, opinions, and feeling dealing with text from social media or customer reviews websites. Sentiments can be positive, negative, or neutral. The Yelp reviews - polarity dataset that we used is constructed from the Yelp data challenge in 2015 by considering stars 1 and 2 as negative polarity, and 3 and 4 as positive with a total number of 560,000 training samples and 38,000 testing samples (see Table 1).

For both datasets, we perform our experiments on a subset of 20,000 balanced examples from the original dataset with 85% of the subset for training and 15% as the validation set. The table 2 describes this details.

---

[1] http://groups.di.unipi.it/~gulli/AG_corpus_of_news_articles.html.

Table 1. Summary of text classification datasets.

Task	Dataset	Train size	Test size	Nb. classes
News classification	AG's news	120,000	7600	4
Sentiment analysis	Yelp reviews - polarity	560,000	38,000	2

Table 2. Summary of AG's news and Yelp datasets.

Dataset	Examples/classes	Train set	Validation set
AG's news	5000	17000	3000
Yelp reviews - polarity	10,000	17000	3000

2.2 BERT Fine-Tuning for Text Classification

BERT is a transformer-based architecture with L transformer layers [19]. Each layer contains A multi-head self-attention layers, and H hidden neurons in the position-wise fully connected feed-forward network. BERT is pretrained and fine-tuned given an input sequence of no more than 512 tokens. For text classification tasks, the special tokens [CLS] and [SEP] delimit each input sequence. The [CLS] token is used at the beginning of the sequence, while the [SEP] is used to separate two segments in the same sequence. As shown in Fig. 1, BERT is linked to a dense layer and takes the last hidden state corresponding to the [CLS] token and outputs the class label representing the input sequence.

We fine-tuned the pretrained BERT-base-uncased model with $L = 12$, $A = 12$, and $H = 768$. We run this experiment using the adamW optimizer [12] with the learning rate $lr = 2e - 5$ during 6 epochs with a batch size of 16 and using 512 as a maximum sequence length for both datasets.

Fine-tuning BERT on Yelp reviews - polarity and AG's news datasets shows overfitting indicators. We regularize our model using the dropout hyper-parameter. First, we apply the dropout for all the attention probabilities in the model, we label it by (*dpAttention*). Second, the dropout for the fully connected layers in the whole model we designate by hidden layer dropout probability (*dpHidden*).

According to Devlin et al. [9], the optimal hyper-parameters during fine-tuning are task-specific. For example, the authors fine-tuned BERT on the SQUAD benchmark [17] for 3 epochs, with 32 as batch size and a learning rate of 5e−5, whereas, for the SWAG [21] they used 3 epochs with a batch size of 16 and a learning rate of 2e−5. Indeed, the recommended configuration for fine-tuning BERT is a batch size of 16 or 32, a number of epochs of 3, 4, or 5, and a learning rate of 5e−5, 3e−2 or 2e−5 using Adam optimizer [9].

We studied the dropout influence using three experiments to reduce the overfitting effect: *dpAttention* only, *dpHidden* only, and both *dpAttention* and *dpHidden*. For the three experiments, the dropout probabilities were between 0.1 to 0.6. A value greater than 0.6 decreases the performance.

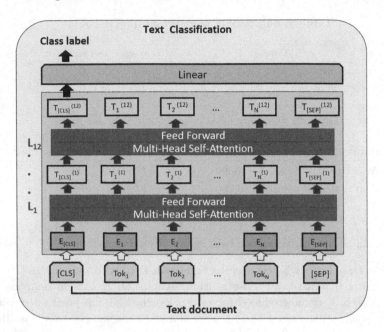

Fig. 1. The BERT model used for text classification. It contains 12 transformer block each contains a multi-head attention mechanism layer and a position-wise feed forward network as defined in [19].

3   Results and Discussion

The Fig. 2 shows the training and validation BERT's accuracies and losses using the dropout probabilities in both *dpAttention* and *dpHidden* for news classification on AG's news. In the Fig. 2a and 2b the maximum of training and validation accuracy was found with minimal dropout probabilities (0.1 and 0.2) in both *dpAttention* and *dpHidden*. However, in 2c the minimum training loss was found for minimum *dpHidden* (0.1) independently of the *dpAttention* dropout, whereas in 2d, minimum validation loss was found with higher dropouts probabilities in both *dpHidden* and *dpAttention*.

Similarly, the Fig. 3 shows the same experiment on Yelp reviews. In the Fig. 3a, the minimum training accuracy was found in the highest *dpHidden* (0.6) and *dpAttention* (0.5, 0.6) dropout probabilities. In contrast, for 3b the minimum validation accuracy was found for higher *dpHidden* probabilities (0.6). However, minimal training loss was found in 3c for lower dropout probabilities. For the Fig. 3d minimal validation loss was found for higher *dpAttention* probabilities (0.5, 0.6) and higher *dpHidden* probabilities (0.5).

The Fig. 4 describes the behaviour of the *dpHidden* and *dpAttention* separately on news classification. In Fig. 4a and Fig. 4b, the maximum training accuracy and validation was found in small *dpHidden* probabilities, after the 0.5 probability the performance diminishes under 90%. However, the maximal

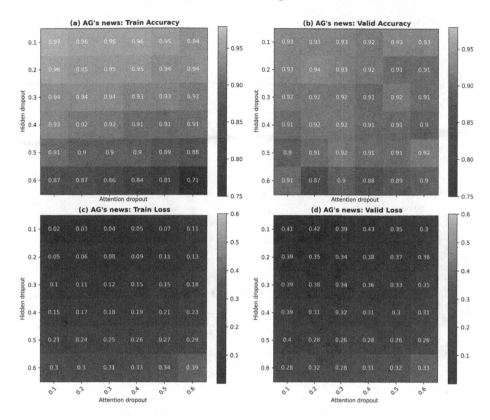

Fig. 2. Training and validation BERT's accuracies and losses using the dropout probabilities in both the hidden and attention layers for news classification on AG's news.

accuracy observed for small *dpAttention* probabilities and starts to diminish for higher *dpAttention* probabilities (0.5, 0.6). In contrast, the training loss augments for higher *dpHidden* (0.5, 0.6), while the training loss for *dpAttention* is limited between 0.0 and 0.05 as shown in the Fig. 4c. Furthermore, minimal validation losses were found for the *dpHidden* in (0.3, 0.4, and 0.6), and for the *dpAttention* in (0.3, 0.5, and 0.6) (Fig. 4d).

Similarly, the Fig. 5 describes the behaviour of the *dpHidden* and *dpAttention* separately on sentiment analysis. In the Fig. 5a and 5b, the maximum accuracy was found for small *dpHidden* probabilities, the accuracy started to reduce for *dpHidden* > 0.5 different from the *dpAttention*, the accuracy is high for all the *dpAttention* probabilities. However, in Fig. 5c and 5d for the *dpAttention* only and *dpHidden* only, the training loss is small, while the validation loss quickly increased.

These detailed experimental results show that *dpAttention* and *dpHidden* impact the performance of the model in many ways. The *dpAttention* only and *dpHidden* only are not sufficient to reduce the overfitting effect and improve

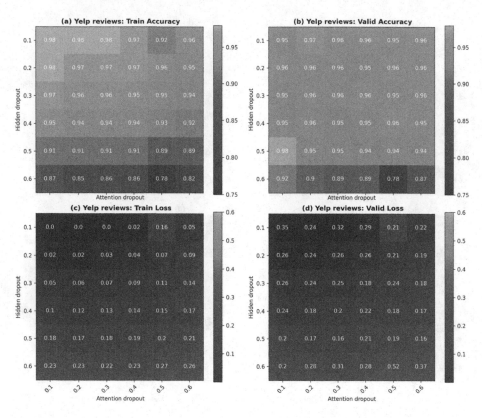

Fig. 3. Training and validation accuracies and losses of the dropout probabilities in both the hidden and attention layers for sentiment analysis on Yelp reviews.

the performance. Our experimental results show that, on one hand, the use of *dpAttention* only causes additional overfitting, and, on the other hand fine-tuning the model using the *dpHidden* only on the validation set degrades the performance of the model. On the contrary, carefully combining both attention and hidden layers dropouts improves the BERT's performance. When combining higher *dpAttention* probabilities (0.5) and lower *dpHidden* probabilities (0.2, 0.3), the overfitting effect is clearly reduced. Still, the accuracy is not completely optimal, we argue however that the training was done with a small size subsets.

4   Conclusion and Future Works

This study was set out to explore the process of fine-tuning BERT on news classification and sentiment analysis using small subsets of AG's news and Yelp reviews datasets. The experiments have shown that overfitting is detected and that tuning of the dropout hyper-parameter of the BERT's hidden and attention layers lead to different model performances. These findings suggest regularizing

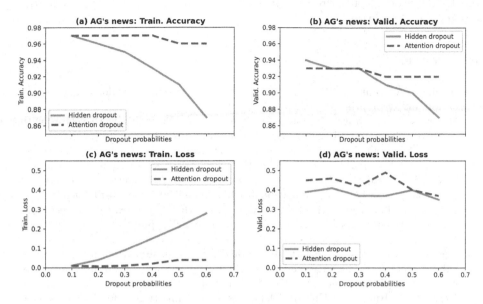

Fig. 4. Training and validation accuracies and losses of AGs news dataset with respect to dropout probabilities of hidden and attention BERT layers.

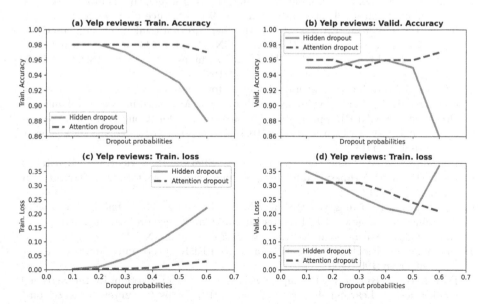

Fig. 5. Training and validation accuracies and losses of Yelp reviews dataset with respect to dropout probabilities of hidden and attention BERT layers.

the BERT model using dropout on both the fully connected layers and the attention probabilities.

References

1. Adhikari, A., Ram, A., Tang, R., Lin, J.: DocBERT: BERT for document classification. arXiv preprint arXiv:190408398 (2019)
2. Bojanowski, P., Grave, E., Joulin, A., Mikolov, T.: Enriching word vectors with subword information. Trans. Assoc. Comput. Linguist. **5**, 135–146 (2017)
3. Dai, A.M., Le, Q.V.: Semi-supervised sequence learning. In: Cortes, C., Lawrence, N., Lee, D., Sugiyama, M., Garnett, R. (eds.) Advances in Neural Information Processing Systems, vol. 28. Curran Associates Inc. (2015). https://proceedings.neurips.cc/paper/2015/file/7137debd45ae4d0ab9aa953017286b20-Paper.pdf
4. Devlin, J., Chang, M.W., Lee, K., Toutanova, K.: BERT: pre-training of deep bidirectional transformers for language understanding. arXiv preprint arXiv:181004805 (2018)
5. Dodge, J., Ilharco, G., Schwartz, R., Farhadi, A., Hajishirzi, H., Smith, N.: Fine-tuning pretrained language models: weight initializations, data orders, and early stopping. arXiv preprint arXiv:200206305 (2020)
6. Howard, J., Ruder, S.: Universal language model fine-tuning for text classification. In: Proceedings of the 56th Annual Meeting of the Association for Computational Linguistics (Volume 1: Long Papers), pp. 328–339. Association for Computational Linguistics, Melbourne (2018). https://doi.org/10.18653/v1/P18-1031. https://www.aclweb.org/anthology/P18-1031
7. Huang, L., Ma, D., Li, S., Zhang, X., Wang, H.: Text level graph neural network for text classification. In: Proceedings of the 2019 Conference on Empirical Methods in Natural Language Processing and the 9th International Joint Conference on Natural Language Processing (EMNLP-IJCNLP), pp. 3444–3450. Association for Computational Linguistics, Hong Kong (2019). https://doi.org/10.18653/v1/D19-1345. https://www.aclweb.org/anthology/D19-1345
8. Kim, Y.: Convolutional neural networks for sentence classification. In: Proceedings of the 2014 Conference on Empirical Methods in Natural Language Processing (EMNLP), pp. 1746–1751. Association for Computational Linguistics, Doha (2014). https://doi.org/10.3115/v1/D14-1181. https://www.aclweb.org/anthology/D14-1181
9. Lan, Z., Chen, M., Goodman, S., Gimpel, K., Sharma, P., Soricut, R.: ALBERT: a lite BERT for self-supervised learning of language representations. arXiv preprint arXiv:190911942 (2019)
10. Lee, C., Cho, K., Kang, W.: Mixout: effective regularization to finetune large-scale pretrained language models. In: International Conference on Learning Representations (2020). https://openreview.net/forum?id=HkgaETNtDB
11. Liu, Y., et al.: RoBERTa: a robustly optimized BERT pretraining approach. arXiv preprint arXiv:190711692 (2019)
12. Loshchilov, I., Hutter, F.: Decoupled weight decay regularization. In: International Conference on Learning Representations (2019). https://openreview.net/forum?id=Bkg6RiCqY7
13. Mikolov, T., Chen, K., Corrado, G., Dean, J.: Efficient estimation of word representations in vector space. arXiv preprint arXiv:13013781 (2013)

14. Pennington, J., Socher, R., Manning, C.D.: GloVe: global vectors for word representation. In: Proceedings of the 2014 Conference on Empirical Methods in Natural Language Processing (EMNLP), pp. 1532–1543 (2014)
15. Peters, M.E., et al.: Deep contextualized word representations. arXiv preprint arXiv:180205365 (2018)
16. Radford, A., Narasimhan, K., Salimans, T., Sutskever, I.: Improving language understanding with unsupervised learning (2018)
17. Rajpurkar, P., Zhang, J., Lopyrev, K., Liang, P.: SQuAD: 100,000+ questions for machine comprehension of text. In: Proceedings of the 2016 Conference on Empirical Methods in Natural Language Processing, pp. 2383–2392. Association for Computational Linguistics, Austin (2016). https://doi.org/10.18653/v1/D16-1264. https://www.aclweb.org/anthology/D16-1264
18. Sun, C., Qiu, X., Xu, Y., Huang, X.: How to fine-tune BERT for text classification? In: Sun, M., Huang, X., Ji, H., Liu, Z., Liu, Y. (eds.) CCL 2019. LNCS (LNAI), vol. 11856, pp. 194–206. Springer, Cham (2019). https://doi.org/10.1007/978-3-030-32381-3_16
19. Vaswani, A., et al.: Attention is all you need. arXiv preprint arXiv:170603762 (2017)
20. Yang, Z., Dai, Z., Yang, Y., Carbonell, J., Salakhutdinov, R.R., Le, Q.V.: Xlnet: generalized autoregressive pretraining for language understanding. In: Wallach, H., Larochelle, H., Beygelzimer, A., d'Alché-Buc, F., Fox, E., Garnett, R. (eds.) Advances in Neural Information Processing Systems, vol. 32. Curran Associates Inc. (2019). https://proceedings.neurips.cc/paper/2019/file/dc6a7e655d7e5840e66733e9ee67cc69-Paper.pdf
21. Zellers, R., Bisk, Y., Schwartz, R., Choi, Y.: SWAG: a large-scale adversarial dataset for grounded commonsense inference. In: Proceedings of the 2018 Conference on Empirical Methods in Natural Language Processing, pp. 93–104. Association for Computational Linguistics, Brussels (2018). https://doi.org/10.18653/v1/D18-1009. https://www.aclweb.org/anthology/D18-1009
22. Zhang, X., Zhao, J., LeCun, Y.: Character-level convolutional networks for text classification. arXiv preprint arXiv:150901626 (2015)

Big Data, Datamining, Web Services and Web Semantics (Full Papers)

Selection of Composite Web Services Based on QoS

Abdellatif Hair[1] and Hicham Ouchitachen[2(✉)]

[1] Faculty of Science and Techniques, Bén-Mellal, Morocco
[2] Polydisciplinary Faculty, Bén-Mellal, Morocco

Abstract. Today, web services and Service-Oriented Architecture (SOA) are causing a lot of ink and research to be reinforced. This architecture makes the composition of web services a necessity. Considering the increase of these in a way that has many proposals for a single customer request. The proposal of other forms of composition based on quality of service (QoS) that will save time and will perfectly meet the user's request given the criteria required by each customer presents a challenge today.

In this article, we propose a new concept of composition of web services based on QoS, this concept is based on dominance in the sense of Pareto.

Keywords: Web services · Quality of service · Pareto dominance · Binary composition · Selection algorithms · Function of fuzzy dominated

1 Introduction

The composition of web services has become a research point in development, customer demand in terms of quality of service is growing and web services are growing and the need to offer several services for a single demand is increasing, leading to the search for new compositional methods that are more efficient and less expensive.

This composition must therefore choose the service providers that obtain a composition that perfectly meets the customer's needs based on the quality of service.

1.1 Example

It is assumed that there is a user who wants to plan a trip, for this he needs to consume 03 types of services (03 domains), a hotel reservation, a plane ticket reservation and a car rental, we also note that we must select a single service (or company) of each category, using QoS criteria (reputation, reliability, costs, turnaround time...) [1].

1.2 Formalizing the Problem

To formalize the problem of composing QoS-based services, we need to design an objective function that takes into account negative and positive QoS criteria; we model the problem as follows:

© Springer Nature Switzerland AG 2021
M. Fakir et al. (Eds.): CBI 2021, LNBIP 416, pp. 143–159, 2021.
https://doi.org/10.1007/978-3-030-76508-8_12

- CD = {D1, ..., Dn}, an abstract composition that represents the user's query, i.e. the n classes (domains) of services to consume Di.
- C = {S1, ..., Sn} a concrete composition, i.e. we replace each Di class with a concrete service Si ∈ Di.

We need to look for a concrete C composition such as:

The objective function (functions) Uk(C) is (are) optimized. Uk represents the objective function of the kth attribute of QoS, if we aggregate the m attributes of QoS into a single value then we will need a single objective function U.

The aim is to develop a web service composition system in such a way that it gives us a better quality of service but that does not affect complexity or execution time.

The rest of the article is organized as follows:

In the next section, we present the related work. In Sect. 3, we provide definitions and relationships related to Pareto dominance and show the proposed approach with examples. Section 4 presents the results of our experiment. Finally, Sect. 5 presents a conclusion and the prospects that we wish to achieve in the future

2   Related Work

In this section, we introduce some related concepts of web service selection. Some categories and QoS attributes of web service are shown in Table 1. The problem of selecting QoS-based web services is solved using Pareto optimality. A set Pareto is a set of achievable solutions that have at least one optimized goal while keeping others unchanged [2]. In [3], Chen and al proposed an algorithm using a partial selection approach. In [4], the authors adopted 4 objective functions: (minimization of cost, minimization of duration, maximizing turnover, maximizing reputation). In [5, 6], the authors used reinforcement learning algorithms to find the set of optimal Pareto solutions that meet multiple QoS factors and user requirements. In [7], the problem of selecting optimal Pareto service compositions is addressed using an AFPTAS (Approximations by Fully Polynomial Time Approximation Scheme) polynomial approximation approach. The 'α − error' is used to estimate the accuracy of the approximation [8]. To calculate this metric, each composition of the Pareto set is associated with the composition of the approximation set with comparable QoS values. In [9] the authors proposed multi-purpose optimization based on particulate swarms, they considered two objective functions (cost and wait time) and neglected global constraints, they adopted the best version of algorithm. In [10] the authors presented the notion of k-dominance that relaxes the notion of Pareto dominance to a subset of k parameters. In [11], the authors proposed the top-k representative Operator Skyline which allows to select a set of points, for which the number of points dominated, by at least one of them, is maximized. In [12], Benouaret and al proposed a variant of Skyline that is based on an extension of Pareto dominance; they offered a fuzzification of Pareto dominance called 'α − dominance'. In [13], the authors proposed an approach based on the notion of skyline to effectively and efficiently select services for composition, reducing the number of candidate services to be considered. In [17], an improved multi-verse optimization algorithm for web service composition that is called IMVO algorithm is proposed to improve QoS while satisfying service level agreement (SLA).

Table 1. Categories and QoS attributes of web service [18].

Categories	QoS attribute
Service provider	Availability, Reputation, Security, Discoverability, Accountability, Interoperability, Throughput, Performance, Dependability, Price, Reputation, Security, Response time, Capacity, Robustness, Exception Handling, Transaction Integrity, Regulatory, Cost, Support standard, Stability, Completeness, Reliability, Time out, Exception duration, Penalty rate, Compensation rate, Scalability, Resource utilization, Dynamic Discoverability, Dynamic Adaptability, Dynamic Composability
Composition	Response time, Throughput, Scalability, Capacity, Availability, Reliability, Security, Reliable message, Integrity, Interoperability, Execution Cost, Reputation, Execution duration
Multi-perspectives	Performance, Security, Relative importance
Domain related	Performance, Availability, Reliability, Failure-semantic, Robustness, Accessibility, Scalability, Capacity, Continuous Availability, Transaction support, Security, Configuration Management, Network and Infrastructure, Routes set, Detail level, Accuracy, Completeness, Validity, Timeliness, Coverage, Correctness, Accuracy, Precision, Input output parameters, Efficiency, Availability Consistency, Load Management
Service Consumer	Price, Response time, Reputation, Successability, Throughput, Availability, Reliability, Latency, Accuracy, Availability, Scalability, Resource utilization, Security, Usability, Composability, Robustness, Security
Unique feature of SOA	Availability, Reliability, Performance, Usability, Discoverability, Adaptability, Composability
Service developer	Maintainability, Reusability, Conformance, Security, Reliable Messaging, Transaction, Accounting, Availability, Response time

In [19], to find the best combinations of atomic web services, S. Fateh formulated the multi-objective QoS-driven web service composition problem (MOQWSCP) as a fuzzy multi-objective optimization problem (FMOQWSCP). In [20], the particular problem of many consumers that are competing to acquire services with same funtionalities but with different QoS, is considered. the authors developed different models based on Discrete Time Markov Chain to implement several policies of the system. In [21], the authors proposed a novel hybrid algorithm to address the personalized recommendation for manufacturing service composition (MSC). The algorithm solves the insufficient individualization defect of MSC optimization by comprehensively considering the QoS objective attributes and customer preference attributes.

3 Proposed Approach

The proposed approach is to break down the overall composition query into sub compositions. This is followed by two repetitive steps of binary compositions and selection such as:

- **Binary composition** allows you to combine two classes of services, or one class and one selection result.
- **Selection** applied to a class or the result of the binary composition, consists of reducing the space of candidates.

Definition 1 (Skyline): The skyline brings together all the vectors of criteria (solutions) that are not Pareto dominated by no other vector, so the skyline represents undominated or undominated optimal solutions.

Definition 2 (Dominance of Pareto): Consider two services X, Y ∈ S, characterized by a set of QoS R values. X dominates Y, (we note X > Y) if and only if:

- X is better or equal to Y in all QoS settings (R values).
- X is at least better compared to Y in QoS. i.e. ∀k∈ QoS settings: Qk(X) ≥ Qk(Y) and ∃k ∈ QoS settings: Qk(X) > Qk(Y) [13].

3.1 Binary Composition Approach AC1

AC1 allows us to start the composing by the first two areas on which the skyline is carried out, then the result of this composition will also undergo the method skyline and then will be added to the next domain, this process is repetitive until the end of all areas.

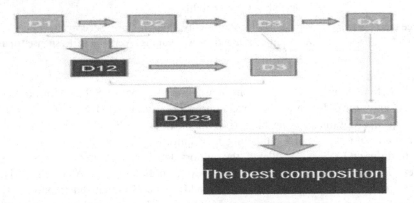

Fig. 1. Model of the AC1 composition approach.

This behavior is described under a model illustrated in Fig. 1 on which four basic domains D1, D2, D3 and D4 are taken and the AC1 method is applied, D12 represents the result of the first composition which is then combined with D3 to give D123, in the end we add D4 to have the final result which is represented by D1234 (Fig. 2).

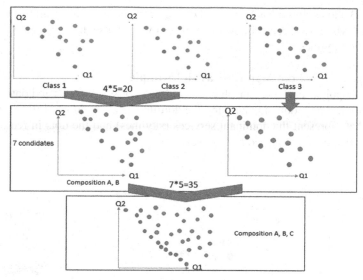

Fig. 2. Example of AC1 application.

3.2 Selection Algorithms

These algorithms were introduced into the composition of the service as a reduction pre-treatment before the composition of the services. It consists of minimizing the space of services in order to reduce the calculation rate for optimal composition.

3.2.1 Dominance Algorithm

The idea of the Dominance Algorithm is to locate all the nominated points in a vector space. The objective of this algorithm is the reduction of the space of the candidate elements with the use of the dominance relationship of Pareto.

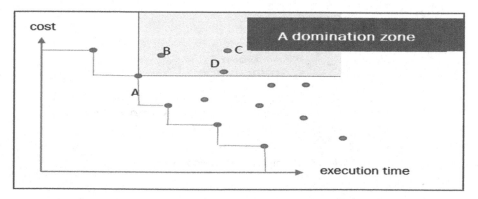

Fig. 3. Example of area of domination.

As Fig. 3 shows the points B, C and D are dominated by point A and so on for the other points, when we precede to the composition we just use the dominant points they are called the Skylines.

3.2.1.1 Application of the Algorithm as an Example

In Fig. 4, all points presented in blue and red are considered all candidate services belonging to the same domain, applying the dominance algorithm and the result is the blue dots that represent the dominant services (skylines) and the dots in red represent the services that can be replaced by skylines.

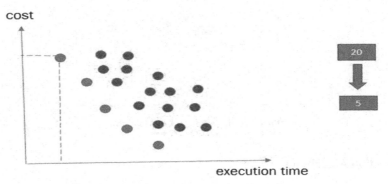

Fig. 4. Example of service reduction using the dominance algorithm.

In this way we were able to minimize the space of the points and the number of services that will be part of the composition.

3.2.1.2 Drawback

The dominance algorithm does not differentiate between web services that are a good compromise and those with a bad compromise (Fig. 5).

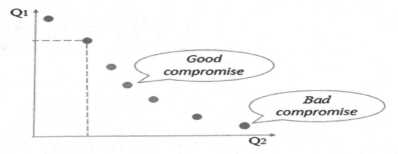

Fig. 5. Example of unrivalled services.

The term 'good compromise' is used to describe services that are balanced in the values of QoS i.e. services that have satisfactory values in all their attributes. In contrast, the term 'bad compromise' is used to describe services that are not balanced in the values of their attributes i.e. incomparable services.

3.2.1.3 Solution

To improve selection results and determine the right services to compromise, it is interesting to fuzzify Pareto dominance relationship.

The aim of fuzzing the dominance relationship is to allow a comparison between incomparable web services with fuzzy dominance.

Definition 3 (fuzzification): It's a process of transforming a numerical size into a blurred subset. It also allows to qualify a numerical value with a language term [14].

Definition 4 (Fuzzy dominance): Given two d-dimensional vectors u and v, the fuzzy dominance function [15] calculates the degree with which u dominates v:

$$FD(u, v) = \left(\frac{1}{d}\right) \sum_{i=1}^{d} EFD(u_i, v_i) \tag{1}$$

Where $FED(u, v)$ is defined as:

$$FED(u_i, v_i) = \begin{cases} 0 & \text{if } u_i - v_i \geq \varepsilon \\ \frac{|u_i - v_i - \varepsilon|}{|\lambda + \varepsilon|} & \text{if } (\lambda + \varepsilon) \leq u_i - v_i \leq \varepsilon \text{ where } \varepsilon, \lambda \in [0, -1] \\ 1 & \text{if } u_i - v_i < \lambda + \varepsilon \end{cases} \tag{2}$$

These parameters are chosen heuristically by an expert or by a machine learning algorithm.

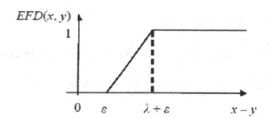

Fig. 6. Elemental fuzzy dominance function.

FED represents the basic fuzzy dominance function. It expresses the extent to which x is more or less superior to y. The function of fuzzy dominance is simply an aggregation of the various basic fuzzy dominance scores. Figure 6 shows the graduated values of x dominance on y, specifically:

– If (x-y) is less than ε, then x is not greater than y.
– If (x-y) is bigger than λ +, then x is much bigger than y.
– If (x-y) is between ε and λ + ε, then x is much larger than y to some extent.

3.2.2 Top-K Ranking Algorithm

3.2.2.1 The Steps of the Algorithm Normalization

- The normalization function consists of giving a score for each data for a result between 0 and 1.
- If we have the G set and the data A, B, C, D, E, so G = {A, B, C, D, E}.
- We find the Min(G) (Suppose the Min(G) = A).
- We find the Max(G) (Suppose the Max (G) = E, A' = 0 and E' = 1).
- To find the values of B, C and D, we use:

$$B' = (B - Min(G))/(Max(G) - Min(G)) \tag{3}$$

Where B' is the new number between 0 and 1 that represents the B.

Example:
G = {A = 2, B = 4, C = 5, D = 8, E = 10}, Min(G) = A = 2, Max(G) = E = 10, A' = 0, E' = 1 and B' = (4-2)/(10-2) = 2/8 = 0,25.

Function of Fuzzy Dominated (FED)
The dominated fuzzy function is denoted FED. It processes two vectors at the same time.

$$FED(U_i, V_i) = \begin{cases} 0 & \text{if } U_i - V_i \geq \varepsilon \\ \frac{|U_i - V_i - \varepsilon|}{|\lambda + \varepsilon|} & \text{if } (\lambda + \varepsilon) \leq (U_i - V_i) \leq \varepsilon \quad \text{where } \varepsilon, \lambda \in [0, -1] \\ 1 & \text{if } U_i - V_i < \lambda + \varepsilon \end{cases} \tag{4}$$

Ranking Function
The ranking function is rated Rang (U).
  It classifies the previously called U vector.
  We have the BASE variable which is the number of vectors or classes or services.

$$Rang(U) = \frac{1}{|BASE - 1|}\left(\sum (U \neq V)AFDS(U, V)\right) \tag{5}$$

Ranking
The ranking allows to give the number of vectors ranked increasingly from the best to the best at least [16], this number is called top-k where k represents the number of the first best vectors. For example: top-5 means that the first 5 best vectors are displayed.

3.2.2.2 Applying the algorithm as an example

This algorithm is applied to a set of 8 services belonging to the SMS class. Each service has 4 QoS settings.

  Table 2 gives the results of the ranking of services according to the top-K algorithm (Table 3).

  We can observe that the S1 service is top-1. Indeed it is better than the others in Q1, Q2 and has good value in other service quality settings.

Table 2. A set of SMS sending web services.

Web service	Provide	Operation	Q1 (ms)	Q2 (hits/sec)	Q3 (%)	Q4 (%)
S1	Acrosscom munication. com	SMS	113.8	5.2	81	84
S2	Sjmillercons ultants.com	SMS	179.2	0.7	65	69
S3	Webservic ex.net	SendSMS	1308	6.3	67	84
S4	Webservic ex.net	SendSMSWorld	3103	5.3	79.3	91
S5	Smsinter. sina.com.cn	SMSWS	751	6.8	64.3	87
S6	Sms.mio.it	SendMessages	291.07	5.2	53.6	84
S7	www.barnal and.is	SMS	436.5	4.5	43.2	84
S8	Emsoap.net	emSoapService	424.54	4.3	11.9	80

Table 3. Ranking services by top-k algorithm.

Web service	AFDetS	Nq1	Nq2	Nq3	Nq4
S1	0.071	1	0.74	1	0.68
S5	0.107	0.79	1	0.76	0.82
S6	0.143	0.94	0.74	0.60	0.68
S3	0.25	0.60	0.92	0.80	0.68
S7	0.286	0.89	0.62	0.45	0.68
S4	0.312	0	0.75	0.98	1
S8	0.393	0.90	0.59	0	0.50
S2	0.571	0.98	0	0.77	0

4   Experimentation

In this section, we describe the experiments that allow the performance analysis of the proposed approach. These experiments are conducted on a platform, having a Core i5 processor with 4 GB of RAM and under the Windows10 operating system.

4.1 Evaluation of Top-K Algorithms Results

To evaluate the effectiveness of this algorithm and observe the different results, several simulations were carried out on a set of 8 services belonging to the SMS class. Each service has 4 QoS settings (Table 2).

For each simulation, we take the top-3 services and vary the following parameters:

- ε, λ.
- d: number of QoS settings.
- n: number of services in the same Class S.

Impact of ε and λ on the Ranking Results

We present below a scenario (Table 4) such that each time by varying ε and λ.

Table 4. Ranking results of the services according to the variation of ε, λ.

λ	ε	Number of top-3 services	Scores
0	0	(5, 4, 3)	(0.10/0.14/0.25)
−0.1	−01	(5, 4, 3)	(0.04/0.06/0.16)
−0.2	−0.2	(4, 5, 3)	(0.004/0.035/0.036)
−0.3	−0.3	(4, 5, 3)	(0/0.035/0.036)
−0.4	−0.4	(4, 3, 5)	(0/0.008/0.017)
−0.5	−0.5	(4, 3, 5)	(0/0.003/0.010)
−0.6	−0.4	(4, 3, 5)	(0/0.007/0.013)
−0.7	−0.3	(4, 3, 5)	(0/0.011/0.006)
−0.2	−0.8	(4, 3, 5)	(0/0/0)
−0.1	−0.2	(4, 5, 3)	(0.05/0.035/0.072)
−0.1	−0.4	(4, 3, 5)	(0/0.035/0.035)
−0.4	−0.1	(4, 5, 3)	(0.014/0.047/0.074)
−0.4	−0.2	(4, 3, 5)	(0.002/0.035/0.043)
−0.2	−0.4	(4, 3, 5)	(0/0.035/0.035)
−0.6	−0.4	(4, 3, 5)	(0/0.007/0.013)
0	−0.1	(4, 5, 3)	(0.07/0.10/0.214)
−0.1	0	(4, 5, 3)	(0.089/0.10/0.24)
0	−0.4	(4, 3, 5)	(0/0.035/0.035)
−0.2	0	(4, 5, 3)	(0.07/0.10/0.21)

We can observe from these results (Table 4) that the ranking given is the same for all the values we took during this evaluation and that the only change made is on the scores assigned to the services.

Impact of Number of Services on the Ranking
We present below a scenario (Table 5) such that each time by varying the number of services.

Table 5. Service ranking results based on changes in the number of services.

Number of services	Number of top-3 services	Scores
8	(4, 5, 3)	(0.07/0.10/0.21)
7	(4, 5, 3)	(0.09/0.11/0.25)
6	(4, 5, 3)	(0.12/0.14/0.30)
5	(4, 3, 5)	(0.12/0.31/0.36)
4	(4, 3, 1)	(0.16/0.16/0.34)
3	(3, 1, 2)	(0/0.28/0.87)

Since we varied the number of services and after the analysis, we found that the ranking results were not changed.

Impact of QoS on the Ranking
We present below a scenario (Table 6) such that each time by varying the number of QoS.

Table 6. Service ranking results by number of QoS.

Number of QoS	Number of top-K	Scores
4	(4, 5, 3)	(0.07/0.10/0.21)
3	(5, 4, 3)	(0.09/0.09/0.20)
2	(3, 4, 5)	(0.10/0.12/0.13)
1	(7, 8, 4)	(0.36/0.36/0)

We have varied the number of QoS here we find that the ranking results have been changed, i.e. the number of parameters can have an impact and be subject to a change in the ranking results.

Discussion
We concluded that the values of ε, λ don't lead to any change in ranking results. So to simplify the evaluation of our approach, we will set $\varepsilon = 0$ and $\lambda = -0.2$. Regarding the QoS parameters that have an impact on the ranking of services, we will only use two QoS parameters.

4.2 Evaluation of the Results of the Proposed Approach

4.2.1 Context

To evaluate the performance of the proposed approach, an application was made with NetBeans. This application is an implementation of a set of classes and functions that will allow us to generate web services and then treat them with the AC1 composition approach based each time on a selection algorithm (the algorithm of dominance or top-k algorithm), which led us to make a comparison between the services skylines found.

These two algorithms take input a set of domains (D1, ..., Dn) and for each Domain Dk a set of web services (S11, ..., Smn) with their quality, whose domains are randomly generated by a function and which constitute our database of data application. After the construction of the database, the database will pass in each Algorithm respectively through the following two steps: selection and composition of web services that will give us at the end the services skylines.

These skylines services obtained by each algorithm respect the scanning problem and their intersections will provide us with the best web services. The performance evaluation criteria for the proposed approach are based on time and accuracy.

Accuracy: this is the ratio between the number of best services obtained on the number of web skylines services provided by one of these two algorithms (P-dominance and P-topk).

Execution Time: This is the period of time between the launch of the user's query and the receipt of the final results (time- dominance and time-topk) (Fig. 7).

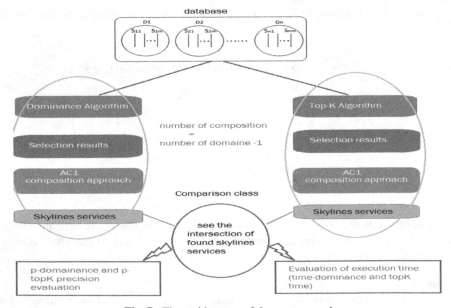

Fig. 7. The architecture of the system used.

4.2.2 Evaluation Results

These results obtained after several simulations such that each time we vary the number of domain and the number of services. After the analysis of the results obtained it was found that the results are segmented according to the number k of the Top-k algorithm. The results are represented by histograms, such as execution time is given per millisecond and precision is a percentage.

4.2.2.1 Evaluation of Execution Time
(See Fig. 8).

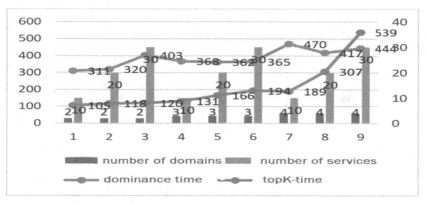

Fig. 8. Average running time associated with top-2 query.

Note that the execution time of the dominance algorithm is greater than that of the top-2 algorithm, but this time knows a large increase from the domain number becomes greater than or equal to 4 (Fig. 9).

Note that the execution time of the top-5 algorithm from domain 3 becomes greater by contributing to the execute on time of the dominance algorithm (Fig. 10).

Note that the execution time of the top-10 algorithm is greater than that of the dominance algorithm as soon as the domain number takes the value 2.

4.2.2.2 Evaluation of Precision
(See Fig. 11).

The accuracy of the top-2 algorithm is much higher than that of the dominance algorithm and knows values close to 1, which means that the number of the best web services obtained is close to that of; k number of top-2 skyline services (Fig. 12).

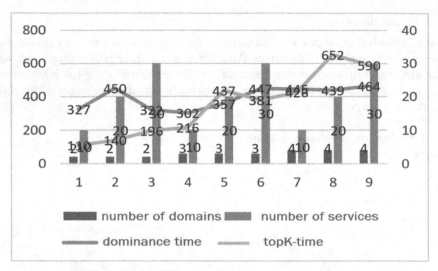

Fig. 9. Average running time associated with top-5 query.

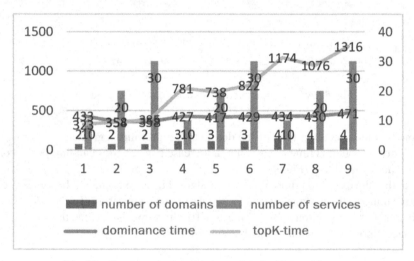

Fig. 10. Average running time associated with top-10 query.

The accuracy of the top-5 algorithm is superior to that of the dominance algorithm (Fig. 13).

The accuracy of the dominance algorithm is far superior to that of top-10 but the results change from the domain number becomes greater than or equal to 4.

Discussion

From top-2 to top-5 the execution time of the top-k algorithm in small data sets is perfect compared to that of the dominance algorithm. Indeed, for k between 2 and 5 the number of combinations between skyline services does not exceed 25 which does not require

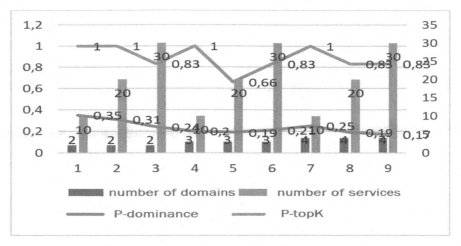

Fig. 11. Average accuracy associated with top-2 query.

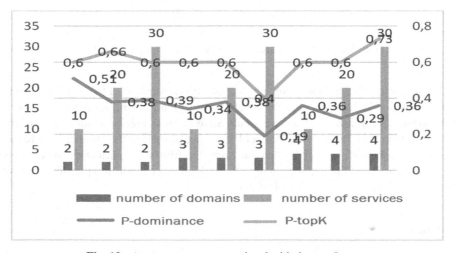

Fig. 12. Average accuracy associated with the top-5 query.

enough execution time for the calculation. When k exceeds the value 5, the number of combinations between the skyline services of the top-k algorithm increases, which will influence directly on the number of computation operations and therefore on the execution time which will know an increase.

It can be concluded that the execution time of the top-k algorithm depends essentially on the number of domains, the number of services and the number k initially fixed.

From top-2 to top-5, the accuracy of the top-K algorithm is perfect compared to the dominance algorithm. Indeed, the number of the best web services found by the two algorithms (in most simulations) is always between 2 and 5 which shows the performance of the accuracy of the top-k algorithm.

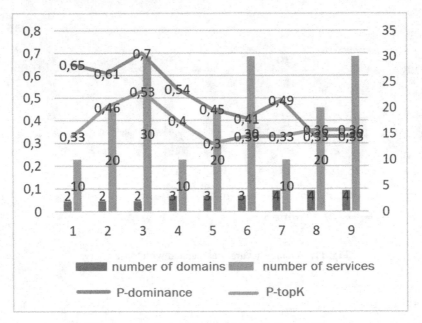

Fig. 13. Average accuracy associated with the top-10 query.

When K exceeds the value 5, the number of skyline services of the top-k algorithm moves away from the number of the best web services found and this results in a precision of the top-k algorithm which becomes weak. So in terms of precision, the results of the top-k algorithm are relative to the number k such that increasing the number k decreases their accuracy. Finally, we can say that the top-k algorithm gives perfect results in terms of execution time and precision compared to the dominance algorithm provided that we work on small data sets and with a k small.

5   Conclusion

In this article, we introduced a new concept of web services composition. This concept can solve the problem of increasing web services in a perfect composition at the level of quality of service and the level of response time. On the one hand, we used a selection algorithm as a preliminary step to exclude web services that are less compatible with the client request and to just keep the web services that perfectly meet the user's requirements. On the other hand we used the AC1 composition approach to compose them.

The results obtained are acceptable and deserve to be extended to improve performance. In this way we can propose a third algorithm that reclassifies the equivalent services and we can take for example the Condorcet based vote or the probabilistic dominance.

References

1. Yu, Q., Bouguettaya, A.: Foundations for Efficient Web Service Selection. Springer, Heidelberg (2010). https://doi.org/10.1007/978-1-4419-0314-3
2. Chankong, V., Haimes, Y.: Multi Objective Decision Making: Theory and Methodology. Courier Dover Publications (2008)
3. Chen, Y., Huang, J., Lin, C., Hu, J.: Partial selection methodology for efficient QoS-aware service composition. IEEE Trans. Serv. Comput. **8**(3), 384–397 (2015)
4. Claro, D.B.: Spoc: un canevas pour la composition automatique de services web dédiés à la réalisation de devis. Thèse de Doctorat, Ecole Doctorale d'Angers (2006)
5. Moustafa, A., Zhang, M.: Multi-objective service composition using reinforcement learning. In: Basu, S., Pautasso, C., Zhang, L., Fu, X. (eds.) ICSOC 2013. LNCS, vol. 8274, pp. 298–312. Springer, Heidelberg (2013). https://doi.org/10.1007/978-3-642-45005-1_21
6. Feng, L.-I., Obayashi, M., Kuremoto, T., Kobayashi, K., Watanabe, S.: QoS optimization for web services composition based on reinforcement learning (2013)
7. Trummer, I., Faltings, B., Binder, W.: Multio-bjective quality-driven service selection a fully polynomial time approximation scheme. IEEE Trans. Softw. Eng. **40**(2), 167–191 (2014)
8. Zitzler, E., Thiele, L., Laumanns, M., Fonseca, C.M., Fa Fonseca, V.G.: Performance assessment of multi-objective optimizers: an analysis and review. IEEE Trans. Evol. Comput. **7**(2), 117–132 (2003)
9. Gonsalves, T., Yamagishi, K., Toh, K.: Service cost and waiting time-a multi-objective optimization scenario. In: Applications of Digital Information and Web Technologies (2009)
10. Chan, C.-Y., Jagadish, H.V., Tan, K.-L., Tung, A.K.H., Zhang, Z.: On high dimensional skylines. In: Ioannidis, Y., et al. (eds.) EDBT 2006. LNCS, vol. 3896, pp. 478–495. Springer, Heidelberg (2006). https://doi.org/10.1007/11687238_30
11. Lin, X., Yuan, Y., Zhang, Q., Zhang, Y.: Selecting stars: the k most representative skyline operator. In: Data Engineering (2007)
12. Benouaret, K., Benslimane, D., Hadjali, A.: On the use of fuzzy dominance for computing service skyline based on QoS. In: Web Services (icws) (2011)
13. Alrifai, M., Skoutas, D., Risse, T.: Selecting skyline services for QoS-based web service composition. ACM (2010)
14. Elkosantini, S.: Introduction à la logique floue (2010)
15. Benouaret, K., Benslimane, D., Hadjali, A., Barhamgi, M., Maamar, Z., Sheng, Q.Z.: Web service compositions with fuzzy preferences: a graded dominance relationship-based approach. ACM Trans. Internet Technol. (toit) **13**(4), 12 (2014)
16. Zahri, A.: Le classement, vol. 1 (2008)
17. Yaghoubia, M., Maroosi, A.: Simulation and modeling of an improved multi-verse optimization algorithm for QoS-aware web service composition with service level agreements in the cloud environments. Simul. Model. Pract. Theory **103**, 102090 (2020)
18. Fariss, M., Asaidi, H., Bellouki, M.: Comparative study of skyline algorithms for selecting web services based on QoS. Proc. Comput. Sci. **127**, 408–415 (2018)
19. Fateh, S.: FDMOABC: fuzzy discrete multi-objective artificial bee colony approach for solving the non-deterministic QoS-driven web service composition problem. Expert Syst. Appl. **167**, 1 (2021)
20. Mokdada, L., Fourneaub, J.-M., Abdellic, A., Othman, J.B.: Performance evaluation of a solution for composite service selection problem with multiple consumers. Simul. Model. Pract. Theory **109**, 102271 (2021)
21. Liu, Z., Wang, L., Li, X., Pang, S.: A multi-attribute personalized recommendation method for manufacturing service composition with combining collaborative filtering and genetic algorithm. J. Manuf. Syst. **58**, Part A (2021)

A MapReduce Improved ID3 Decision Tree for Classifying Twitter Data

Fatima Es-sabery[1](✉) ⓘD, Khadija Es-sabery[2]ⓘD, and Abdellatif Hair[1]ⓘD

[1] Department of computer science, Faculty of Sciences and Technology,
Sultan Moulay Slimane University, 23000 Beni Mellal, Morocco
[2] Department of computer science, National School of Applied Sciences,
Cadi Ayyad University, 40000 Marrakech, Morocco

Abstract. In this contribution, we introduce an innovative classification approach for opinion mining. We have used the feature extractor Fast Text to detect and capture the given tweets' relevant data efficiently. Then, we have applied the feature selector Information Gain to reduce the dimensionality of the high feature. Finally, we have employed the obtained features to carry out the classification task using our improved ID3 decision tree classifier, which aims to calculate the weighted information gain instead of information gain used in traditional ID3. In other words, to measure the weighted information gain for the current conditioned feature, we follow two steps: First, we compute the weighted correlation function of the current conditioned feature. Second, we multiply the obtained weighted correlation function by the information gain of this current conditioned feature. This work is implemented in a distributed environment using the Hadoop framework, with its programming framework MapReduce and its distributed file system HDFS. Its primary goal is to enhance the performance of a well-known ID3 classifier in terms of accuracy, execution time, and ability to handle massive datasets. We have performed several experiments that aim to evaluate our suggested classifier's effectiveness compared to some other contributions chosen from the literature.

Keywords: ID3 decision tree · Opinion mining · Hadoop · HDFS · MapReduce · Fast text · Information gain

1 Introduction

Opinion mining (also recognized as sentiment analysis) is an intense research issue in the domain of NLP. It pursues at classifying, studying, selecting, and evaluating the opinions, attitudes, emotions, and reactions from user-posted texts in social media platforms towards entities, such as organization, service, individual, product, topic, event, and issue [1]. With the spread of user-generated text in social networks and microblogging websites such as YouTube, Trip Advisor, Tiktok, Instagram, Twitter, Facebook, Amazon, and WhatsApp; sentiment analysis in web sites and social networks has acquired rising popularity amongst many scientific research and industry communities [2].

© Springer Nature Switzerland AG 2021
M. Fakir et al. (Eds.): CBI 2021, LNBIP 416, pp. 160–182, 2021.
https://doi.org/10.1007/978-3-030-76508-8_13

Twitter is the most communal social network in real-time to express an individual or group of individuals' opinions and ideas about a specific topic through short messages of 280 characters called tweets. Due to the limitation in terms of length of a tweet, the users tend to use Slang, Informal language, many abbreviations, URLs, short forms, and heavy use of emoticons along with Twitter specified expressions like hashtags and user mentions [2], and that pose considerable defies for Twitter sentiment classification. Therefore, it is mandatory to apply intelligent mechanisms to extract useful information from tweets. In this proposal, we use a parallel supervised approach based on Hadoop Big data framework to handle the tweets in such a manner as to overcome the above defies.

Machine learning techniques can be utilized to reveal valuable knowledge which is hidden in such noisy social media data created on daily basis [3]. There are numerous algorithms of machine learning that support learning, such as K-Nearest Neighbors (KNN) [4], Support Vector Machines (SVM) [5], ID3 decision tree [6], Logistic Regression (LR) [7], Naive Bayesian (NB) [8], or Random Forest (RF) [9]. These machine learning techniques are divide into two types, which are supervised and unsupervised learning algorithms [3]. Supervised learning algorithms have widely been utilized for OM [10–14].

In this work, we will use our improved ID3 supervised learning algorithm, because it is executed only on labeled data. So, our work described in this paper carries out opinion mining for English language content applying an improved ID3 decision tree implemented by Hadoop technology. Therefore, the principal proposals of our study can be summed as follows:

1. Our work classifies the collected tweets into three class labels: negative, positive, or neutral.
2. Data preprocessing techniques like lemmatization operation, stemming process, and effect of negation technique are applied to improve the tweets quality by removing the noise and unwanted data.
3. Representation approach FastText is applied on the used dataset of the tweets to convert text-based data (tweets) into text-based numerical.
4. Feature selector like Information Gain is employed to reduce the high feature dimensionality.
5. Our Improved ID3 decision tree algorithm [15] with numerous parameters is used for classifying the analyzed tweet into a neutral, negative or positive class label.
6. Our introduced model is implemented on parallel manner using Hadoop Big data framework to prevent the long execution time problem and improve our proposal's ability to deal with the massive dataset.
7. Comparing the performance of the suggested procedure with other chosen methods from the literature.
8. Our proposed model outperforms baseline approaches by a significant margin in terms of Recall, specificity, error rate, false-positive rate, execution time, kappa statistic, F1-score, false negative rate, accuracy and precision rate.

The remainder of our work is constructed as follows: previous literature researches are described in the "Previous research" section; the "Methodology of our proposed approach" section introduces our developed approach in detail; in "Experiment and results", we introduce the empirical setup and obtained results. And the "Conclusions" section summarizes the proposed method and recommends future work.

2   Previous Research

Here are some works that employ different machine learning techniques for carrying out opinion mining in diverse languages, such as in Sharma and Dey [14]; Patil et al. [16]; Shein and Nyunt [17]; Gamallo et al. [18]; Anjaria and Guddeti [19]; Duwairi [20]; Soni and Pawar [21]; Ngoc et al. [22]. This paper [14] presents a comparative study between five feature selectors, and it deduces that the Gain ratio performs better than other feature selectors. Also, this paper carries out a comparative study between eight classifiers, and it deduces that SVM gives good accuracy for the sentiment analysis. At the same time, the NB approach demonstrates a better classification rate when applied with fewer feature vector spaces. The authors in Patil et al. [16]; proposed a new hybrid approach that integrates the TF-IDF vectorization method with the SVM algorithm to find out the polarity of analyzed textual data. This work proved that SVM could extract high dimensional feature space from textual data, which cancels the need for other feature selection techniques. This suggested contribution in [17] integrates formal concept analysis ontology and SVM for labelling the collected software reviews negative, neutral, or positive. The proposed system in this work [18] combines the Naive-Bayes classifier with unigrams+bigrams to detect Spanish tweets' polarity. Experimental results display that this work achieved an accuracy equal to 67%. This suggested approach [19] developed a new feature extractor to detect the terms influencing the event, it also evolved a method to find the impact factor produced with re-post over Twitter, and it achieved the classification rate equal to 88% in the case of US Presidential Elections 2012. This contribution [20] has performed the sentiment analysis on the Arabic Twitter dataset in dialectical words. In [21], an improved ID3 technique is proposed to avoid the shortcomings of classical ID3. Then this novel ID3 is used to perform the sentiment analysis on the English dataset. The author of the paper [22] have used the C4.5 decision tree algorithm for classifying polarities (positive, negative, neutral) of the English sentences.

The newest innovative research papers for the OM are Lakshmi et al. [23]; Guerreiro et al. [24]; Mehta et al. [25]; Zhang [26]; Lopez-Chau et al. [27]; Addi et al. [28], Patel and Passi [29], and Wang et al. [30]. In [23], the authors proposed a new contribution that intends to classify reviews using two classifiers and determine which of the both perform better performance. Both classifiers are DT and NB. Guerreiro et al. [24] designed a new text mining approach, which is applied to online collected reviews to extract the drivers behind each explicit recommendations. In [25], the author implemented nine separate algorithms, which are: Long short-term memory (LSTM), NB, NLP, Multi-layer perceptron (MLP), Convolutional neural network (CNN), RF, XGBoost, Max entropy, DT, SVM to classify tweets and compare their accuracy on preprocessing data. The experimental result proved that the convolutional neural network outperforms other applied approaches by achieving 79% accuracy. Zhang in [26] evolved a new approach that used the Term Frequency Inverse Document Frequency (TF-IDF) as the feature. It employed Chi-Square and Mutual Information as feature selectors. Then the extracted features are fed into Logistic Regression, Linear SVC, and Multinomial Naive Bayes classifiers for performing the sentiment classification. The authors of the research paper Lopez-Chau *et al.* [27] analyzed the data sets collected from Twitter about the earthquake topic on September 19, 2017, applying OM tools and supervised machine learning. They built three classifiers to find out the sentiment of tweets that appear on the same topic. The experimental result proved that the SVM and NB achieved the best classification rate of classifying the emotions. The author of [28] suggested an innovative method based on a three-level binary tree structure for multi-class hierarchical opinion mining in Arabic text. Empirical outcomes show that their proposed method obtains considerable improvements over other literature approaches. In [29], authors carried out a sentiment analysis on Twitter data for World Cup soccer 2014 held in Brazil to capture the sentimental polarity of the people everywhere employing SVM, KNN, and NB machine learning techniques. The NB achieves good classification equals 88.17%. Wang *et al.* [30] used gradual machine learning process for performing sentiment analysis.

3   Materials and Methods

In the subsequent sections of this section, we will present why we are motivated to propose and evolve this approach. Generally, this proposed paradigm's fundamental structure is constituted of six-phase; the first stage is the data collection in which we used COVID-19_Sentiments massive dataset to evaluate our suggested classifier. The second stage called data pre-processing which aims to remove unwanted and noisy data. The third phase is the data representation phase which converts the tweets into numerical data. The fourth step is the feature selection stage for reducing the dimensionality of extracted features. Finally, we have applied our improved ID3 [15] to classify each tweet sentence into the negative, or neutral, or positive label.

3.1   Motivation

Opinion mining is an important field of research that endeavours to design computational techniques to detect, capture, and evaluate people's ideas and opinions about an entity and its diverse sides and extract the emotions expressed in those ideas and opinions. These expressed sentiments about a product, event, or service have a significant commercial worth. For example, the user-written re-views about products help new users in decision-making, such as buying this new product or no. And are extremely valuable for big companies/organizations in the supervision of their products/events/services, consolidating best relationships with their clients, designing and evolving efficient marketing strategies, enhancing and devising their products/events/services. Motivated by the significant importance of sentiment analysis in different domains of our daily life. We evolved in this work a novel method that aims to carry out the sentiment analysis. This approach combines NLP techniques for performing the text preprocessing, vectorization techniques for converting tweets into numerical values, selecting methods for capturing the essential features, and improved ID3 for performing the classification [15], and the Hadoop framework to parallelize our work.

The first stage of this contribution is the text preprocessing task. Therefore, preprocessing plays an essential role in the text classification process, as introduced in the paper [34]. The authors provided us with a comparative study that aims to evaluate the influence of text preprocessing steps on text classification in terms of accuracy. The experimental result proved that the implementation of the text preprocessing step on linguistic data performed a considerable improvement in classification performance. Many literature studies [35–39] proved that the preprocessing steps positively impact text classification task in terms of accuracy. Thence we motivated by the presented positive result about the text preprocessing steps, and we have applied these preprocessing techniques in this work.

After the preprocessing phase, the next phase is the representation stage that turns out the tweets into numerical values. In this step of our work, we used FastText. The next stage of our work is the feature selection phase, which uses the filter approach like Information Gain. Also, in this stage, we did a comparative study to find an efficient feature selection method.

Finally, in the classification stage, we used our improved ID3. So why we use our improved ID3 proposed in [15] as a classifier in this work? We used our proposed improved ID3 because of two reasons. Firstly it has been achieved good performance as presented in the paper [15]. Secondly, rule-based opinion mining (ID3) often has many advantages, such as the structure of the rule-based algorithm is very simple, which makes it easy to understand by scientific researchers. The ID3 rule-based algorithm is pertinent in large-scale datasets. It often has a high predictive classification rate, and the rules are widespread in the Data Mining field. In the literature, many scientific researchers have endeavoured to discover several manners to apply Data Mining generated rules in opinion mining and find out several diverse relationships between NLP and Data Mining field. The ID3 decision tree is the most common and essential algorithm in the Data Mining field; thus, the produced rules using the ID3 decision tree technique are very accurate. This will result in various scientific researches, hence the motivation for this work.

As a summarized conclusion, this study's goal is to raise the classification performance of opinion mining by incorporate the strength points of text preprocessing techniques in improving the data quality by removing the unwanted and noisy data, feature extraction method in converting the text data into numerical values and extracting the most relevant feature, feature selection method in reducing the high dimensionality of extracted feature in the preceding step and selecting the most interesting feature of and our improved ID3 decision tree algorithm in classifying the input sentence and improving the classification accuracy. As illustrated in Fig. 1, our suggested hybrid approach's overall architecture consists of five stages: data collection step, text preprocessing phase, data representation stage, feature selection methods and our improved ID3 decision tree algorithm.

3.2   Data Collection Stage

In this work, we have been selected two massive datasets to assess the effectiveness of our suggested classifier for opinion mining. The first dataset is termed COVID-19_Sentiments, which is downloaded from this link https://www.kaggle.com/abhaydhiman/covid19-sentiments employing Twitter application programming interfaces (API). It contains 637978 of collected tweets. It classified collected tweets into three class labels which are neutral that takes the value 0, negative that takes a value in the interval $[-1, 0[$, and positive that takes a value in the interval $]0, 1]$. It comprises five features which are introduced below:

- Target: determine the class label of each tweet, where neutral class label takes the value 0, negative class label takes a value in the interval $[-1, 0[$, and positive class label takes a value in the interval $]0, 1]$.
- Ids: is a unique number (520442) that identify each tweet.
- Date: indicates the exact data when the tweet is posted (Sat June 26 05:43:52 +0101 2019).
- Location: indicates the exact location where the tweet is posted (India Country, New Delhi City).
- Text: introduces the text of each tweet, for example "Delhi government will pay salaries to all contract workers, daily wage laborers, guest teachers a cause of Corona Virus".

The essential features in this contribution are the Target and Text features. Consequently, all other features were eliminated. Besides, the "Target" feature divided the data of this given dataset in an inequitable apportionment, where the number of neutral tweets equal to 259458, the number of negative tweets equal to 120646, and the number of positive tweets is equal to 257874. Figure 2 shows the number of neural, negative and positive posted tweets inside of training and testing subset. In the training operation, we have used the number of tweets equal to 574182 tweets. In the testing operation, we have used the number of tweets equal to 63796 tweets. In other terms, the testing subset represents 10% of the total number of tweets in this dataset.

The second dataset is named sentiment140, which is downloaded from this link https://www.kaggle.com/kazanova/sentiment140 employing Twitter application programming programming interfaces (API). This dataset contains

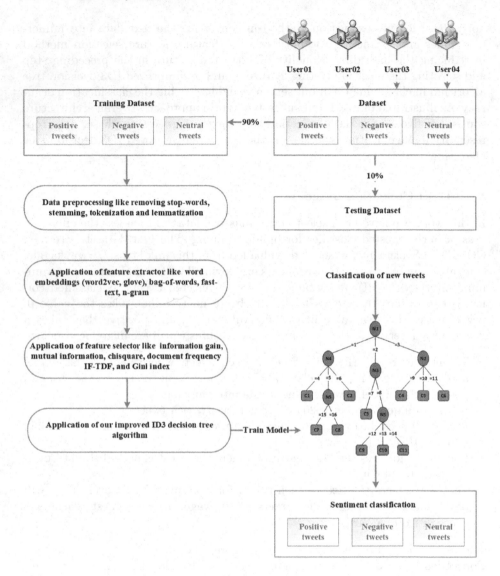

Fig. 1. Universal architecture of the suggested approach.

1600000 collected tweets and all emoticons in this dataset were eliminated. Each tweet have been classified using two class labels positive, and negative, where the value 4 indicates the positive label and the value 0 indicates the negative label. It comprises six features which are presented below:

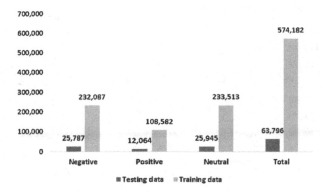

Fig. 2. Number of neutral, negative, and positive tweets for the COVID-19_Sentiments dataset.

- **Target:** determine the class label of each tweet, where the value 4 indicates the positive label and the value 0 indicates the negative label.
- **Ids:** is a unique number (2356221408) that identify each tweet.
- **Date:** for example this feature takes this expression (Mon April 22 19:15:33 PDT 2005) as value which indicates the exact time when the tweet is posted.
- **Flag:** describes the text of the user-query. The 'NO_QUERY' value is assigned to this feature in the case the user does not posted any query.
- **User:** indicates the username that posted the tweet (username:Merissa).
- **Text:** introduces the text of each tweet, for example "thought sleeping in was an option tomorrow but realizing that it now is not. evaluations in the morn..." is the text tweeted by the user Merissa.

In our contribution, we are focused on opinion mining. That is means, obtain every opinion communicated by the Twitter-user in every posted tweet. In consequence, the other features in this dataset have not any impact on the training operation. Consequently, we eliminated the "Date", "Flag", "Ids", "User" features, and we saved the "Target" and "Text" features from the dataset. The "Target" feature apportionment of the dataset is equitable apportionment, because 50% of the data being labelled negative class label, ranging from the row number 0 to 799999*th* row, and another 50% of the data labelled positive class label, which are ranging from the row number 800000 to 1600000*th* row. In this work, the given dataset is splitted into two subsets which are training and testing subsets. Hence, we have utilized these two subsets to demonstrate our proposed method's classification effectiveness compared to other suggested techniques picked from the literature. Figure 3 presents the number of user tweets in each training and testing subsets, where a total of 160000 posted tweets were utilized in the testing operation and 1440000 posted tweets were utilized in the training operation.

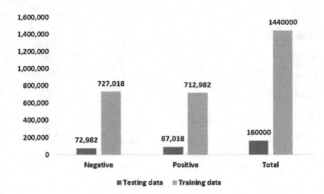

Fig. 3. Number of positive and negative tweets for the sentiment140 dataset.

3.3 Data Preprocessing Stage

To examine the semantic data, noise or abnormal data must be purified, so that the accurate emotions and meaning can be procured from the examined linguistic text. At this point, the text preprocessing techniques play a primary role. In this contribution, the type of analyzed data is the tweets extracted from the Twitter platform. These tweets often are in an unregulated form of text and contain inefficient, noise, and undesired knowledge. Data preprocessing techniques normalize the semantic data, purifier noise, eliminate unwanted information, and clarify the vocabulary employed to analyze emotions from the tweets. There are numerous steps to be carried out to reappraise the data accurately and extract the real sense behind it via data processing. Thence these followed up preprocessing steps in this study are introduced below:

Remove Usernames, Numbers, Hashtags, and URLs: It is a common technique to remove the usernames, numbers, hashtags, and URLs from the tweet since they do not express any sentiments. Eliminate white-spaces, special characters, and punctuation: The foremost stage to perform is removing all occurred white-spaces in the tweet, then we delete exclamation, stop and question punctuation marks. Finally, all existed special characters are eliminated since they do not hold any negative or positive effect on communicated sentiment in the given tweet. After the application of all text preprocessing tasks introduced formerly, we preserved only the uppercase and lowercase letters.

Lower-Casing: After the application of both formerly presented stages, all special characters have been removed and only the letters have been preserved. Therefore the next stage is the application of lower casing process. That means all the uppercase letters in the given tweet were converted to lower case, and that minimize the dimensionality of each terms.

Replace Elongated Words: This process aims to remove the letter, which is appeared at least more than two times in the word like the elongated term "Saaaad". So, after effecting this technique, the elongated word "Saaaad" becomes "Saad" and stamped with at most two characters.

Remove Stop-Words: It is the technique that removes the words which are frequently confronted in texts without dependence on a specific topic (e.g., "a", "an", "of", "on", "the", "in", "I", "she", "it", "your", "and" etc.). These frequent use words are often ineffective and meaningless for the data classification task and deleted beforehand of the classification. Stop-words are particular to the studied language, as in the situation of stemming technique.

Stemming: It is a technique that aims to reduce the size of each word by converting the derived word into its root, or stem form. Since the stem word is semantically analogous to its derived word. The stemming process is ordinarily referred to as stemming algorithms or stemmers. A straightforward stemming algorithm looks up the stem form of the derived word in a lookup table. This type of procedure is fast and easy to understand. Its shortcoming the lookup table does not contain all inflected forms of the stem word. For example, the words "using", "usage", "used" must be reduced to only one root form, which is "use".

Lemmatization: it is a technique that acts as a Stemming algorithm. Its main goal is to find the root or stem form of words in the analyzed sentences. The only difference between these similar techniques is in the followed algorithm to capture the lemma or stem words, since the lemmatization takes into consideration the morphological analysis of the derived words.

Tokenization: it is a technique that divides each input tweet into a set of meaningful words since each word is named a token. For example, a piece of text is split into sentences or words. In this contribution, we employed an NLTK tokenizer developed by Python.

In this work, the stage that follows up the preprocessing data phase is the data representation phase. In other terms, the obtained text after the utilization of all text preprocessing tasks previously described will be the input of the data representation methods.

3.4   Data Representation Stage

Most machine learning classifiers can only deal with numerical data; the kind of data we have in this work is text-based data. Therefore, to address our contribution for handling the obtained textual data after applying the preprocessing data stage employing machine learning algorithms, these textual data needs to be turned out into numerical values. This process is termed text vectorization or feature extraction phase. It is one of the fundamental issues in NLP that allowing machine learning techniques for examining text-based data. As we said previously, Feature extraction is the operation of turning out the text-based input data into a set of numerical features. The effectiveness of each machine learning classifier is depended significantly on the extracted features. Therefore, it is critical to select the better features that have a positive impact on classification accuracy. There are several diverse feature extractors to represent the textual data in numerical data, including bag-of-words, n-grams, GloVe, word2vec, Fast-Text, and TF-IDF. These different feature extractors will lead to different analysis results and influence differently the classification performance. This stage's primary purpose is to summarize and convert text-based input data into a set of feature vectors that operate agreeably to the used machine learning classifier. On the other hand, our principal purpose is to apply FastText as feature extractor be-cause it is more efficient than other previously cited extractors due to this study [34].

FastText: Facebook AI Research Team proposed new word embedding approach which is named Fast Text in order to improve the learning effectiveness of word representation. This technique is considered as a new version of the Word2vec word embedding technique; So, instead of learning the word embedding vector of a set of words immediately as in the Word2vec approach, Fast Text approach learns the word embedding vector of a set of n-gram of characters. For example, the vectorization of the term "Happy" employing the Fast Text approach with n-gram = 2 is (H, Ha, ap, pp, py, y), and the brackets indicate the starting and ending of the vectorized term. The advantage of the Fast Text approach permits to capture the meaning of shorter terms and aids the word embedding process to pick up the prefixes and suffixes of the learned term. Therefore, the next stage after dividing the inputted term by using the character n-grams is the application of either continuous bag-of-words or skip-gram in order to generate term embed-ding. Fast Text approach performs better with unseen terms and the terms out-of-lexicon than other word embedding techniques. Therefore, even if the term is out-of-lexicon in the former learning stages, this approach is broken down it to multiple n-gram characters to learn its word embedding vector. GloVe and Word2vec approaches both be unsuccessful to generate word embedding vector for the out-of-lexicon terms, unlike Fast Text approach that has the ability to generate the word embedding vector of the unseen terms. This is the powerful advantage of this technique in comparing with other approaches.

After the Word embedding stage that endeavors to turn out the text into numerical values, which it inputs a preprocessed text, and it outcomes a term em-bedding vector. The next phase is the feature selection, as outlined in the next subsection.

3.5   Feature Selection

After the features extraction stage, in which the features are derived from the text corpus through the application of the Fast text word embedding approach. It is worth taking into account the dimensionality of features. The reason is that the high feature dimensionality requires more costly computational resources. And also may probably decrease the accuracy and effectiveness of the applied learning algorithm. That is named "The Curse of Dimensionality".

The cause why high features dimensionality may lead to the decreasing in the accuracy of applied patterns is because when the vector space of the features expands bigger and bigger by appending further and further features, the vector space of the features becomes further and further sparse. The further sparse in the vector space of the features causes further overfitting in the applied machine learning algorithm on the training dataset. Therefore, suppose the vector space of the features is overly sparse. In that case, applied learning algorithms will over-fit the training dataset and thus became inefficient in classifying the unknown instances in the testing dataset. On the other hand, the try of decreasing dimensions of the features vector space may lead to decreasing in the performance as well, as this might eliminate some relevant features and cause to underfitting of the trained algorithms to the data.

In the literature, there are two techniques for decreasing the dimensionality of the feature space. The first technique is feature projection, and the second technique is the feature selection. The distinction between these techniques is that the former technique shrinks the dimensionality of the vector space of the features by projecting the features in high-dimensional vector space upon low-dimensional vector space. While the latter technique eliminates irrelevant features. A new subset of features is created with the former techniques, while a subset of features is kept with the latter techniques. In this paper, Information Gain is employed as feature selection technique for reducing the dimensionality of feature vector spaces for text opinion mining in order to ameliorate the classification rate and diminish the execution time of used machine learning algorithms.

Information Gain (IG): has been utilized commonly as a term (feature) goodness criterion in the text classification based on machine learning techniques such as C4.5 and ID3. It gauges the information needed in bits for class label prediction of a given sentence (or document) by measuring the entropy that is computed based on the absence or existence of a feature word in that sentence. IG is obtained by measuring the effect of the feature word's inclusion on diminishing overall entropy. The predictable information required to predict the class label of an example in partition PA is recognized as entropy and is computed as follows:

$$Entropy(PA) = -\sum_{i=1}^{C}(P_i) \times log_2(P_i) \tag{1}$$

Where:

- C indicates the overall number of the class labels in the corpus.
- $P_i = \frac{|C_{i,PA}|}{|PA|}$ is the probability that an arbitrary example sentence in partition PA has the class label C_i.

To predict the class label of an example sentence in PA on some feature $F = \{f_1, f_2, .., f_v\}$, and the partition PA is divided into k sub-partitions $\{PA_1, PA_2, ..., PA_k\}$. Therefore, the information required to get an exact prediction is calculated by:

$$Entropy_F(PA) = -\sum_{j=1}^{k}\frac{|PA_j|}{|PA|} \times Entropy(PA_j) \tag{2}$$

Where:

- $\frac{|PA_j|}{|PA|}$ is the weight of the jth partition PA_j.
- $Entropy(PA_j)$ is the entropy of jth partition.

Finally, IG by dividing up on feature F is measured by the following equation:

$$IG(F) = Entropy(PA) - Entropy_F(PA) \tag{3}$$

In this research, before reducing the feature vector space dimension, each feature within the sentence is ordered depending on their relevance for the text classification in decrescent order employing the IG technique. Thus, in the learning process of text classification, features that have lesser importance are neglected, and dimension decrease techniques are exercised to the features that have higher importance (i.e. IG).

3.6   Classifier

In this work, and after the feature selection phase, the next phase is the classification task, in which we used improved ID3 in order to classify each input sentence into the negative or neutral or positive class label. ID3 decision tree technique is a classification paradigm that pursues to measure the IG criterion described by the equation for creating a decision tree. This decision tree technique computes the IG to pick out the better splitting feature in every algorithm's round. The computing IG process considers only a present conditional feature and class label feature, and the other conditional features cannot be employed to gauge the feature importance. Thus, we proposed a novel enhanced ID3 based on weighted adjusted IG. The weighted modified IG considers the association between the current conditional feature, the feature decision, and the other conditions features in the learning process's progress. The principal purpose of our enhanced ID3 based on weighted adjusted IG is to gauge the influence of the other conditional features on the IG criterion for the present conditional features in the training operations. Or rather, our enhanced ID3 approach takes into account the association between the present analyzed feature and the other features and the association between each conditional feature and the class label feature in the progress of the training operation. The weighted modified IG is computed using weighted attribute and the weighted correlation function.

3.7   Weighted Attribute

Let $F = \{F_1, F_2, ..., F_k\}$ be a be a collection of K conditional features. We suppose the occurrence number of feature $F_i (i = 1, 2, \ldots, K)$ is K_i. Thus, the frequency of F_i can be described as:

$$TF_i = \frac{K_i}{K} \tag{4}$$

Then the weight of the feature F_i can be calculated as:

$$WF_i = \frac{TF_i}{\sum_{i=1}^{K} TF_i} \tag{5}$$

Due to the conditions of weight theory, the total importance of all features satisfy the equation described below:

$$WF_i = 1 \tag{6}$$

3.8 Weighted Correlation Function

Let $F = \{F_1, F_2, ..., F_k\}$ be a collection of K conditional features with an interval of values $\{RV_1, RV_2, ..., RV_k\}$, respectively. Let C be a class label feature with an interval of values RV_C. $F_{i \in (i=1,2,...,K)}$ is one of conditional features of the collection and has N values, whereas $RV_{i \in (i=1,2,...,K)} = f_1, f_2, ..., f_N$. C is the class label feature and has Y values $RV_C = \{C_1, C_2, ..., C_Y\}$. Thus, the weighted correlation function (WCF) between the conditional feature A_k and the class label feature Y is computed using the following formula:

$$CF(F_i, C) = \frac{\sum_{v=1}^{N} \left| |F_{vj}| - \sum_{j=2}^{Y} |F_{vj}| \right|}{N} \tag{7}$$

Where N is the number of values of F_i; $|F_{vj}|$ is the examples' number of that vTh value of F_i be associated with the jTh category of class label feature C. Then WCF of the feature F_i can be calculated as:

$$WCF(F_i, C) = \frac{CF(F_i, C)}{\sum_{i=1}^{K} CF(F_i, C)} \tag{8}$$

3.9 Weighted Modified Information Gain

Weighted modified information gain ($WMIG$) is measured using the calculated IG in the Eq. (3) and the computed WCF in the Eq. (8):

$$WMIG(F) = IG(F) \times \sum_{i=1}^{N} WCF(F). \tag{9}$$

Subsequently, due to the IG criterion and the weight representation theory, the particular implementation of our improved ID3 decision tree technique based on $WMIG$ is introduced in detail in the Algorithm 1.

4   Experiment and Results

The previous sections were consecrated to the general introduction, previous research, materials, and proposed classifier methods. As we said early,

---

Algorithm 1: OUR ENHANCED WEIGHTED ALGORITHM ID3

Input:

. **Training feature set :** $F = \{F_1, F_2, ..., F_K\}$ be a collection of K conditional features with an interval of values $\{RV_1, RV_2, ..., RV_N\}$, respectively. And C is a class label feature and has RV_Y values $RV_Y = \{C_1, C_2, ..., C_Y\}$.

. **Training example set :** $E = \{(e_i, c_i) \| e_i \in RV_1 \times RV_2 \times ... \times RV_n, c_i \in RV_Y\}$ is an example picked from an unknown distribution, where e_i has an outcome c_i related with it.

. **Ending Condition:** is the condition to end the training operation.

   1 All samples in the training dataset relate to a unique value of c.

   2 All feature values of E are the same or the feature set F is empty.

Output: DT (Decision tree)

TreeCreate(E, F) : generate a novel decision tree DT with a single root

if EndingCondition(E) = 1 **then**

   Tick the root node as a leaf node labelling the class C.

   return

else if EndingCondition(E) = 2 **then**

   Tick DT as a leaf node with the most prevalent value of class feature C.

   return

else

   for $F_i \in F$ **do**

     Compute the IG employing the next equation:

     $IG(F_i)[i] \leftarrow Entropy_{F_i}(C) - Entropy_{F_i}(F_i, E)$

   end for

   $SImp \leftarrow 0$

   for featureValues $FV_i \in RV_i$ **do**

     $PImp \leftarrow 1$

     Compute the importance of each feature employing the training set E_i of each value FV_i of feature F_i

     $S \leftarrow 0$

     for featureValues $F_k \in F \setminus \{F_i\}$ **do**

       compute the frequency employing the following equation

       $\bullet CF_{(F_k, C)}[k] \leftarrow \frac{\sum_{v=1}^{|FV_i|} \|F_{vy}\| - \sum_{c=2}^{Y} \|F_{vy}\|}{V_k}$ $\bullet S \leftarrow S + CF_{(F_k, C)}[k]$

     end for

     for featureValues $F_k \in F \setminus \{F_i\}$ **do**

       compute the importance employing the following equation

       $\bullet WCF_{(F_j, C)}[k] \leftarrow \frac{CF_{(F_k, C)}[k]}{S}$ $\bullet PImp \leftarrow PImp \times WCF_{(F_j, C)}[k]$

     end for

     $SImp \leftarrow SImp + \frac{|E_i|}{|E|} \times PImp$

   end for

   $IG(F_i)[i] \leftarrow IG(F_i)[i] \times SImp$

end if

Determining the better dividing feature F_{better} that has the maximum weighted adjusted $IG(F_i)[i]$

$F_{better} \leftarrow argmax_F IG(F_i)[i]$; Attach F_{better} into DT

for featureValues $v \in F_{better}$ **do**

   create an edge that links the new node with the previous created DT, and E_v describes the subset of the examples in E of which the F_{better} feature is v

   if E_v = null **then**

     Tick the edges of the new node as the values labels of the preceding leaf node, and its class label is ticked as the class label that contains the highest number of examples in E

     return

   else

     Recursion of **TreeCreate(E_v,F\\{F_{better}})** continues

   end if

end for

---

our proposed classifier consists of five steps; data collection in which we have chosen Sentiment140 (https://www.kaggle.com/kazanova/sentiment140 and COVID-19_ Sentiments (https://www.kaggle.com/abhaydhiman/covid19-sentiments) datasets to apply our contribution to them. After the data collection phase, the next phase is the text preprocessing phase. We have applied several preprocessing tasks to reduce the noisy data and remove unwanted information for improving the data quality. The data representation stage follows up the text-preprocessing stage, which is converting the text-based input data into a set of numerical features. We have applied several techniques in this phase, including N-grams, Bag-of-words, TF-IDF, GloVe, Word2vec, and FastText. After the data representation stage, the next phase is the feature selection phase to reduce the high feature dimensionality. Also, in this phase, we have applied many feature selectors such as Chi-square, IG, Gain Ratio, and Gini Index. Finally, and after the feature selection phase, the next phase is applying our improved ID3 as presented in the subsection "Classifier". This section performs five experiments to demonstrate our suggested classifier's correctness and effectiveness compared to other literature's methods. And to assess its effectiveness, we have selected nine evaluation criterion as presented in subsection "Evaluation metrics". And to assess our text classification method, we principally compute ten assessment metrics: *True Positive Rate* (TPR), *True Negative Rate* (TNR), *Kappa Statistic* (KS), *False Positive Rate* (FPR), *Precision* (PR), *False Negative Rate* (FNR), *Classification Rate or Accuracy* (AC), *Error Rate* (ER), *Time Consumption* (TC), and *F1-score* (FS) [31,32].

In this experiment, we are going to introduce the empirical outcomes of our proposed classifier. These empirical results are achieved by practicing our proposed classifier and other methods such as ID3, C4.5, Soni and Pawar [21], Ngoc *et al.* [22], Lakshmi *et al.* [23], Addi *et al.* [28], Patel and Passi [29], and Wang *et al.* [30], on both chosen datasets as described in subsection "Data Collection Stage". To demonstrate which of these methods is more efficient and has better performance, we measure nine evaluation metrics as outlined previously. Our classifier will be performed parallelly, employing the Hadoop framework with the Hadoop distributed file system and the MapReduce programming paradigm [33]. The Hadoop cluster comprises four salve nodes and one master node.

Figure 4, depicts the experimental result achieved for the classification rate applying our suggested classifier on Sentiment140 and COVID-19_Sentiments datasets, and we compare the empirical outcome reached with other approaches like ID3, C4.5, Soni and Pawar [21], Ngoc *et al.* [22], Lakshmi *et al.* [23], Addi *et al.* [28], Patel and Passi [29], and Wang *et al.* [30]. As shown in Fig. 4, we remark that our proposed classifier outperforms the other implemented approaches in terms of classification rate. As Fig. 4 depicts, our proposed classifier achieved a higher accuracy equal to 86.53%, and 88.82% when it applied on Sentiment140 and COVID-19_Sentiments datasets, respectively. And the Addi *et al.* [28] has the lower accuracy equal to 43.02%, and 31.08% in this experiment. Our suggested approach achieved a good performance compared to other methods because we have applied several preprocessing techniques to improve data quality. And according to an experiment we have performed for choosing the most efficient vectorization method amongst n-gram, bag-of-words, IF-IDF, GloVe, Word2Vec, and FastText approaches. The experimental results proved that n-gram, bag-of-words, IF-IDF, GloVe, Word2Vec, and FastText achieved an accuracy equal to 51.76%, 64.24%, 71.05%, 72.23%, 77.43%, 87.13%, respectively. Therefore, we have used the word embedding technique FastText to vectorize the data because it achieved higher accuracy, which equals 87.13%. Then, we have carried out a second experiment to compare Chi-square, Information Gain, Gain Ratio, and Gini Index performance, and the experimental result demonstrated that Chi-square, Information Gain, Gain Ratio, and Gini Index reached an accuracy equal to 73.35%, 80.16%, 86.53%, 60.62% respectively. Hence, we have applied the information gain as a feature selector to reduce the high dimensionality of extracted features. Furthermore, we have used our improved ID3 as a classifier due to its reached good result presented in the paper [15]. Finally, we have implemented our proposed approach on the Hadoop cluster of five computer nodes to avoid long processing times.

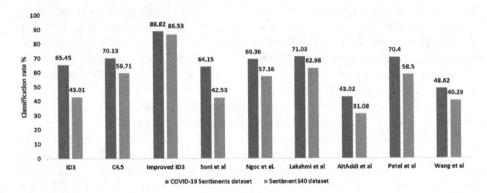

Fig. 4. Classification rate obtained by implementing our classifier and other approaches.

Figure 5, illustrates the empirical result obtained for the error rate applying our suggested classifier and other previously chosen approaches on Sentiment140 and COVID-19_Sentiments datasets. As shown in Fig. 5, we notice that our proposed classifier has a lower error rate than other implemented approaches. And if we compared our classifier with Addi et al. [28] approach, we note that our classifier reduce the error rate from 56.98%, and 68.92% (Addi et al. [28]) to 11.18%, and 13.47% (our improved ID3) for the Sentiment140, and COVID-19_Sentiments datasets, respectively. Soni and Pawar [21] has an error rate that equals 57.47% when it applied on Sentiment140. Its error rate is very higher than our approach because in the [21] method; the authors do not apply all preprocessing techniques to improve data quality, perform the data vectorization manually, and do not apply any feature selector. Ngoc et al. [22] approach also has a higher error rate equals 42.84% in the case of Sentiment140 data due to the application of C4.5 decision tree algorithm without the applicability of data preprocessing, data vectorization, and data selection phases. Lakshmi *et al.* [23] method gives an error rate equals to 37.02%, which is lower compared to Soni and Pawar [21], and Ngoc *et al.* [22] approaches because the authors of this work [23] combines the Naïve Bayes and decision tree as the classifier. However, it is still higher than our proposed approach because they do not use data preprocessing, data vectorization, and feature selection phases. Patel and Passi [29], and Wang *et al.* [30] achieved an error rate equals 41.5%, and 59.71% respectively, and they are also higher values compared to our proposed approach because the authors of these both works do not follow the appropriate steps for performing the sentiment analysis correctly.

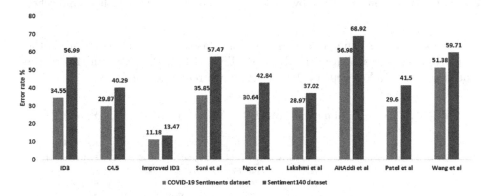

Fig. 5. Error rate obtained by implementing our classifier and other approaches.

Another experiment is conducted to examine the execution time between our classifier and the other chosen methods. Figure 6 displays the experimental result after applying all proposals on both elected datasets. Without forgetting that our classifier is executed in a parallel mode on five computers utilizing the framework Hadoop. As shown in Fig. 6, we observe that our developed model has a lower execution time rate than other implemented methods. And let's compared our classifier with the Patel and Passi [29] method. We remark that our classifier reduces the execution time from 5402.15 s (Patel and Passi approach) to 45.21 s (our improved ID3) for the Sentiment140 dataset and from 842.91 s (Patel and Passi approach) to 15.95 s (our improved ID3) for the COVID-19_Sentiments dataset. Soni and Pawar [21], Ngoc *et al.* [22], Lakshmi *et al.* [23], Addi *et al.* [28], Patel and Passi [29], and Wang *et al.* [30] approaches consumed 987.48 s, 1587.34 s, 4250.98 s, 2161.83 s, 5402.15 s, and 1730.1 s for Sentiment140 dataset,respectively. We deduced that all approaches have a higher processing time than our proposed approach, which confirms that the implementation of our classifier using the Hadoop cluster of five machines is a more efficient tool to reduce the consumption time.

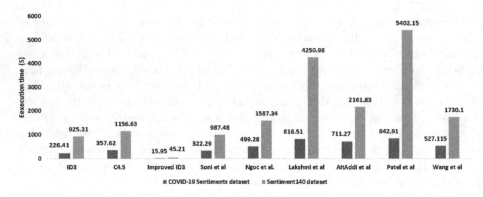

Fig. 6. Execution time obtained by implementing our classifier and other approaches.

For more proving the performance of our classifier, we have computed other evaluation measures such as TPR, FNR, TNR, FPR, PR, KS, and FS as presented earlier. Table 1 depicts the experimental result reached.

Table 1. Experimental outcome of TPR, FNR, TNR, FPR, PR, KS, and FS.

		TPR	FNR	TNR	FPR	PR	KS	FS
COVID-19_Sentiments dataset	Our classifier	85.72	14.28	86.51	13.49	86.67	87.69	85.54
	[21]	64.57	35.43	59.12	40.88	60.76	61.89	60.48
	[22]	69.08	30.92	67.15	32.85	66.51	65.70	66.84
	[23]	71.46	28.54	70.89	29.11	72.32	73.51	70.54
	[28]	43.54	56.46	44.62	55.38	42.27	43.25	42.92
	[29]	71.92	28.08	70.52	29.48	72.69	70.56	71.96
	[30]	49.11	50.89	48.17	51.83	47.92	49.43	48.69
Sentiment140 dataset	Our classifier	81.41	18.59	82.33	17.67	83.04	84.58	83.87
	[21]	43.56	56.44	42.71	57.29	40.19	41.79	42.64
	[22]	56.24	43.76	57.19	42.81	56.92	54.61	55.24
	[23]	63.78	36.22	64.50	35.5	63.29	62.26	64.73
	[28]	32.91	67.09	33.13	66.87	32.92	34.51	33.85
	[29]	58.27	41.73	59.62	40.38	57.92	58.41	57.86
	[30]	40.91	59.09	43.78	56.22	41.92	40.54	42.81

As shown in Table 1, we deduce that our classifier outperforms all other classifiers applied on both used datasets COVID-19_Sentiments and Sentiment140. Our classifier achieved higher values (COVID-19_Sentiments dataset: 85.72%, 14.28%, 86.51%, 13.49%, 86.67%, 87.69%, 85.54%) and (Sentiment140 dataset: 81.41%, 18.59%, 82.33%, 17.67%, 83.04%, 84.58%, 83.87%) at the level of all computed evaluation measures (TPR, FNR, TNR, FPR, PR, KS, FS). The good performance achieved by our suggested classifier is due to the followed steps in this work, such as the applied preprocessing techniques in the data preprocessing stage like "Remove usernames, numbers, hashtags, and URLs", "Lower-casing", "Replace elongated words", "Remove stop-words", "Stemming", "Lemmatization", and "Tokenization". The used FastText word embedding method in the data representation phase. The application of Information Gain in the feature selection phase. Then, the use of our improved ID3 in the classification phase. Finally, the implementation of our approach using the Hadoop framework.

5   Conclusions

This study has proposed an innovative approach to classify tweets into positive, negative, or neutral based on social media big data. The suggested system consists of five parts: data collection, data preprocessing, data representation, data classification, feature selection, and application of the Hadoop framework. In the data collection phase, we have chosen Sentiment140 and COVID-19_Sentiments datasets to evaluate our proposal. For the data preprocessing phase, we have applied multiple preprocessing tasks on both chosen datasets. Then, we have applied the FastText technique to convert the textual data into numerical data in data representation. After the feature extraction, the next phase is the feature selection, in which we have applied the Information Gain method in order to reduce the dimensionality of the feature space. Finally, we have applied our improved ID3 classifier on both used datasets, and we obtained an accuracy equal to 86.53% and 88.82% for Sentiment140 and COVID-19_Sentiments, respectively. Our proposal is parallelized using the Hadoop Framework (MapReduce+HDFS).

Our future work is to merge the fuzzy logic theory and our proposal in this paper in order to handle with continuous-valued features, taking into consideration various parameters concerning the feature extractors and feature selectors. Utilization of Mamdani fuzzy system as a classifier for sentiment analysis in order to deal with uncertainty and vagueness sentiments held in the data expressed by social media users. Integration of fuzzy rule-based model with our MapReduce improved ID3 decision tree for inferring the sentiment expressed in speech cues on the social media networks.

References

1. Tang, D., Qin, B., Liu, T.: Deep learning for sentiment analysis: successful approaches and future challenges. Wiley Interdiscip. Rev. Data Min. Knowl. Discov. **5**(6), 292–303 (2015). https://doi.org/10.1002/widm.1171
2. Severyn, A., Moschitti, A.: Twitter sentiment analysis with deep convolutional neural networks. In: Proceedings of the 38th International ACM SIGIR Conference on Research and Development in Information Retrieval, pp. 959–962. Association for Computing Machinery, New York (2015). https://doi.org/10.1145/2766462.2767830
3. Han, J., Kamber, M., Pei, J.: Getting to know your data. In: Data Mining: Concepts and Techniques. Morgan Kaufmann, San Francisco (2006)
4. Daeli, N.O.F., Adiwijaya, A.: Sentiment analysis on movie reviews using information gain and K-nearest neighbor. J. Data Sci. Appl. **3**(1), 1–7 (2020). https://doi.org/10.34818/jdsa.2020.3.22

5. Zainuddin, N., Selamat, A.: Sentiment analysis using support vector machine. In: 2014 International Conference on Computer, Communications, and Control Technology (I4CT), Langkawi, Malaysia, pp. 333–337 (2014). https://doi.org/10.1109/I4CT.2014.6914200

6. Rathi, M., Malik, A., Varshney, D., Sharma, R., Mendiratta, S.: Sentiment analysis of tweets using machine learning approach. In: 2018 Eleventh International Conference on Contemporary Computing (IC3), Noida, India, pp. 1–3 (2018). https://doi.org/10.1109/IC3.2018.8530517

7. Ramadhan, W.P., Novianty, S.T.M.T.A., Setianingsih, S.T.M.T.C.: Sentiment analysis using multinomial logistic regression. In: 2017 International Conference on Control, Electronics, Renewable Energy and Communications (ICCREC), Yogyakarta, Indonesia, pp. 46–49 (2017). https://doi.org/10.1109/ICCEREC.2017.8226700

8. Troussas, C., Virvou, M., Espinosa, K.J., Llaguno, K., Caro, J.: Sentiment analysis of Facebook statuses using Naive Bayes classifier for language learning. In: 2013 International Conference on Information, Intelligence, Systems and Applications (IISA), Piraeus, Greece, pp. 1–6 (2013). https://doi.org/10.1109/IISA.2013.6623713

9. Al Amrani, Y., Lazaar, M., El Kadiri, K.E.: Random forest and support vector machine based hybrid approach to sentiment analysis. Proc. Comput. Sci. **127**, 511–520 (2018). https://doi.org/10.1016/j.procs.2018.01.150

10. Neethu, M.S., Rajasree, R.: Sentiment analysis in twitter using machine learning techniques. In: 2013 Fourth International Conference on Computing, Communications and Networking Technologies (ICCCNT), Tiruchengode, India, pp. 1–5 (2013). https://doi.org/10.1109/ICCCNT.2013.6726818

11. Boiy, E., Moens, M.-F.: A machine learning approach to sentiment analysis in multilingual web texts. Inf. Retr. Boston **12**(5), 526–558 (2009). https://doi.org/10.1007/s10791-008-9070-z

12. Gautam, G., Yadav, D.: Sentiment analysis of twitter data using machine learning approaches and semantic analysis. In: 2014 Seventh International Conference on Contemporary Computing (IC3), Noida, India, pp. 437–442 (2014).https://doi.org/10.1109/IC3.2014.6897213

13. Hasan, A., Moin, S., Karim, A., Shamshirband, S.: Machine learning-based sentiment analysis for Twitter accounts. Math. Comput. Appl. **23**(1), 11 (2018). https://doi.org/10.3390/mca23010011

14. Sharma, A., Dey, S.: A comparative study of feature selection and machine learning techniques for sentiment analysis. In: Proceedings of the 2012 ACM Research in Applied Computation Symposium, San Antonio, Texas, pp. 1–7 (2012). https://doi.org/10.1145/2401603.2401605

15. Es-Sabery, F., Hair, A.: An improved ID3 classification algorithm based on correlation function and weighted attribute*. In: 2019 International Conference on Intelligent Systems and Advanced Computing Sciences (ISACS), Taza, Morocco, pp. 1–8 (2019). https://doi.org/10.1109/ISACS48493.2019.9068891

16. Patil, G., Galande, V., Kekan, M.V., Dange, K.: Sentiment analysis using support vector. Int. J. Innov. Res. Comput. Commun. Eng. **2**(1), 2607–2612 (2014). https://citeseerx.ist.psu.edu/viewdoc/download?doi=10.1.1.1084.6923&rep=rep1&type=pdf

17. Shein, K.P.P., Nyunt, T.T.S.: Sentiment classification based on ontology and SVM classifier. In: 2010 Second International Conference on Communication Software and Networks, Singapore, pp. 169–172 (2010). https://doi.org/10.1109/ICCSN.2010.35

18. Gamallo, P., Garcia, M., Fernandez-Lanza, S.: TASS: a Naive-Bayes strategy for sentiment analysis on Spanish tweets. In: XXIX Congreso de la Sociedad Española de Procesamiento de Lenguaje Natural, Spain, pp. 126–132 (2013). https://gramatica.usc.es/~gamallo/artigos-web/TASS2013.pdf
19. Anjaria, M., Guddeti, R.M.R.: A novel sentiment analysis of social networks using supervised learning. Soc. Netw. Anal. Min. **4**, 181 (2014). https://doi.org/10.1007/s13278-014-0181-9
20. Duwairi, R.M.: Sentiment analysis for dialectical Arabic. In: 2015 6th International Conference on Information and Communication Systems (ICICS), Amman, Jordan, pp. 166–170 (2015). https://doi.org/10.1109/IACS.2015.7103221
21. Soni, V.K., Pawar, S.: Emotion based social media text classification using optimized improved ID3 classifier. In: 2017 International Conference on Energy, Communication, Data Analytics and Soft Computing (ICECDS), Chennai, India, pp. 1500–1505 (2017). https://doi.org/10.1109/ICECDS.2017.8389696
22. Ngoc, P.V., Ngoc, C.V.T., Ngoc, T.V.T., Duy, D.N.: A C4.5 algorithm for English emotional classification. Evol. Syst. **10**(3), 425–451 (2019). https://doi.org/10.1007/s12530-017-9180-1
23. Lakshmi Devi, B., Varaswathi Bai, V., Ramasubbareddy, S., Govinda, K.: Sentiment analysis on movie reviews. In: Venkata Krishna, P., Obaidat, M.S. (eds.) Emerging Research in Data Engineering Systems and Computer Communications. AISC, vol. 1054, pp. 321–328. Springer, Singapore (2020). https://doi.org/10.1007/978-981-15-0135-7_31
24. Guerreiro, J., Rita, P.: How to predict explicit recommendations in online reviews using text mining and sentiment analysis. J. Hosp. Tour. Manag. **43**, 269–272 (2020). https://doi.org/10.1016/j.jhtm.2019.07.001
25. Mehta, R.P., Sanghvi, M.A., Shah, D.K., Singh, A.: Sentiment analysis of tweets using supervised learning algorithms. In: Luhach, A.K., Kosa, J.A., Poonia, R.C., Gao, X.-Z., Singh, D. (eds.) First International Conference on Sustainable Technologies for Computational Intelligence. AISC, vol. 1045, pp. 323–338. Springer, Singapore (2020). https://doi.org/10.1007/978-981-15-0029-9_26
26. Zhang, J.: Sentiment analysis of movie reviews in Chinese. Uppsala University (2020). https://www.diva-portal.org/smash/get/diva2:1438431/FULLTEXT01.pdf
27. López-Chau, A., Valle-Cruz, D., Sandoval-Almazán, R.: Sentiment analysis of Twitter data through machine learning techniques. In: Ramachandran, M., Mahmood, Z. (eds.) Software Engineering in the Era of Cloud Computing. CCN, pp. 185–209. Springer, Cham (2020). https://doi.org/10.1007/978-3-030-33624-0_8
28. Addi, H.A., Ezzahir, R., Mahmoudi, A.: Three-level binary tree structure for sentiment classification in Arabic text. In: Proceedings of the 3rd International Conference on Networking, Information Systems & Security (NISS2020), Marrakech, Moroccopp. 1–8 (2020). https://doi.org/10.1145/3386723.3387844
29. Patel, R., Passi, K.: Sentiment analysis on twitter data of world cup soccer tournament using machine learning. IoT **1**(2), 218–239 (2020). https://doi.org/10.3390/iot1020014
30. Wang, Y., Chen, Q., Shen, J., Hou, B., Ahmed, M., Li, Z.: Aspect-level sentiment analysis based on gradual machine learning. Knowl.-Based Syst. **212**, 106509–106521 (2021). https://doi.org/10.1016/j.knosys.2020.106509
31. Es-sabery, F., Hair, A.: A MapReduce C4.5 decision tree algorithm based on fuzzy rule-based system. Fuzzy Inf. Eng. 1–28 (2020). https://doi.org/10.1080/16168658.2020.1756099

32. Es-Sabery, F., Hair, A., Qadir, J., Sainz-De-Abajo, B., García-Zapirain, B., Torre-Díez, I.D.L.: Sentence-level classification using parallel fuzzy deep learning classifier. IEEE Access **9**, 17943–17985 (2021). https://doi.org/10.1109/ACCESS.2021.3053917

33. Es-Sabery, F., Hair, A.: Big data solutions proposed for cluster computing systems challenges: a survey. In: Proceedings of the 3rd International Conference on Networking, Information Systems & Security, Marrakech, Morocco, pp. 1–7 (2020). https://doi.org/10.1145/3386723.3387826

34. Uysal, A.K., Gunal, S.: The impact of preprocessing on text classification. Inf. Process. Manag. **50**(1), 104–112 (2014). https://doi.org/10.1016/j.ipm.2013.08.006

35. Vijayarani, S., Ilamathi, J., Nithya: Preprocessing techniques for text mining - an overview. Int. J. Comput. Sci. Commun. Netw. **5**(1), 7–16 (2015)

36. Abel, J., Teahan, W.: Universal text preprocessing for data compression. IEEE Trans. Comput. **54**(5), 497–507 (2005). https://doi.org/10.1109/TC.2005.85

37. Yao, Z., Ze-wen, C.: Research on the construction and filter method of stop-word list in text preprocessing. In: Proceedings 2011 Fourth International Conference on Intelligent Computation Technology and Automation, March 2011, vol. 1, pp. 217–221 (2011). https://doi.org/10.1109/ICICTA.2011.64

38. Kruse, H., Mukherjee, A.: Preprocessing text to improve compression ratios. In: Proceedings DCC 1998 Data Compression Conference (Cat. No. 98TB100225), March 1998. https://doi.org/10.1109/DCC.1998.672295

39. Anandarajan, M., Hill, C., Nolan, T.: Text preprocessing. In: Anandarajan, M., Hill, C., Nolan, T. (eds.) Practical Text Analytics. AADS, vol. 2, pp. 45–59. Springer, Cham (2019). https://doi.org/10.1007/978-3-319-95663-3_4

Clustering Techniques for Big Data Mining

Youssef Fakir[✉] and Jihane El Iklil

Laboratory of Information Processing and Decision Support, Sultan Moulay Slimane University,
Beni Mellal, Morocco

Abstract. This paper introduces the Clustering method as an unsupervised machine learning where the input and the output data are unlabeled. Many algorithms are designed to solve clustering problems and many approaches were developed to enhance deficiency or to seek efficiency and effectiveness. These approaches are partitioning-based, hierarchical-based, density-based, grid-based, and model-based. With the evolution of data amounts in every second, we become faced to deal with what is called big data that compelled researchers to develop the algorithms based on these approaches in order to adjust them to manage warehouses in a fast way. Our main purpose is the comparative of representative algorithms of each approach that respect most of the big data criterions which are called the 4Vs. The comparison aims to figure out which algorithms could mine efficiently information by clustering big data. The studied algorithms are FCM, CURE, OPTICS, BANG, and EM respectively from each approach aforementioned. Assessing these algorithms based on the 4Vs big data criterions which are Volume, Variety, Velocity and Value shows some deficiency in some of them. All trained algorithms clusters well large datasets but exclusively FCM and OPTICS algorithms suffer from the curse of dimensionality. FCM and EM algorithms are very sensitive to outliers which affect badly the results. FCM, CURE, and EM algorithms require the number of clusters as input which plays a deficiency if the optimal one wasn't chosen. FCM and EM algorithms give spherical shapes of clusters unlike CURE, OPTICS, and BANG algorithms which give arbitrary ones that play an advantage for cluster quality. FCM algorithm is the fastest in performing big data, unlike EM algorithm that takes the longest time in training. For diversity in types of data CURE algorithm trains both numerical and categorical data types. Consequently, the analysis leads us to conclude that both CURE and BANG are efficient in clustering big data but we noticed that CURE lacks a bit of accuracy in data assignment. Therefore we infer to qualify the BANG algorithm to be the appropriate one to cluster a large dataset with high dimensionality and noise within it. BANG algorithm is based on a grid structure but comprises implicitly partitioning, hierarchical and density approaches the reason behind its efficiency in giving good accurate results. But even so, the ultimate accuracy in clustering isn't reached yet but almost close. The conclusion we observe from the BANG algorithm should be applied to more algorithms by mixing approaches in order to attain the ultimate accuracy and effectiveness that lead consequently to accurate future decisions.

Keywords: Datamining · Clustering · Clustering approaches · Big data · 4Vs · Machine learning · FCM · CURE · OPTICS · BANG · EM

© Springer Nature Switzerland AG 2021
M. Fakir et al. (Eds.): CBI 2021, LNBIP 416, pp. 183–200, 2021.
https://doi.org/10.1007/978-3-030-76508-8_14

1 Introduction

Clustering [1, 2] is one of the crucial methods for segmentation in any field whatever it's about individuals or objects. The main thing is how to catch similarities within them constituting homogeneous groups in order to mine information that lead to major decisions. Clustering is considered as unsupervised machine learning therefore the data are not necessarily labeled to be trained. Many algorithms are designed to cluster data some of them deal with small data only, others could also deal with large data but some of them suffer from the curse of dimensionality. This discrimination was the idea of our contribution that aims to do a comparison of some algorithms as representatives of each clustering approach to figure out the ones that fit well with big data. The choice of these algorithms was made based on previous analysis [3–6]. In this paper, we represent the different approaches invented to cluster data and the algorithms chosen from each one. Then we figure out the big data characteristics allowing the evaluation of studied algorithms. Then we introduce the method used to assess the accuracy in clustering big data. Finally, we exhibit the experimental results leading us to make decisions on these algorithms.

2 Clustering Approaches

Many approaches were invented for the clustering method trying every time to reach the effectiveness in different cases. So far there are five main approaches [5] (see Fig. 1) figured out and developed to cure the dysfunctionality of some algorithms and some are combined to give an efficient algorithm.

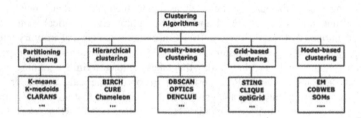

Fig. 1. An overview of clustering techniques taxonomy

Partitioning-Based Approach: Partitioning based is a method that the well-known algo-rithm K-means [7] is based on. This approach proposes dividing data into partitions in which the number is specified by the user as the number of clusters. The procedure begins with an initial assignment of data constituting prior clusters. Subsequently, the enhancement of quality of these partitions is based on updating centroids in different ways depending on the algorithm until they become immutable. Through this process, the algorithms depending on this approach seek to optimize an objective function also known as the sum of squared errors (SSE). As much as the value of SSE is lower as much as the data within clusters are correlated and the clusters are distanced from each other. This type of approach forms clusters with only spherical shapes which make it sensitive to outliers. Many algorithms were designed based on this approach such as the well-known K-means and its variations that deal with categorical types of data, PAM [8], CLARA [9] and CLARANS [10] which all are hard clustering and for the fuzzy logic, we can cite FC-Means [11] algorithm.

Hierarchical-Based Approach: The hierarchical-based method is about building a hierarchy throughout making clusters. There are two types of hierarchical clustering: agglomerative clustering and divisive clustering which are the opposite of one another. For agglomerative hierarchical clustering, each data point is considered as an initial cluster and the procedure begins with computations of the distances between these clusters before mentioned. Subsequently, the nearest clusters which mean the ones with the lower distance are picked to eventually construct a cluster. Afterwards the recalculation occurs again with this new cluster lately merged to choose the nearest clusters among all clusters and so on. The procedure ends up with all the data points gathered in one big cluster building a bottom-up hierarchy of clusters drawn as a dendrogram. In contrast, for divisive hierarchical clustering all the data points are considered as one cluster then gradually this big cluster is split into smaller ones. The procedure stops when it reaches the initial clusters then a top-down hierarchy is built. This type of clustering forms clusters in non-convex, arbitrary, hyper and rectangular shapes, unlike the partitioning method that forms only spherical shapes of clusters. Hierarchical clustering gives birth to such algorithms as BIRCH [12], CURE [13], ROCK [14], and many more developed for a better hierarchical clustering.

Density-Based Approach: Density-based approach differs from the last approaches it doesn't neither require a number of clusters to begin clustering or follow a hierarchy structure while clustering it focuses on the density around each data point. For this reason it defines two required parameters which are minimal points as minimal data's neighbors and the radius ε which determines the circle limiting the neighboring density. By these parameters this method categorizes data points into tree types: Core, Border and noise point. A Core point is a point which has at least the required minimal number of points within the limits defined by the radius. A border point is a Core point's neighbor but doesn't reach the required number of neighbors to become a core point. And finally the Noise point is neither a Core point nor a Border point which becomes easily detected by this method. Algorithms based on this approach form arbitrary shapes of clusters for instance the well-known algorithm DBSCAN [15], its variation the OPTICS [16] algorithm and many more like DBCLASD [17] and DENCLU [18].

Grid-Based Approach: Grid based approach is another proposed approach that has a great advantage of reducing the computational complexity in clustering big data. This approach proposes a grid structure that adapts itself to the distribution of data inserted in. the procedure of clustering is based on identifying the dense cells of the grid to become the clusters center. Then it proceeds by seeking neighbors of these centers and so on until it ends up with final clusters. Many algorithms are designed based on this approach for instance there is OptiGrid [19], BANG [20], ENCLUS [21], PROCLUS [22], ORCLUS [23] and many more that all known as very fast processing algorithms.

Model-Based Approach: This approach is based on building a model supposed to be a mixture of prior clusters with fixed number. Then the algorithm based on this approach tries to train the model in order to fit with the distribution of data and catch the final clusters. The model is either based on a statistical approach which relay on a mixture of probability distributions or based on neural approach which relay on neural network concept. For instance we cite the well-known algorithm SOM [24] which follows the neural approach. In the other hand we cite the well-known algorithm EM [25] which follows the statistical approach.

3 Big Data Characteristics

The growth of massive amounts of data every day leads to many different observations and studies about how to store these quantities, how to manage these large data and how to extract accurate information concealed behind in order to developing the field and the society for more growth. Therefore, researchers come to the so-called data warehouses to store these large databases, invent tools and software to deal with it and design algorithms to extract main information for better future decisions using data mining methods. Clustering as one of the main data mining methods knew a development to handle more with the big data. Gartner [26] has defined tree characteristics for big data which are Volume, Velocity and Variety which well-known as Gartner's commentary or 3Vs. After that a new attribute is added which is Veracity by the scientists of IBM to become 4Vs then it adds another one which is Value [27] to become 5Vs. Many more dimensions were added by researchers but regarding to our study we consider only 4Vs described as follows:

Volume: This criterion refers to the size of data set, high dimensionality and the capability to avoid or deal with noise. The target is to find the algorithms that deal with large data in size, with high dimensionality if there is many attributes assigned to the data. They also need to handle with outliers the way they shouldn't make groups or accurately assigning them to approximate clusters.

Variety: For variety criterion the algorithms are examined by their ability to support many types of data like both numerical and categorical types. In addition to the capability of forming arbitrary shapes of clusters to have more efficiency in terms of clusters quality.

Velocity: Velocity denotes the speed of the algorithm which refers to the complexity time based on big O notation which is a way of showing how the runtime change influenced by the input size which is whether linear, constant or quadratic...etc.

Value: The value characteristic is about the input parameters of the algorithm which differs from one another depending on the kind of methods and the less it is the less the computations are made.

4 Clustering Assessment

Since measuring distance within data determines whether they are close or not to group them in a cluster then this distance could be an efficient assessment for a good clustering or not. There are many different assessment methods designed for clustering where we choose *silhouette method* [28]. It is efficient in searching the optimal number of clusters that gives good clustering. This method is relaying on two primarily measurements one is measuring the mean (a) of the distances between a data point (i) and the rest of data points within its cluster C. The other is measuring the mean (b) of the distances between the previous data point (i) and the data points of the closest cluster (*min*) with the following expressions

$$a(i) = \frac{1}{|C_i| - 1} \sum\nolimits_{j \in C_i, i \neq j} d(i,j) \ And \ b(i) = min_{k \neq i} \frac{1}{|C_i|} \sum\nolimits_{j \in C_k} d(i,j) \qquad (1)$$

The silhouette score of one data point is measured using the previous measurements as follows:

$$s(i) = \frac{b_i - a_i}{max\{a_i, b_i\}} \in [-1, \ 1] \qquad (2)$$

Then the final result over the clustered dataset is measured by the mean of silhouette score of all the data points. The best value is 1 which indicates the optimal number of clusters. The worst is -1 which denotes that there are some data points that are wrongly assigned to their clusters. The 0 value indicates that clusters are close to each other.

5 Studied Algorithms

Clustering big data requires standards to respect in order to reach the ultimate effectiveness in results. Many studies invent algorithms based on approaches or evolve previous ones to fix their deficiency or bring new approaches to fulfill more efficiency or why not design the perfect algorithm. After the analysis of many algorithms, we decided to study the following algorithms that satisfy the majority of big data characteristics aforementioned. These algorithms according to different approaches are: for the partitioning-based approach we decide to work with a fuzzy logic algorithm which is Fuzzy-C-means (FCM). For the hierarchical-based approach and density-based approach respectively we choose Clustering Using Representatives (CURE) algorithm as a divisive clustering algorithm and Ordering Points To Identify Clustering Structure (OPTICS) algorithm. Finally, for the grid-based and model-based approaches, we choose respectively BANG algorithm and Expectation-Maximization (EM) algorithm. These algorithms are chosen based on their dealing with most of the big data's features that after experiment we could obtain accurate results and figure out the deficiency about them.

Fuzzy-C-Means [11] is an algorithm for clustering data following the so-called *fuzzy logic* that deals with fuzziness. As a *soft clustering* gender, FCM assigns data points to their clusters with a probability of belongingness to each one. FCM is a kind of K-means algorithm except for the assignment of data to clusters which is probabilistic in FCM and asserted for K-means that's why it is categorized as a soft clustering and the other as a hard clustering. This algorithm demands one input which is the number of clusters since we are doing a partitioning of data and pursues the following procedure defined in 4 steps:

Step 1: Randomly assign values to all *memberships* ω_{ij}, which is the degree of belongingness of a data i to a cluster j, with the condition that $\sum_{j=1}^{k} \omega_{ij} = 1$ where k is the number of clusters.

Step 2: Calculation of the centroids which are the centers of the clusters depending on the memberships assigned before with the flowing expression:

$$C_j = \frac{\sum_{i=1}^{n} \omega_{ij}^p \times x_i}{\sum_{i=1}^{n} \omega_{ij}^p} \tag{3}$$

- p is a parameter called *fuzzifier* which is suggested to be bounded between [1.5, 2.5] [29]. A recent analysis [30] suggests a large value (p = 4) when the data contain outliers in order to make the FCM more robust against them.
- n is the number of data points.

Step 3: Updating the memberships after calculating the distance between points and centroids $dist(x_i, C_j)$ by the following expression where i is the data and j is the cluster:

$$\omega_{ij} = \frac{\left(\frac{1}{dist(x_i, C_j)}\right)^{\frac{1}{p-1}}}{\sum_{q=1}^{k} \left(\frac{1}{dist(x_i, C_q)}\right)^{\frac{1}{p-1}}} \tag{4}$$

Step 4: Repeat updating by Step 2 and Step 3 until the centroids become immutable.

These steps aforementioned are the major steps of the Fuzzy-C-Means algorithm that try to optimize an objective function like K-means by adding the influence of membership under the following expression:

$$SSE = \sum_{j=1}^{k} \sum_{i=1}^{n} \omega_{ij}^p \times dist(x_i, C_j) \tag{5}$$

Where SSE is the Sum of Squared Errors, k is the number of clusters, n is the number of data and p is the fuzzifier.

Clustering Using Representatives [13] is a clustering algorithm, unlike the partitioning approach, which focuses on centroids to represent the cluster. This algorithm suggests the well-scattered points in a cluster as *representatives*. As this algorithm follows the divisive hierarchical approach it primarily proceeds by splitting the dataset into random samples then it works on one by one. Working on a random sample the algorithm starts by its partitioning into partitions then it partially clusters them. After this initial clustering, the algorithm shrinks the representative points into the cluster's center with an angle of α which is suggested to be big in order to be more close to the center. This way the algorithm partitions the clusters in order to eliminate the outliers which become visible due to the shrinking of most scattered points of the clusters. Subsequently, the algorithm proceeds by a final clustering of the partial clusters by merging them. This merge is depending on the minimal distance between the representative points and their cluster's center. This distance is used to be compared with distances between the representative points of adjacent clusters. In case the distances between representatives are equal or less than the minimal distance then the merge is applied for these adjacent clusters. After having final clusters the algorithm labels data on disk by storing the representative points of clusters. Afterwards it iterates these steps until it ends up with the required number of clusters entered as initial input.

Ordering Points To Identify Clustering Structure [16] is a clustering algorithm evolved in order to fix the deficiency observed on DBSCAN algorithm requiring both *minimal points* and *radius* as inputs. The deficiency was in accurately choosing the appropriate radius therefore they propose an ordering point to discover the radius given only the minimal points. The idea of *ordering points* relay on higher density points then lower density points. With a given minimal number of points, the algorithm will try to find the radius that covers this number using a distance called *Core-distance*. This distance is defined as the smallest value of ε that comprises the minimal number of points as neighbors of a point a ($N_\varepsilon(a)$) then its value is as follows:

$$\textbf{Core} - \textbf{Distance}_{\varepsilon,\textbf{minPts}}(a) = \textbf{undefined; if card}(\textbf{N}_\varepsilon(a)) < \textbf{MinPts}$$
$$= \textbf{MinPts}^{\textbf{th}}\textbf{smallest distanceinN}_\varepsilon(a), \textbf{otherwise}$$

Subsequently, the algorithm discovers a big radius that covers the same number using a distance called *Reachability-distance*. This distance is the maximum distance among the distances of nearest neighbors that are density reachable from a core point a as expressed below:

$$\textit{Reachability} - \textit{Distance}_{\varepsilon,minPts}(b,a) = \textit{undefined; if a is not a core point}$$
$$= \textbf{max}\{\textbf{Core} - \textbf{Distance}(a), \textbf{distance}(a,\textbf{b})\}, \textbf{otherwise}$$

When we say that a point p is *density reachable* from q if there is a chain of points $p_0 \cdots p_n$ as $p_0 = q$ and $p_n = p$ such that p_{i+1} is *directly density-reachable* from p_i. Also, a point p is *directly density-reachable* from a point q if it respects the following conditions: p should belong to the ε-neighborhood of q ($N_\varepsilon(q)$) and q is a core point.

Both *Core & Reachability distances* build a structure of clustering applied on all points. When the *Core-distance* is undefined it means this point is an outlier. Therefore OPTICS algorithm is stronger to detect outliers based on the minimal points around it and this is a privilege of density-based algorithms.

BANG [20] is a clustering algorithm based on a grid structure called *Bang structure*. Bang structure is composed of a *grid directory* split into blocks containing a maximum number of data points. The algorithm dynamically generates this grid directory by *scaling* the features of data in order to be adapted to the distribution of data inserted in. After inserting data the algorithm splits the blocks into two subspaces called *grid regions*. Each block is a grid region but subsequently, after the splitting, any block could contain more than one grid region constituting what is called *block region*. The splitting is actually a *binary split* generated by a cyclic sequence of dimensions entry. This binary split produces a *hierarchy* of grid regions storing information which are the *number* of the region and its *level* in order to keep track of the root of the split. All the sub-regions of the grid directory belong to region number 0 in level 0. Subsequently, after splitting blocks the algorithm starts clustering by calculating the *Density Index* of each block by the following expression where P_B is the number of patterns (*pattern* means a data point described by its features) within the block and V_B is its spatial volume (*volume* is the Cartesian product of the extents of this block in each dimension i).

$$V_B = \prod e_{B_i} \quad and \quad D_B = \frac{P_B}{V_B} \quad where \ i = 1, \ldots k \tag{6}$$

Afterwards all the density indexes are sorted decreasingly in order to assign the ones with the higher density to be the centers of the clusters. Then the algorithm starts seeking *neighbors* of these blocks as centers by measuring the dimensionality of touching areas between regions supported by the hierarchy of regions early constructed. Since each region is at such a level then in order to have an equitable comparison between regions defined in different levels the algorithm proceeds by transforming the lower level to the higher level. After finding neighbors, Bang algorithm sorts them decreasingly by their density index. Starting with the ones with higher density it seeks its following neighbors and the same procedure is applied on each found neighbor constituting a dendrogram each time. The algorithm applies these steps concerning the search of neighbors on the rest of the centers to end up with the final clusters.

Expectation Maximization [25] is an algorithm that clusters data based on a *model* called the *Gaussian Mixture Model*. First, the idea is based on the so-called *Gaussian distribution* that shows the distribution of x values in one dimension or more. The distribution falls from the *mean* μ to both sides evenly following a specific *standard deviation* σ. It produces a probability of each point x in this dimension called *Probability Density Function* (PDF) which refers to the density of this point in this dimension. PDF is calculated in the case of *one dimension* by the following expression:

$$Y = \frac{1}{\sigma\sqrt{2\pi}} e^{-\frac{(x-\mu)^2}{2\sigma^2}} \tag{7}$$

Where μ is the Mean, σ^2 is the Variance and σ is the Standard deviation.

For a data point defined in *d dimensions,* $x = \left(x^1, x^2, \ldots, x^d\right)^T$ we obtain its Probability Density Function by the following expression:

$$N(x|\mu, \Sigma) = \frac{1}{(2\pi)^{\frac{d}{2}}\sqrt{|\Sigma|}} exp\left(-\frac{1}{2}(x-\mu)^T \Sigma^{-1}(x-\mu)\right) \tag{8}$$

Where μ is the Mean (vector) and Σ is the Covariance (matrix)

The Gaussian Mixture Model (*GMM*) then is a model composed of multiple Gaussians mixed together where the probability of each point in this model is obtained by the following expression:

$$P(x) = \sum_{j=1}^{K} \omega_j . N\left(x|\mu_j, \Sigma_j\right) \tag{9}$$

Where: ω_j is a prior probability of the jth Gaussian with regards to $\sum_{j=1}^{k} \omega_j = 1$ & $0 \leq \omega_j \leq 1$.

EM algorithm aims to maximize the likelihood function of each point by updating the estimated parameters θ initially assigned to each Gaussian distribution where this method is known as the *Maximum likelihood Estimation* method.

$$\theta^* = \arg \max P(x|\theta) = \arg \max \prod_{i=1}^{N} P(x_i|\theta) \tag{10}$$

For this purpose, the EM algorithm goes through two major processes. The first one is the phase of expectation followed by the phase of maximization giving existence to a *latent variable* z which unobserved and defined as a joint (related) distribution on x:

Initialization Step: the algorithm requires as input the number of Gaussians, which refers to randomly prior clusters as K-means perform. It also requires the prior probabilities of choosing each Gaussian. Besides the parameters of Gaussians which are the *mean* μ and *standard deviation* σ in case of one dimension or *covariance* Σ in case of multi-dimensions which constitute the values of the prior parameters of θ.

Expectation Step: This step is about computing probabilities of the latent variable z given current parameter values of θ by the following expression using *Bayes Rules* and the *Law of Total Probability*:

$$P(z|x, \theta) = \frac{P(x|\theta, z)P(z)}{P(x, \theta)} \Rightarrow P(z_i = k|x_i, \theta) = \frac{P(x_i|\theta, z_i = k)P(z_i = k)}{\sum_{j=1}^{k} P(x_i|z_i = j, \theta)P(z_i = j)} \tag{11}$$

Where $P(z_i = k) = \omega_k$ and $P(x_i|\theta, z_i = k)$ is *conditional Probability Mass Function* (PMF). It's the probability that a discrete random variable is exactly equal to some value which means it describes a discrete probability distribution unlike, PDF which describes a continuous probability distribution.

Maximization Step: Subsequently trying to maximize the likelihood estimated using the probabilities above to estimate the new parameters that fit the data gradually by maximizing the *loglikelihood* of the data and the hidden variable under the following expression:

$$Q(\theta, \theta^*) = \sum_{i=1}^{n} \sum_{j=1}^{k} P(z_i = k|x_i, \theta^*) \log(P(z_i = j|\theta)P(x_i|z_i = j, \theta)) \tag{12}$$

Iteratively we repeat the *E-step* and *M-step* until reaching the ultimate likelihood estimation by improving the model parameters by evaluating the following expression:

$$\theta^* = \arg \max Q(\theta, \theta^*) \tag{13}$$

Since the EM algorithm follows almost the principle of K-means algorithm then it tries to minimize an objective function, which is $\sum_{j=1}^{k} \frac{(x-\mu_k)^2}{\sigma^2}$. Unlikely K-means tries to minimize $\sum_{j=1}^{k} (x - \mu_k)^2$. After a set of iterations, we become able to fit the GMM into our data to end up with the final clusters.

6 Experimental Results

Since we aim to compare the algorithms' ability of mining big data adopting different clustering methods, we choose a dataset with 100000 rows and 10 columns that gathers the two features of big data that are large data and high dimensionality. The dataset contains 100000 sales records [31] with 6% of noise with two categorical features and eight numerical ones.

After preprocessing of data we used python's libraries Pyclustering and sickit-learn to implement our five algorithms with easiness using methods they provide that allows us to harvest the following results. As it well known most of the clustering algorithms require the number of clusters, unlike some developed ones that do not. Therefore, in our work, we split the five algorithms into types: three ones that rely on the number of clusters to construct which are FCM, CURE and EM. In the other hand, we have two algorithms left OPTICS and BANG which don't require it. Starting with the first type, which requires the number k of clusters which in most cases is ambiguous to define the efficient one that will give accurate results. For this reason, we choose to perform the algorithms for different numbers of k. We analyzed their performance by measuring their execution time using a Python's library called *time* and their accuracy in finding results by measuring the silhouette score aforementioned. The performance of an algorithm depends on the amount of data and the input it requires which is evaluated based on its complexity in time and space. In our case, we obtained the results shown in Table 1, which represents the performance of each algorithm depending on a different number of clusters.

Table 1. Runtime processing for FCM, CURE and EM

k	Runtime/minutes		
	FCM	CURE	EM
2	0,08	69,7	140.31
3	0,09	51,92	154
4	0,1	41,62	157.41
5	0,17	41,21	272.49

The results show that the performance of FCM and EM algorithms which are soft clustering algorithms becomes greater with the growth of clusters' number because of their complexity time O(n) which is a linear time. Unlikely the CURE algorithm which is a divisive hierarchical algorithm its performance becomes lower with the increasing number of clusters due to its logarithmic and quadratic time complexity O(n2 log n) called super-quasi quadratic time. We notice that the CURE algorithm takes a long time which is many times the time that takes the FCM algorithm. The EM algorithm is influenced by the number of iteration it took to complete fitting the model to data. If complete training requires much iteration like in our case of big data it is 2000 iterations, the algorithm takes a long time bigger than the CURE algorithm due to computations complexity.

FCM algorithm is very fast and its performance is not influenced by iterations whether it is small or big it shows the same performance.

The accuracy of an algorithm denotes its ability to provide accurate and efficient results. For clustering algorithms, it's about the rate of correlation between the data points within the final formed clusters and how much they are assigned to the right ones. Therefore, the accuracy leads to decide about the optimal number of clusters to choose in case of this type of clustering algorithms that demand it and to assess the assignment. Table 2 represents the accuracy obtained for each algorithm according to different input k number of clusters. The results below show that the three algorithms give significant values of accuracy for 2 as an optimal number of final clusters for better clustering. Even the good accuracy that the algorithms show plots visualize some data points that are wrongly assigned to their appropriate clusters which got negative values of silhouette score.

Table 2. Silhouette score for FCM, CURE and EM algorithms

k	Accuracy		
	FCM	CURE	EM
2	0,43	0,35	0.168
3	0,24	0,098	0,168
4	0,19	0,11	0,168
5	0,14	0,056	0,168

The plotting of silhouette score grouped by final clusters (see Fig. 4) shows that the soft clustering EM algorithm gives the same number of final clusters even with different inputs of k number of clusters. Therefore it doesn't make any effect on the clustering process in order to give the optimal number of final clusters. But it's crucial to take attention to unassigned data points which makes the difference between them. On the other hand, the divisive hierarchical algorithm CURE which processes clustering until it attends the required number of clusters demands an accurate number of clusters. Observing the plots resulting in every entry of k clusters (see Fig. 3), the CURE algorithm gives exactly the same number of clusters required. The same with FCM algorithm (see Fig. 2) which iterates clustering until the centers' values becomes unchangeable conserve the number of clusters required.

The clustering algorithms left don't require the number of clusters but each one of them had one indispensable input that affects results. Tables 3 and 4 show the performance and accuracy results for both algorithms BANG and OPTICS respectively.

For the grid-based algorithm BANG, the performance grows as the number of levels assigned grows due to its complexity time O(n) which is a linear time. Accuracy shown by this algorithm is good with two clusters. The plot (see Fig. 6) visualizes that all the data points were perfectly assigned to their clusters, which refer to a well-performed clustering besides a fast performance.

Table 3. Performance and accuracy of OPTICS algorithm

Table 4. Performance and accuracy of BANG algorithm

OPTICS			
minPts	Accuracy	Runtime	Clusters
900	0.032	34,13	3/−1
1000	−0.034	30,16	2/−1
1200	0.13	30,43	2/−1
1500	0,13	25,91	2/−1
1520	0,13	24,84	2/−1
1560	0,067	21,94	1/−1

BANG		
Levels	Accuracy	Runtime/min
15	0,07	4,34
16	0,07	11,88
20	0,07	14,11
21	−0,029	17,33
22	−0,029	20,17
25	−0.065	30,93

For density-based algorithm OPTICS, the performance grows as the number of minimal points assigned is low with a complexity time $O(n \log n)$ which is a quasi-linear time. Experimenting with different inputs of a minimal number of points as the results show in Table 3 we deduce that two clusters show a good clustering accuracy too. (-1) in Table 3 denotes that outliers are detected and construct a group. We visualize (see Fig. 5) that the data points grouped in both clusters are well assigned to their appropriate clusters with no negative silhouette score detected.

After analyzing both the performance and accuracy of these algorithms for clustering our big dataset, we deduce that two clusters is the optimal number of clusters for its good clustering. Now we observe the final clusters shape depicted in Fig. 6 in two and three dimensions for two clusters.

As observed (see Fig. 7), the clusters formed by FCM are not suitable due to their sensitivity to outliers since they choose the mean of data points to assign the centers of clusters. Consequently, they could lean toward the outlier and gives erroneous centers. In addition, FCM is sensitive to high dimensionality and lean to give spherical shapes that affect badly the clusters formed above. Clusters formed by EM algorithm are a little bit confusing because it lacks a bit of accuracy due to sensitivity to outliers too. The CURE algorithm shows well-separated clusters since it could avoid outliers based on the shrinking of representative points of clusters beginning with splitting and merging of partial clusters until obtaining the final clusters. The final clusters of OPTICS algorithm are well too since within the clusters formed there is one of the noisy points that are easily detected thanks to the low density around these points. However, the results are untrusted because OPTICS suffers from the curse of dimensionality. FCM and OPTICS algorithms based more specifically on distance functions where for data defined in multiple dimensions with different scales could provide similar distances. Consequently, algorithms cannot differentiate between the nearest neighbor and the farthest one, which affects badly results.

Finally, the clusters constructed by the BANG algorithm shows well-separated clusters in terms of shape. BANG algorithm is based on a grid structure that adapts with the dimensionality of data points besides the density of grid blocks that allows centers to be efficiently detected.

7 Comparison

Many analyses and the results acquired above clarify most of the things about these chosen algorithms. Our contribution concludes some insights that make us choose the more appropriate algorithm(s) to deal with clustering big data and capable to extract the essence information within it with accuracy and efficiency. We briefly develop these insights in the following lines:

For Volume criterion: Partitioning-based algorithms are known by their sensitivity to outliers that deviates the results towards them. In other words, they deviate the cluster centers towards them every time the mean is calculated. Model-based algorithms also have the same problem because they almost follow the same concept of the K-means algorithm. Consequently, FCM and EM show poor results in terms of accuracy in clustering data with noise generally. In the contrary, a density-based approach endows its algorithms with robustness to detect easily outliers by noisy points. Based on the density of neighbors around a point, noise cannot cluster the required minimal points as neighbors.

For Value Criterion: When we talk about clustering, the number of clusters formed it's whether arbitrary obtained by clustering algorithm processing or demanded as input for other algorithms. This number requirement plays a deficiency for these algorithms if it is not the optimal number to give accurate and efficient results. Partitioning based algorithms are centroid-based then the number of clusters is indispensable which plays a disadvantage for FCM algorithm. CURE algorithm demands the number of clusters too because its clustering process trains data until the number of clusters required is obtained. The same for the EM algorithm that requires the number of clusters which is the number of Gaussians in order to initialize their parameters to proceed clustering. In the case of OPTICS and BANG algorithms, they require specialized parameters, which are respectively the number of minimal points as neighbors and the number of levels as the grid height. These last parameters need experiments with different values until accuracy occurred in final clusters.

For Variety Criterion: Clusters' shape plays a crucial factor in deciding about the quality of the algorithm. Partitioning approach and model approach lean to give spherical shapes of final clusters that influence badly the results in some cases. In the contrast, the density approach and the grid approach play an advantage by giving arbitrary shapes. For instance, the data correlation doesn't follow a circle shape but an arbitrary one then the algorithm loses absolutely effectiveness. It's the case of FCM and EM algorithm that produce spherical shapes unlike OPTICS, BANG and CURE where arbitrary shapes are formed.

- The diversity of data types makes a huge problem for algorithms designed for clustering. Especially categorical data which is rarely supported by algorithms where we can site for instance our algorithm CURE that deals with both categorical and numerical data types. Most of the algorithms support only numerical data type as FCM, BANG and OPTICS algorithms do. Another type of data that is supported by the EM algorithm is spatial data, which is more supported by algorithms based on the grid approach, which emphasizes dealing with high dimensions.

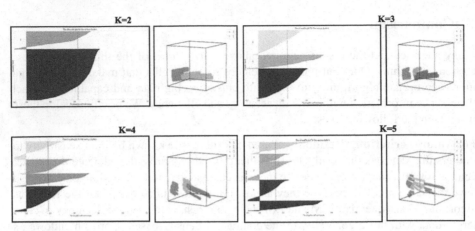

Fig. 2. Plotting of Silhouette score grouped by clusters and visualization of clustered data by FCM algorithm

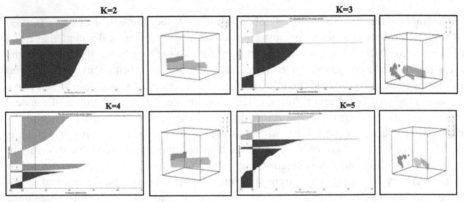

Fig. 3. Plotting of Silhouette score grouped by clusters and visualization of clustered data by CURE algorithm

For Velocity Criterion: The partitioning approach based only on the distance to the centroids does not take much time in computations. It is one of the FCM algorithm pros that it shows the great velocity in clustering large datasets. The grid-based approach gives an advantage with fast performance due to slight computations for clustering. The performance of the BANG algorithm shows that despite the large datasets and high dimensionality the results were rapid to appear. In the contrary, we find the model-based approach that demands complex computations necessitates a long time to run an algorithm based on it. EM algorithm even dealing with spatial data the mathematical computations it takes gives it a long performance especially for a large dataset with high dimensions. Hierarchical based approach either divisive or agglomerative had complexity in proceeding clustering. In the case of large datasets with high dimensions, the procedure becomes more complex that affects the performance of the algorithm. The CURE algorithm that requires random sampling, partitioning and clustering, shrinking,

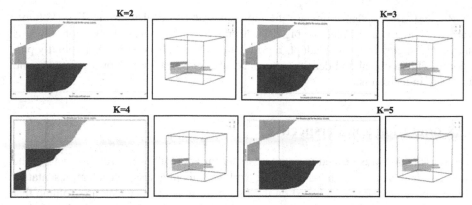

Fig. 4. Plotting of Silhouette score grouped by clusters and visualization of clustered data by EM algorithm

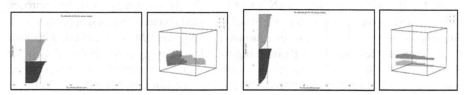

Fig. 5. Plot of silhouette score grouped by clusters visualization of clustered data by OPTICS algorithm

Fig. 6. Plot of silhouette score grouped by clusters and visualization of clustered data by BANG algorithm

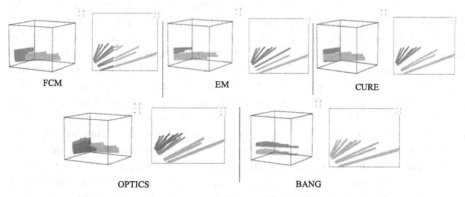

Fig. 7. Plotting of final clusters of each algorithm in 2D and 3D

calculating distances between representative points, merge of partitions and labeling shows that iteration of this sequence of steps takes a long time trying to train a large dataset with high dimensions. The density-based algorithm takes time seeking neighbors of each point within a large dataset with high dimensions. But as much as the required density around a point becomes greater the velocity decreases as shown in OPTICS algorithm performance. After discussing all the properties above, we conclude that both

CURE and BANG algorithms are the most qualified to deal with the 4Vs (Volume, Variety, Velocity, and Value) of big data. Moreover, the privilege leans more towards the BANG algorithm, which adopts a grid approach but also comprises implicitly partitioning, hierarchical and density approaches the reason why it shows efficiency and accuracy in its results.

8   Conclusion and Perspective

Our contribution was headed to compare different approaches in order to find out the ones that deal the most with big data. Our work was about assessing each representative algorithm of each approach based on the four criterions (4Vs) which are Volume that comprises the size of the dataset, high dimensionality and outlier's sensitivity. Variety in terms of dataset type and the shape of the constructed clusters, Velocity refers to the time complexity of the algorithm and finally Value which is the parameters required as inputs. FCM, CURE, OPTICS, BANG and EM algorithms were chosen by approach and evaluated by the last criterions. Partitioning-based and model-based approaches are sensitive to outliers. Therefore results shown by FCM and EM were untrusted because of the mean deviates the results. Partitioning-based and density-based approaches suffer from the curse of dimensionality because they focus on distance in their clustering. As consequence all algorithms gives good results when clustering large dataset but only CURE, BANG and EM could handle large dataset with high dimensions. Clustering quality was measured by the shape of clusters where the arbitrary shape is required. CURE, OPTICS and BANG algorithms give arbitrary shapes unlike FCM and EM that lean to give spherical ones. Partitioning-based and model-based approaches require the number of clusters as input which is a deficiency for an algorithm to choose the optimal one. The divisive hierarchical algorithm CURE also requires it. In terms of velocity, FCM shows great rapidity, unlike EM and CURE which takes a long time due to complex computations and procedure. The results shown by these algorithms in terms of approaches they are based on lead us to decide that CURE and BANG algorithms are the more qualified to respect most of criterions. However, exclusively BANG is the winner in all the criterions and shows efficiency and accuracy in clustering big data. We infer that future works should be evolved in terms of mixing different approaches as the BANG algorithm does. BANG algorithm is based on a grid structure but implicitly involves partitioning, hierarchical and density approaches the reason why it deals with big data and shows more accuracy and efficiency. The ultimate result with one hundred percent of accuracy isn't reached yet. The perspective is clear for future algorithms by not only fixing the previous algorithms but furthermore a merging of algorithms is needed for more accurate results and right future decisions.

References

1. Tufféry, S.: Data mining et statistique décisionnelle l'intelligence dans les bases de données. TECHNIP ed., Paris (2005)
2. Gan, G., Ma, C., Wu, J.: Data Clustering: Theory, Algorithms, and Applications. SIAM, Philadelphia (2007)

3. Abbas, A.: Comparisons between data clustering algorithms. Int. Arab J. Inf. Technol. **5**(3), 320–325 (2008)
4. Fahad, A., et al.: A survey of clustering algorithms for big data: taxonomy and empirical analysis. IEEE Trans. Emerg. Top. Comput. **2**(3), 267–279 (2014)
5. Sajana, T., Rani, C.M.S., Narayana, K.V.: A survey on clustering techniques for big data mining. Indian J. Sci. Technol. **9**(3), 1–12 (2016)
6. Nayyar, A., Puri, V.: Comprehensive analysis & performance comparison of clustering algorithms for big data. Rev. Comput. Eng. Res. **4**(2), 54–80 (2017)
7. Macqueen, J.: Some methods for classification and analysis of multivariate observations. In: Proceedings 5th Berkley Symposium on Mathematical Statistics Probability, pp. 281–297 (1967)
8. Kaufman, L., Rousseau, P.: Finding Groups in Data: An Introduction to Cluster Analysis. Wiley Series in Probability and Statistics, pp. 68–125 (1990)
9. Kaufman, L., Rousseau,P.: Finding Groups in Data: An Introduction to Cluster Analysis. Wiley Series in Probability and Statistics, pp. 126–163 (1990)
10. Ng, R., Han, J.: Efficient and effective clustering methods for spatial data mining. In: Proceedings International Conference on Very Large Data Bases (VLDB), pp. 144–155 (1994)
11. Bezdek, J., Ehrlich, R., Full, W.: FCM: the fuzzy c-means clustering algorithm. Comput. Geosci. **10**(2–3), 191–203 (1984)
12. Zhang, T., Ramakrishma, R., Livny, M.: BIRCH: an efficient data clustering method for very large data bases. In: Proceedings of the ACM SIGMOD International Conference on Management of Data, vol. 25, no. 2, pp. 103–114 (1996)
13. Guha, S., Rastogi, R., Shim, K.: CURE: an efficient clustering algorithm for large databases. In: ACM SIGMOD International Conference on Management of Data, vol. 27, no. 2, pp. 73–84 (1998)
14. Guha, S., Rastogi, R., Shim, K.: Rock: a robust clustering algorithm for categorical attributes. In: 15th International Conference on Data Engineering, pp. 512–521 (1999)
15. Ester, M., Kriegel, H., Sander, J., Xu, X.: A density-based algorithm for discovering clusters in large spatial databases with noise. In: Proceedings ACM SIGMOD Conference on Knowledge Discovery and Data Mining (KDD), pp. 226–231 (1996)
16. Ankerst, M., Breuning, M., Kriegel, H., Sander, J.: OPTICS: ordering points to identify the clustering structure. In: Proceedings ACM SIGKDD International Conference on Management of Data, vol. 28, no. 2, pp. 49–60 (1999)
17. Xu, X., Ester, M., Kriegel, H., Sander, J.: A distribution-based clustering algorithm for mining in large sptial databases. In: Proceedings 14th IEEE International Conference on Data Engineering (ICDE), pp. 324–331 (1998)
18. Hinneburg, A., Keim, D.: An efficient approach to clustering in large multimedia databases with noise. In: Proceedings ACM SIGKDD Conference on Knowledge Discovery Ad data Mining (KDD), pp. 58–65 (1998)
19. Hinneburg, A., Keim, D.: Optimal grid-clustering: towards breaking the curse of dimensionality in high-dimensional clustering. In: Proceedings 25th International Conference on Very Large Data Bases (VLDB), pp. 506–517 (1999)
20. Schikuta, E., Erhart, M.: The BANG-clustering system: grid-based data analysis. In: Liu, X., Cohen, P., Berthold, M. (eds.) IDA 1997. LNCS, vol. 1280, pp. 513–524. Springer, Heidelberg (1997). https://doi.org/10.1007/BFb0052867
21. Cheng, C., Fu, A., Zhang, Y.: Entropy based sub space clustering for mining numerical data. In: Proceedings of the fifth ACM SIGMOD International Conference on Knowledge Discovery and Data Mining, pp. 84–93 (1999)
22. Aggarwal, C., Wolf, J., Yu, P., Procopiuc, C., Jong, S.: Fast algorithms for projected clustering. ACM SIGMOD Rec. **28**(2), 61–72 (1999)

23. Aggrawal, C., Yu, P.: Finding generalized projected clusters in high dimensional spaces. In: Proceedings of the 2000 ACM SIGMOD International Conference on Management of Data, vol. 29, no. 2, pp. 70–81 (2000)
24. Kohonen, T.: The self-organizing map. Neurocomputing **21**(1–3), 1–6 (1998)
25. Dempster, A., Laird, N., Rdin, D.: Maximum likelihood from incomplete data via the EM algorithm. J. Roy. Stat. Soc.: Ser. B (Methodol.) **39**(1), 1–38 (1977)
26. Laney, D.: 3-D data management: controlling data volume, velocity and variety. META Group Res. Note **6**, 1 (2001)
27. Demchenko, Y., Membrey, P., Grosso, P., de Laat, C.: Addressing big data issues in scientific data infrastructure. In: First International Symposium on Big Data and Data Analytics in Collaboration (BDDAC 2013), May 2013
28. Rousseeuw, P.J.: Silhouettes: a graphical aid to the interpretation and validation of cluster analysis. J. Comput. Appl. Math. **20**, 53–65 (1987)
29. Pal, N., Bezdek, J.: On cluster validity for the fuzzy C-means model. IEEE Trans. Fuzzy Syst. **3**(3), 370–379 (1995)
30. Wu, K.-L.: Analysis of parameter selections for fuzzy c-means. Pattern Recogn. **45**(1), 407–415 (2012)
31. Datasets (2020). https://www.kaggle.com/jahina/100000-scales-records. Accessed 25 Oct 2020

Data Mining Approach for Intrusion Detection

Mohamed Azalmad and Youssef Fakir$^{(\boxtimes)}$

Laboratory of Information Processing and Decision Support, Faculty of Sciences and Technics,
Sultan Moulay Slimane University, Beni-Mellal, Morocco

Abstract. Intrusion detection systems are simply a security layer that aims to
detect malware and unusual events in a network where they have been installed
and notify the system administrator with an alarm. Intrusion detection systems
divided into several types depending on the configuration that they have, these
configurations are linked to the following components, which are defined as (1)
the information source of the IDS, (2) the detection approach, and (3) the architecture of the IDS itself. The component that interests us is the detection approach, in general, two major detection approaches can be used within an IDS,
a signature-based approach and a behavioural or anomaly-based approach, these
two detection approaches can be treated with different techniques, one of these
techniques is the technique of Data Mining. An intrusion is an activity that differs
from the usual events, while an anomaly is an observation that differs so much
from other observations. The intrusion and anomaly arouse suspicion that a different mechanism generated them. The objective is to understand these mechanisms
behind intrusions and anomalies. Based on this idea, we say that the analogy of
intrusion detection systems in Data Mining detects anomalies. The objective of
intrusion detection systems is to detect attacks in a network, while the objective of
the Data mining anomaly detection approach is to detect anomalies in a dataset.
The anomaly detection approach is divided into three main techniques, supervised
detection, unsupervised detection, and semi-supervised detection. The selection
of the anomaly detection technique is based on the availability of class labels in the
dataset. In our research study, we are using two real datasets, which are the KDD-
Cup99 dataset and the NSL-KDD dataset. These two datasets contain thousands
of normal network events and real attacks. Our goal behind this research study is
to evaluate supervised and unsupervised detection techniques and compare each
technique's performance based on the results obtained.

Keywords: Data mining · Clustering · Intrusion detection · SVM · K-means ·
KNN · C4.5 · ANN

1 Introduction

There are countless examples of where attackers have successfully compromised networks and severely damaged organizations. Nowadays, there are many examples of
intrusive activities on the internet. They fall into a few separate classes. It involves
guessing or cracking passwords, copying or viewing sensitive data or databases, and
running a packet analyzer. Basic attack methodologies consist of acquiring targets and

© Springer Nature Switzerland AG 2021
M. Fakir et al. (Eds.): CBI 2021, LNBIP 416, pp. 201–219, 2021.
https://doi.org/10.1007/978-3-030-76508-8_15

gathering information, taking control of a system, and using system vulnerabilities. No system can be 100% secure, and security breaches cannot be avoided completely. Intrusion detection systems are one of the security methods used to increase the security of network systems, and these intrusion detection systems are becoming very important in any network system these days. In intrusion detection systems, there are different detection approaches and techniques. Each technique has its strengths and weaknesses, but the technique that is highly recommended for intrusion detection systems in recent years is Data Mining, many ideas and techniques have been discussed and proposed in the field of Data Mining to be used as detection approaches within intrusion detection systems [28]. The key to the popularity of the data mining technique lies behind the word "data". Every action that occurs in a network has its traceability or indicator when we talk about indicators at the network level, we are talking about things like traffic volumes, ports and protocols, IP addresses, etc. these activities can produce thousands and thousands of indicators, and these indicators are just data. This great availability of data in networks gives data mining the power over traditional techniques to find patterns and hidden data among all the available data. During this study, we focus only on supervised and unsupervised detection techniques, for the supervised technique, we build four learning models, which are C4.5 [11], SVM [13], ANN [14], and KNN [12], while for unsupervised detection technique, we use the K-means clustering technique. Our goal is to evaluate the performance of each detection technique on real data sets.

2   Intrusion Detection Systems

Intrusion detection is the process of monitoring events occurring on a network or a host machine and analyzing them for signs of intrusion or any abnormal behavior that attempts to compromise information security of a system. An intrusion detection system (IDS) is to look for suspected events and triggers a response about these events to the administration console [1]. Intrusion detection systems are classified into several types depending on the configuration that they have, and these configurations are linked to the following components, which are defined as: (1) the information source of the IDS, (2) the detection approach, and (3) the architecture of the IDS itself [2].

2.1   Types of IDS Based on Information Sources

Intrusion detection systems can be classified into two different groups depending on the information sources. The first group is for host-based intrusion detection systems, and the second group is for network-based intrusion detection systems. On the one hand, network-based intrusion detection systems detect attacks by capturing and analyzing network packets by listening on a network segment or switch, these types of IDS are beneficial for understanding what is happening on the network. On the other hand, Host-based intrusion detection systems are located in an individual host machine monitoring and analyzing only the activities and information collected from that individual host machine. This point allows a host-based intrusion detection system to analyze activities with high reliability and precision determining exactly which processes and users are involved in case of attacks against the operating system. However, the Host-based intrusion detection system needs more management effort to install and configure this system on multiple host machines [1, 3–5].

2.2 Types of IDS Based on Detection Approaches

An event is an action performed by a user on an online system or a network if the user is allowed to perform certain actions on the system, these actions are normal; otherwise, if the user is not allowed to perform these actions, the actions are therefore abnormal. There are two main approaches to determining the nature of events on the network and classifying them as normal or abnormal. The first group concerns intrusion detection methods or approaches based on signatures and well-known abnormal events and activities, signature-based intrusion detection systems based on a database of known signatures, and each signature represents a specific attack pattern that has already been detected in the past, that exploits known system vulnerabilities, or that violates information security. A signature-based intrusion detection system's main function is to monitor and analyze all current events occurring on the network to find a match with one of the predefined signatures in the database and report the match found to the administration console. This approach's weakness is the detection only of known attacks that exist in the database of the intrusion detection system. It is unable to detect new invented attacks. The system considers these attacks to be normal events or activities, making the network compromised [6, 7]. The second group concerns behavioral approaches, events or activities are generally divided into two categories, normal events that indicate authorized actions, which appear normal to the system, and abnormal events that indicate unauthorized actions, which compromise the system's security. The following example can summarize the idea behind the behavioral approach or the anomaly detection approach, and we consider as an example that we have So far, we have seen the different types of intrusion detection system depending on the detection approach and information sources. The architecture is how to organize the response alerts after detecting a strange object or character on which we cannot determine its nature, this strange object walks like a human, acts like a human, and speaks like a human, so it is probably a human. In the context of intrusion detection, if we have a new event and we cannot determine its nature, the idea is to compare this new event with the set of a predefined model of normal events. If the new event is very similar to normal events or does not deviate from the model that describes these normal events, then it is a normal event. Otherwise, we consider it as an abnormal event. The dark side of this approach is the poor detection of normal events, which leads to the detection and reporting of certain normal events such as intrusions [6, 8].

2.3 IDS Architecture

An intrusion in the case of several intrusion detection systems of different types and installed on the same internal network. The architecture is a structure for configuring and controlling multiple intrusion detection systems to monitor, detect and respond in groups or individually against intrusions. There are three main architectures, the first architecture is centralized, the second architecture is partially distributed, and the third architecture is fully distributed. For the centralized architecture, all intrusion detection systems in different places of the network are linked to a central intrusion detection system, all controls, analyzes, and responses come from this central unit. With this architecture, intrusion detection systems work in cooperation, and the central unit is aware of any intrusion that may occur at any location in the network. There are local central units for the partially distributed architecture, and there is a global central unit, intrusion detection systems in different places are linked to their local unit, which means monitoring, analysis, and response is controlled through these local central units. These local units themselves are linked to a single main unit. Control is hierarchically distributed from one or more global central units to local central units and finally to intrusion detection systems. For the fully distributed architecture, each intrusion detection system operates individually in the network, and each intrusion detection system is responsible at its specific location for monitoring, analyzing and responding to an intrusion, this architecture is not cooperative [1].

3 Data Mining Approach

In intrusion detection systems, the primary goal is to analyze network traffic and then detect intrusions or abnormal events precisely among all the normal activities on the network. Its analogy in the Data Mining approach is the detection of anomalies, and the goal is to find objects that are different from most other objects, often an anomaly object or an outlier is an observation that differs so much from other observations it raises the suspicion that a different mechanism generated it. The intrusion detection subject is treated in the field of Data Mining as anomaly detection. The objective is to find anomalies (intrusions) through a huge dataset containing the events data collected from a network. The question that is assumed now is how to detect these intrusions? To answer this question, multiple techniques can be used to detect intrusions in a dataset. However, these techniques can be organized according to three main approaches based on the availability of class label in the dataset: (1) supervised detection and (2) semi-supervised detection, and unsupervised detection. In this research study, we focus only on supervised and unsupervised detection techniques. In the supervised detection, each instance of the dataset is labelled as 'normal' or 'attack', and a learning algorithm is trained on the training dataset to build a classification model. Our work in the supervised detection approach is mainly focused on detecting intrusions in a training dataset using four classifiers which are decision tree (C4.5), k nearest neighbors (KNN), support vector machine (SVM), and artificial neural network (ANN), The built models will be used to assign new given events to one of the predefined class labels. On the other hand, in the unsupervised detection technique, we use the K-means clustering. The K-means defines a prototype in terms of a centroid, which is generally the average of a group of points. The idea is to find K clusters, each cluster defined by a center. These clusters represent the instances' new distribution, without using any prior knowledge on the class labels in the training dataset. When we need to decide the nature of a new event we calculate the distance between this new event and the K clusters, then we attribute this new event to the closest cluster [9, 18].

3.1 Supervised Detection

Classification is the process of building a model from an initial dataset called a training dataset. The model built will be used to assign new given objects characterized by a set of attributes to one of the predefined class labels, this task is called classification or prediction, which means predicting a class label of new data. The supervised detection problem is viewed as a classification problem. This section will discuss the four classifiers that we choose to use to predict the nature of network events and classify each event on the network as a normal event or an intrusive event [10, 27]. This section only gives an overview of each classifier:

- **C4.5 Decision Tree Classifier:** Decision tree algorithms use an initial training dataset characterized by a dimension, m designed to the number of rows in the training dataset, and n designed to the number of attributes that belongs to a record. The decision tree algorithms run through that search space recursively, and this method called tree growing, during each recursive step, the tree growing process must select an attribute test condition to divide the records into smaller subsets. To implement this step, the algorithm used the measure of the information gain that we saw previously. This means that the decision tree algorithms calculate the information gain for each attribute, and this process called the attribute test condition, and choose the attribute test condition that maximizes the gain. A child node was then created for each outcome of the attribute test condition and the sub-records distributed to the children node based on the outcomes. The tree growing process repeated recursively to each child node until all the records in a set belong to the same class or all the records have identical attribute values. However, both conditions are sufficient to stop any decision tree algorithm [11].
- **K-nearest Neighbor Classifier:** The k-nearest neighbor's classifier does not need an initial model that must already be built from a training dataset, such as decision tree classifiers. The training dataset modelling process is delayed until it is necessary to classify the new given records. A k nearest neighbors classifier represents each record in the training set as a data point in d-dimensional space, where d is the number of attributes. Given a new record, we calculate its proximity to all data space points, using proximity measures such as Euclidean distance and Manhattan distance. The k-neighbors closest to a given z-record refer to the k points closest to z. The z record is classified according to the class labels of its neighbors. In case the neighbors have more than one class label, the record is assigned to the majority class of its nearest neighbors [12].
- **Support Vector Machine:** SVM classifier work by learning hyperplanes that optimally separate the set of objects presented in an n-dimensional space into two or more groups (classes) to minimize a well-defined penalty function. This penalty function penalizes training data points for being on the wrong side of the decision limit and even being close to the right side's decision limit. The notion of "proximity" to the decision frontier is defined by the margin, which is simultaneously maximized, the goal is to find the optimal hyperplane, and we will say that it is the one that produces the highest margin among all the hyperplanes possible. Thus, an SVM classifier can

be modelled as an optimization problem, in which the goal is to solve this optimization problem to find the coefficients of the optimal hyperplane [13].

- **Artificial Neural Network:** If we connect a series of artificial neurons in a network, we have an artificial neural network or neural network. We can represent a neural network as a graph made up of nodes for each neuron and edges for each connection. An artificial neural network is made up of an input layer, where the source data is fed in, an output layer, which produces a prediction, and we have one or more layers hidden in between. To train the weights of a neural network to make precise predictions. First, we have a forward propagation stage. We use the network with its current settings to compute a prediction for each example in our training dataset when propagating forward. We use the known correct answer provided to determine whether the network made a correct prediction or not. An incorrect prediction, which we call a prediction error, will teach the network to change the weight of its connections to avoid making prediction errors in the future. Second, we have a backward propagation step, too. In this step, we use the prediction error that we calculated in the last step to correctly update the weights of the connections between each neuron to help the network make better future predictions. This is where the entire complex math happens. We use a gradient descent technique to help us decide whether to increase or decrease each connection's weights and then we also use something called a training rate to determine how much to increase or decrease the weights for each connection during each stage of the training. Essentially, we need to increase the strength of the connections that helped predict the correct answers and decrease the connections that led to incorrect predictions. We repeat this process for each training sample in the training dataset, and then we repeat the entire process several times until the weights of the network become stable. When we are done, we have a network that is set to make accurate predictions based on all training data it has seen. However, we represent this network of nodes and edges using much more computationally efficient data structures, such as vectors and matrices. A vector is a one-dimensional array of values. A matrix is a two-dimensional array of values. A deep neural network is a neural network with several hidden layers. Adding more hidden layers allows the network to model increasingly complex functions. The ability to model more complex functions gives deep neural networks their power [14].

3.2 Unsupervised Detection

For intrusion detection with unsupervised methods, we choose to work with the cluster-based approach. The clustering approach we used for intrusion detection is prototype-based. The objective is to maximize intra-cluster similarity and minimize inter-cluster similarity, and that is exactly what the K-means clustering algorithm is trying to achieve. To understand how K-means clustering works, let us imagine two-dimensional data. To perform K-means clustering, we first initialize K centroids. These K values can be any randomly chosen value, or some algorithms can be used to choose those K centroids we start with. Once we have these K centers, we go through all the remaining points and assign each point to a cluster. Measure the distance between the centroid and all the points, and the points that are close to a particular centroid will be assigned to the cluster represented by the centroid. Once we have all the clusters, we will recalculate the mean or centroid for all of these clusters. At this point, the center of gravity or the average values may move a little. Once we have the new average values, we will reassign the points to the groups closest to those points. It is the same process that we do over and over again. Iterate until all the points are in their final groups and the means no longer move. This is where the K-means algorithm achieves convergence. K-means clustering is an example of a centroid-based clustering algorithm, where each cluster can be represented using a centroid [12, 15, 19–21].

4 Intrusion Detection Datasets: Presentation and Preprocessing

Datasets play an important role in assessing the performance of learner models. A well-constructed dataset that presents the real situation of the problem being studied will help assess the models' performance under study. In our case, the datasets must contain real network events, and it must be well constructed to be useful for constructing, testing and validating the methods for detecting intrusions on a network. A good-quality dataset allows us to identify the models' ability and measure their accuracy and effectiveness to detect abnormal activities. In this section, we discuss two publicly available datasets, which are the KDDcup99 dataset and the NSL-KDD dataset. Generated using simulated environments in large networks, by running different attack scenarios [16, 17, 22].

4.1 KDDcup99 Dataset

The version of the KDDcup99 dataset we worked on consists of 494,020 records and 41 attributes. The 23 class labels presented in the KDDcup99 dataset are organized into five main classes, which are, normal class label, denial of service (DOS), remote to local (R2L), user to root (U2R), and probing attack (PROB). The list of all class labels in this dataset and the number of samples related to each class label shown in Table 1 below.

The dataset is organized into five main classes as shown above in Table 1, each row in the dataset represents an event on the network, but the dataset can be reorganized into two main categories: the normal class and the intrusion class. The total number of samples linked to the "normal" class is equal to 97,277. In return, the total number of samples linked to the "intrusion" class is equal to 396,743. As we can see, there is a huge difference between the numbers of samples for each class, which can cause performance issues with the training algorithms that will be applied to that dataset. In the next few paragraphs, we will cover some of the preprocessing techniques we applied to the dataset to prepare for the training and testing phases.

The first preprocessing we adopted is to clean the dataset of duplicates. In this version of the KDDCup99, we found about 348,435 duplicate data. In Table 2, we show the new distribution of samples for each attack after removing all redundancies.

The table above summarizes the number of samples for each class before and after removing all redundancies. The dataset's size has been reduced by 80%, resulting in a change in the number of samples for each class label. The data distribution for each class label has been changed. The total number of samples linked to the normal class becomes equal to 87,831. On the other hand, the total number of samples linked to the

Table 1. The distribution of instances according to each class label in the KDDcup99 dataset.

Attack	Number of samples	Class label
NORMAL	97277	Normal
SMURF	280790	DOS
NEPTUNE	107201	DOS
BACK	2203	DOS
TEARDROP	979	DOS
POD	264	DOS
LAND	21	DOS
WAREZCLIENT	1020	R2L
GUESS_PASSWD	53	R2L
WAREZMASTER	20	R2L
IMAP	12	R2L
FTP_WRITE	8	R2L
MULTIHOP	7	R2L
PHF	4	R2L
SPY	2	R2L
SATAN	1589	PROB
IPSWEEP	1247	PROB
PORTSWEEP	1040	PROB
NMAP	231	PROB
BUFFER_OVERFLOW	30	U2R
ROOTKIT	10	U2R
LOADEMODULE	9	U2R
PERL	3	U2R

intrusion classes becomes equal to 57,754, which means that the normal class represents approximately 60% of the total dataset and that the intrusion category represents approximately 40% of the total dataset. The big difference between the two classes (intrusion and normal) has been reduced, which will help produce results that are more practical in the future when working on this dataset.

The second preprocessing we used is converting categorical attributes to numeric attributes, the set of algorithms we used such as KNN, SVM, ANN, and clustering only support numeric values, while C4.5 is the only algorithm that supports categorical and numeric attributes. This version of the dataset that we used contains only three categorical attributes, which are "protocol_type", "service" and "flag". We take all possible unique

Table 2. According to each class label, the distribution of instances before and after removing all the redundancy data in the KDDcup99 dataset.

Attack	Number of samples before redundancy is removed	Number of samples after removing redundancy	Class
NORMAL	97277	87831	Normal
SMURF	280790	641	DOS
NEPTUNE	107201	51820	DOS
BACK	2203	968	DOS
TEARDROP	979	918	DOS
POD	264	206	DOS
LAND	21	19	DOS
WAREZCLIENT	1020	893	R2L
GUESS_PASSWD	53	53	R2L
WAREZMASTER	20	20	R2L
IMAP	12	12	R2L
FTP_WRITE	8	8	R2L
MULTIHOP	7	7	R2L
PHF	4	4	R2L
SPY	2	2	R2L
SATAN	1589	906	PROB
IPSWEEP	1247	651	PROB
PORTSWEEP	1040	416	PROB
NMAP	231	158	PROB
BUFFER_OVERFLOW	30	30	U2R
ROOTKIT	10	10	U2R
LOADEMODULE	9	9	U2R
PERL	3	3	U2R

values for each categorical attribute, and for each categorical value, we give it a unique integer value.

Until now, we have the training dataset with many numeric attributes. The range of values for each attribute will be different. Because they represent different types of information, they will have different maximum and minimum values. The distribution of data is also different. The mean of each attribute in the dataset will be different, as well as its dispersion, or how it moves around the mean value. Common terms for this are variance and standard deviation. They will also be different. In real datasets like the one we have, classification and clustering algorithms will not work very well when we train it with numeric data at different scales. When the attribute ranges are completely different, classification models tend not to learn very well from this data. In addition, that is why when we use the KDDcup99 dataset, which contains numeric data to train machine-learning models as (C4.5, ANN, KNN, SVM, and clustering) we do something called attribute scaling and attribute normalization.

We will talk about scaling first and see how exactly we scale the numeric attributes. Each attribute has n values, so there are n rows or n records in the dataset. The numeric attributes in the actual dataset can represent anything, which means that each of these attributes' minimum and maximum values can be very different. We talked about the fact that classification and clustering algorithms do not work well when the numeric data have different scales, this is why we perform the scaling technique, we scale all the attributes to be in the same range with the same minimum and maximum values. For each attribute in the dataset, we scale the values in that attribute between 0 and 1. For each attribute, its minimum value will be 0, and the maximum value will be 1. This technique is useful for optimization algorithms such as SVM. It is also useful for algorithms that weight inputs like Neural Networks (ANN) and algorithms that use distance measurements like KNN and K-means. The second method we used is normalization, which refers to rescaling each row line to have a length of 1 called the unit norm or a vector of length 1 in linear algebra. This technique is crucial for algorithms that weight input values such as neural networks (ANN) and algorithms that use distance measures such as KNN and K-means.

The last thing that remains is to label the class attribute as a numeric type, starting with the normal class, which is labelled as 1, DOS labelled as 0, PROB labelled as 2, R2L labelled as 3, and U2R labelled as 4.

4.2 NSL-KDD Dataset

The NSL-KDD dataset consists of two parts: (i) KDD Train + and (ii) KDD Test + . The distribution of attacks and normal instances in the NSL-KDD dataset is shown in Table 3. Table 3 summarizes the distribution of attacks and normal instances for each category in the two datasets. The training dataset contains approximately 125,972 records while in the test dataset the number of records is less than the training dataset equal to 22 543. The size of the NSL-KDD dataset is smaller than the size of KDDCup99, which contains redundant records. The NSL-KDD training dataset does not include redundant records. The learning phase is performed on the NSL-KDD Train + dataset, which contains 22 types of attack, and the testing phases are performed on NSL-KDD Test + dataset, which contains 17 additional types of attack. These additional attacks fall into the same four classes that we discussed earlier in the KddCup99 dataset which are, denial of service (DOS), remote to local (R2L), a user to Root (U2R), and probing attack (PROB).

The NSL-KDD dataset contains numeric attributes and three categorical attributes. The categorical attributes "protocol_type", "service" and "flag" must be converted into numeric attributes as we did previously with the KDDCup99 dataset. Likewise, as we did before with the KddCup99 dataset, we apply the scaling and normalization techniques to make the data more convenient to use, and finally, we convert the categorical class labels into numeric class labels, as we did before, start with the normal class labelled 1, DOS labelled 0, PROBE labelled 2, R2L labelled 3, and U2R labelled 5.

5 Experimental Results

Performance evaluation is a major part of any intrusion detection technique or model. Without a proper assessment, it is not easy to demonstrate that a detection mechanism is accurate and can be deployed in real environments. It is the determining factor to compare the quality of several models or systems if there is more than one model and all do the same job [23, 25]. This section presents the performance evaluation metrics that we used to evaluate the intrusion detection models that we used on the two datasets (KDDCup99 and NSL-KDD) that we previously prepared. These primary metrics include detection rate, specificity, false-negative rate, false-positive rate, accuracy, and error rate [24, 26].

The intrusion detection systems are responsible for predicting the current events occurring on the network and classifying it as normal or intrusion. The evaluation of an intrusion detection system's performance is based on the number of events correctly and incorrectly predicted by the system.

5.1 Evaluation and Comparison of Supervised Detection Methods

This section measures the performance of C4.5, SVM, ANN, and KNN models with KDDCup99 and NSL-KDD datasets. We start by applying these training models on the KDDCup99 dataset, discuss and compare the results obtained. Then we move to the second NSL-KDD dataset, we apply the same training models, and then we discuss and compare the results for that dataset. Finally, we evaluate these models using the six measures discussed above based on the obtained results, and then we compare the performance of these models based on each dataset.

Before presenting and discussing the results obtained from each model on the two datasets, first, we need to clarify some parameters that we used in each training model:

- **C4.5:** We used the entropy function to measure the quality of a split, and we use the "best" strategy to choose the best split at each node.
- **SVM:** There is a parameter that we have to take into consideration to build this model, which is called "C". The parameter C gives control of how the SVM will handle errors. The goal is now not to make zero classification mistakes, but to make as few mistakes as possible. In our practical case, we have assigned the parameter C the value 1.
- **ANN:** For this neural network model, we have specified exactly three hidden layers with eight neurons for each layer, and the activation function we have specified here is RELU activation. In addition to the RELU activation, we specify that the learning rate, which moderates the changes' strength, is set to 0.0001.
- **KNN:** The number of neighbors K that we have chosen for our model is equal to seven, and only the points located in this radius are considered neighbors. We chose K = 7 based on the best score produced, due to we run the KNN algorithm through several values of K and found that KNN produces the best score when k is between 1 and 7.

We divide the KDDCup99 dataset into two datasets. The first is the training dataset, which takes 80% of the base dataset instances, and the second is the test dataset that takes 20% of the KDDCup99 base dataset instances. As we saw earlier in the previous section, the KDDCup99 dataset has around 145,585 instances after removing duplicate data, so the training dataset will have 116,468 records, while the test dataset will have 29,117 records. We used these datasets to train and test the C4.5, SVM, ANN, and KNN models, and we obtain the following results as highlighted below in Table 4.

The evaluation measures according to the prediction results obtained for the C4.5, SVM, ANN, and KNN models are summarized in Table 5.

Table 3. According to each class label, the distribution of instances and the categorization of class labels into five categories in the NSL-KDD Train + and NSL-KDD Test + datasets.

Attack	Number of samples in NSL-KDD Train +	Number of samples in NSL-KDD Test +	Class
NORMAL	67342	9711	Normal
SMURF	2646	665	DOS
NEPTUNE	41214	4656	DOS
BACK	956	359	DOS
TEARDROP	892	12	DOS
POD	201	41	DOS
LAND	18	7	DOS
MAILBOMB	-	293	DOS
PROCESSTABLE	-	685	DOS
UDPSTORM	-	2	DOS
APACHE2	-	737	DOS
WAREZCLIENT	890	-	R2L
GUESS_PASSWD	53	1231	R2L
WAREZMASTER	20	944	R2L
IMAP	11	1	R2L
FTP_WRITE	8	3	R2L
MULTIHOP	7	18	R2L
PHF	4	2	R2L
SPY	2	-	R2L
HTTPTUNNEL	-	133	R2L
NAMED	-	17	R2L
SENDMAIL	-	14	R2L
SNMPGETATTACK	-	178	R2L
XLOCK	-	9	R2L
XSNOOP	-	4	R2L
SATAN	3633	735	PROB
IPSWEEP	3599	141	PROB
PORTSWEEP	2931	157	PROB
NMAP	1493	73	PROB
MSCAN	-	996	PROB
SAINT	-	319	PROB
BUFFER_OVERFLOW	30	20	U2R
ROOTKIT	10	13	U2R
LOADEMODULE	9	2	U2R
PERL	3	2	U2R
PS	-	15	U2R
SNMPGUESS	-	331	U2R
SQLATTACK	-	2	U2R
WORM	-	2	U2R
XTERM	-	13	U2R

Table 4. The confusion matrices summarize the prediction results by the C4.5, SVM, KNN, and KNN models on the KDDCup99.

			Predicted	
			Normal	Intrusion
Actual	C4.5	Normal	17569	24
		Intrusion	22	11502
	SVM	Normal	17621	67
		Intrusion	312	11117
	ANN	Normal	17536	27
		Intrusion	43	11466
	KNN	Normal	17593	25
		Intrusion	43	11456

Table 5. Assessment metrics for the four models on the KDDCup99dataset.

	Detection rate	Specificity	FN/rate	FP/rate	Accuracy	Error rate
C4.5	0.998	0. 998	0.001	0.001	0.998	0.001
SVM	0.972	0.996	0.027	0.003	0.986	0.013
ANN	0.997	0.997	0.002	0.002	0.997	0.002
KNN	0.996	0.998	0.003	0.001	0.997	0.002

Table 5 shows that there is no huge difference in each model's performance; there is no accurate model than the others; all models behave the same. The models are exact in detecting normal events and intrusions as demonstrated by the measurements: detection rate, false-negative rate (FN/rate), false-positive rate (FP/rate). Therefore, the misclassification of normal events as intrusions is very low, as evidenced by the specificity measure. These huge performance values are related to the nature of the KDDCup99 dataset, although we have divided the KDDCup99 dataset into two datasets, one for training and one for testing. However, all the training dataset attacks also exist in the test dataset, which makes this dataset unrealistic, which is why the classification models are very accurate using the KDDCup99 dataset. On the other hand, there are always new attacks in real environments, to be more realistic, and to demonstrate how these models will react in the presence of new unprecedented attacks, we use the NSL-KDD dataset. We proposed to train and test these models with the NSL-KDD dataset, and this dataset has two versions: NSL-KDD Train + and NSL-KDD Test + . The main difference between these two versions of datasets is that the test dataset has more attacks than the training dataset. These new attacks do not exist in the training dataset, making it more realistic than the KDDCup99. In the next step, we will train the four models and test them using the new dataset. The steps are well known now. First, we will train and test the models then take the prediction results for each model, second, we will create the confusion matrices from these initial predictive results, and finally, we calculate the evaluation measures for each model.

As expected, all models' performance metric values have been reduced when working with the NSL-KDD dataset because the NSL-KDD dataset is more realistic and better than the KDD dataset. The NSL-KDD dataset tests the models with additional 17 attacks that it does not exist in the training dataset, so the models have never seen these new attacks in the training phase. This situation is close to real situations, where there are always new attacks that can occur in the network where the detection system intrusion is installed. Table 6 shows the evaluation metrics for each model; we can see that the ANN model is the best but with small margins, followed by the SVM model, while the C4.5 and KNN models are slightly the smallest.

Table 6. Assessment metrics for the four models on the KDDCup99dataset.

	Detection rate	Specificity	FN/rate	FP/rate	Accuracy	Error rate
C4.5	0.608	0.969	0.391	0.030	0.763	0.236
SVM	0.612	0.977	0.378	0.022	0.769	0.230
ANN	0.632	0.968	0.367	0.031	0.777	0.222
KNN	0.601	0.977	0.398	0.022	0.763	0.236

5.2 Evaluation and Comparison of Unsupervised Detection Method

In this section, we have focused on the unsupervised detection method (K-means). We train and test the k-means using only the NSL-KDD dataset because this dataset increasingly reduces, the performance of supervised models compared to KDDCup99 and shows the weaknesses of the supervised models in the presence of new attacks.

Before demonstrating the results obtained when applying the K-means algorithm, we must first visualize the studied datasets. For this reason, we present the NSL-KDD training and test datasets in a 2-dimensional space. The initial datasets contain 41 attributes, well, so it is hard to visualize. To visualize these datasets, we need to reduce dimensionality; for this reason, we use principal component analysis technique. It is generally referred to as a dimension reduction technique, and it is a linear algebra technique that can be used to perform dimensionality reduction automatically. It reduces dimensionality by projecting the data into a lower-dimensional subspace that preserves the most important structure or relationships between the data variables. Figure 1 and Fig. 2 below show the training and test datasets divided into five groups in a 2-dimensional space, after reducing the data space's dimensions into two main components found by analyzing the main components. Each group is identified by a specific color, red for DOS class, green for normal class, blue for PROB class, black for R2L class, and yellow for U2R class.

We applied the K-means clustering algorithm on the NSL-KDD Train + dataset presented in Fig. 2, with K equal to 5, which is the number of classes. As a result, we

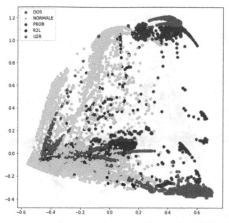

Fig. 1. The distribution of points in a two-dimensional space according to 5 groups of the training dataset NSL-KDD Train +.

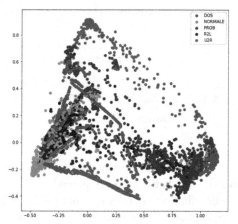

Fig. 2. The distribution of points in two-dimensional space according to 5 groups of the NSL-KDD Test + dataset.

get the centers of each cluster, and then we use these centers along with the test dataset presented in Fig. 3 to predict the index of the cluster for each data point or record in the NSL-KDD Test + dataset.

Figure 3 below shows the new distribution of data in the NSL-KDD Test + dataset, after assigning each data point to a cluster center; we got in the first step when we applied the K-means algorithm on the NSL-KDD Train + dataset.

Cluster 1 represents instances of the normal class, cluster 2 represents class DOS, cluster 3 represents class PROB, cluster 4 represents class R2L, and cluster 5 represents class U2R. Table 7 summarizes the results obtained by the k-means algorithm. It counts the number of samples for the two classes "normal" and "intrusion".

The detection rate using the unsupervised K-means method is greatly increased while the specificity means the number of normal events that are predicted as normal events

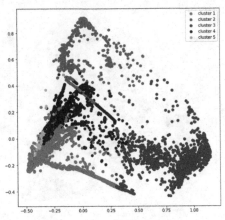

Fig. 3. The new distribution of points in two-dimensional space into 5 groups in the NSL-KDD Test + dataset.

Table 7. The confusion matrix summarizes the results of the classifications made by the K-means algorithm

		Predicted	
		Normal	Intrusion
Actual	Normal	7312	2399
	Intrusion	2352	10480

is greatly reduced compared to the supervised methods. The accuracy and error rate are almost the same for supervised and unsupervised methods.

After determining the distribution values of the "normal" and "intrusion" classes on the NSL-KDD dataset, we then calculate all the performance measures and present the results in Table 8.

Table 8. The confusion matrix summarizes the results of the classifications made by the K-means algorithm

	Detection rate	Specificity	FN/rate	FP/rate	Accuracy	Error rate
K-Means	0.816	0.752	0.183	0.247	0.789	0.210

6   Discussion

We obtain the experimental results for the supervised anomaly detection models (C4.5, ANN, SVM, and KNN) and the unsupervised anomaly detection (K-means clustering)

(Fig. 4). We conclude that the supervised anomaly technique is very accurate when the attacks already exist in the training dataset. When the attacks are already seen and learned by the models, the detection rate and accuracy are very high to predict these future attacks. Nevertheless, in the case of new attacks, attacks that do not exist in the training dataset, attacks never seen before, the detection rate and accuracy of these classification models have been decreased. On the other hand, the clustering technique works well for new attacks that do not exist in the training dataset, giving to the unsupervised anomaly detection techniques the priority over the supervised techniques in the case of unknown attacks. However, the problem with unsupervised techniques is the decrease in the specificity measure, which represents the number of normal events that are predicted as normal, and it means that the rate of misclassification of normal instances has been increased; this is the inconvenient of the unsupervised method, which increases of the false alarm rate.

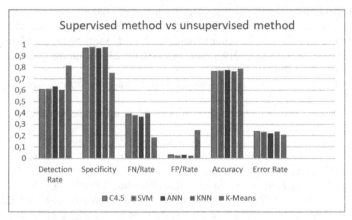

Fig. 4. Performance comparison between supervised and unsupervised methods based on the NSL-KDD dataset.

7 Conclusion

In this research study, we provide an in-depth study of Data Mining's role in intrusion detection. IDS can be divided into two categories depending on the detection approaches: signature-based detection and behavioral detection. The signature-based approach's weakness is as follows: detection only of known attacks, and it is unable to detect new invented attacks, IDS considers these attacks to be normal events or activities. While the behavioral approach's weakness is the poor detection of normal events, which leads to the detection and notification of some normal events as intrusions, this disadvantage is called a false alarm rate. In our experimental study, on the one hand, we prove that the supervised detection technique is very accurate with the attacks seen previously. However, this method's downside is the inability to detect invisible attacks in the training dataset, which used to learn classification models. On the other side, we prove that the unsupervised technique is very accurate with new attacks or attacks not exist in the training dataset, but the downside of this method is the increase in the false alarm rate. From the results obtained, we can see that supervised detection is closely related to the signature-based approach, while unsupervised detection is closely related to the behavior-based approach. These two approaches are complementary in nature; supervised detection can be used to reduce the high value of the false alarm rate, while unsupervised detection can be used to greatly increase the detection rate in the case of new attacks. With this combination, IDS will be more accurate and performant in real situations on the network.

References

1. Rebecca, B., Peter, M.M.: Intrusion Detection Systems. National Institute of Standards and Technology. (2001). https://doi.org/10.6028/NIST.SP.800-94
2. Debar, H.: An Introduction to Intrusion-Detection Systems (2009)
3. Rahul-Vigneswaran, K., Poornachandran, Prabaharan, Soman, K.P.: A compendium on network and host based intrusion detection systems. In: Kumar, Amit, Paprzycki, Marcin, Gunjan, Vinit Kumar (eds.) ICDSMLA 2019. LNEE, vol. 601, pp. 23–30. Springer, Singapore (2020). https://doi.org/10.1007/978-981-15-1420-3_3
4. Kabiri, P., Ghorbani, A.: Research on intrusion detection and response: a survey. Int. J. Netw. Secur. 1(2), 84–102 (2005)
5. Sarkar, T., Das, N.: Survey on host and network based intrusion detection system. Int. J. Adv. Netw. Appl. 6(2), 2266–2269 (2014)
6. Axelsson, S.: Intrusion Detection Systems: A Survey and Taxonomy (2000)
7. Landge, R.S., Wadhe, A.: PMisuse detection system using various techniques: a review. Int. J. Adv. Res. Comput. Sci. 4(6) (2013)
8. Agrawal, S., Agrawal, J.: Survey on anomaly detection using data mining techniques. Procedia Comput. Sci. 60(01), 708–713 (2015). https://doi.org/10.1016/j.procs.2015.08.220
9. Dahima, S., Shitlani, D.: A survey on various data mining technique in intrusion detection system. IOSR J.Comput. Eng. 19(01), 65–72 (2017). https://doi.org/10.9790/0661-190101 6572
10. Lee, W., Stolfo, S.: Data Mining Approaches for Intrusion Detection. 7 (1998)
11. Tom, M.M.: Machine Learning. McGraw-Hill, Maidenhead, U.K. (1997). https://doi.org/10. 1002/(SICI)1099-1689(199909)9:3<191::AID-STVR184>3.0.CO;2-E
12. Tan, P.-N., Steinbach, M., Kumar, V.: Introduction to Data Mining. Pearson Education (2006)
13. Alexandre, K.: Support Vector Machines Succinctly. Syncfusion, Inc. (2017)
14. Tariq, R.: Make Your Own Neural Network. CreateSpace Independent Publishing Platform (2016)
15. Leung, K., Leckie, C.: Unsupervised anomaly detection in network intrusion detection using clusters. In: Proceedings of the Twenty-Eighth Australasian Conference on Computer Science, pp. 333–342 (2005)
16. Bhuyan, M., Bhattacharyya, D.K., Kalita, J.: Network anomaly detection: methods, systems and tools. Commun. Surv. Tutorials IEEE 16, 303–336 (2014). https://doi.org/10.1109/SURV. 2013.052213.00046
17. Parsazad, S., Saboori, E., Allahyar, A.: Fast Feature Reduction in intrusion detection datasets (2013)
18. Antti, J.: Intrusion detection applications using knowledge discovery and data mining (2014)
19. Mahini, R., Zhou, T., Li, P., Nandi, A., Li, H., Li, H., Cong, F.: Cluster Aggregation for Analyzing Event-Related. Potentials. (2017). https://doi.org/10.1007/978-3-319-59081-3_59

20. Irani, J., Pise, N., Phatak, M.: Clustering techniques and the similarity measures used in clustering: a survey. Int. J. Comput. Appl. **134**, 9–14 (2016). https://doi.org/10.5120/ijca20 16907841
21. Syarif, I., Prugel-Bennett, A., Wills, G.: Unsupervised Clustering Approach for Network Anomaly Detection. 293 (2012). https://doi.org/10.1007/978-3-642-30507-8_7
22. Ring, M., Wunderlich, S., Scheuring, D., Landes, D., Hotho, A.: A Survey of Network-based Intrusion Detection Data Sets (2019)
23. Abdulrazaq: AImproving intrusion detection systems using data mining techniques. Doctoral thesis, Loughborough University (2016)
24. Hindy, H., et al.: A Taxonomy and Survey of Intrusion Detection System Design Techniques, Network Threats and Datasets (2018)
25. Jalil, K., Kamarudin, M.H., Masrek, M.: Comparison of machine learning algorithms performance in detecting network intrusion. In: ICNIT 2010 - 2010 International Conference on Networking and Information Technology, pp. 221–226 (2010). https://doi.org/10.1109/ICNIT.2010.5508526
26. Juvonen, A., Sipola, T.: Anomaly Detection Framework Using Rule Extraction for Efficient Intrusion Detection (2014). https://arxiv.org/pdf/1410.7709.pdf
27. Laskov, P., Düssel, P., Schäfer, C., Rieck, K.: Learning Intrusion Detection: Supervised or Unsupervised? **3617**, 50–57 (2005). https://doi.org/10.1007/11553595_6
28. Salo, F., Injadat, M., Nassif, A., Shami, A., Essex, A.: Data mining techniques in intrusion detection systems: a systematic literature review. IEEE Access **PP**, 1 (2018). https://doi.org/10.1109/ACCESS.2018.2872784. Author, F., Author, S.: Title of a proceedings paper. In: Editor, F., Editor, S. (eds.) CONFERENCE 2016, LNCS, vol. 9999, pp. 1–13. Springer, Heidelberg (2016)

Optimization and Decision Support
(Full Papers)

Markov Decision Processes with Discounted Rewards: New Action Elimination Procedure

Abdellatif Semmouri[1]([⊠])[iD], Mostafa Jourhmane[1][iD],
and Bahaa Eddine Elbaghazaoui[2][iD]

[1] Faculty of Sciences and Techniques, Laboratory TIAD, Sultan Moulay Slimane University, Beni Mellal, Morocco
abd_semmouri@yahoo.fr, jourhman@hotmail.com
[2] Faculty of Sciences, Laboratory of Computer Sciences, Ibn Tofail University, Kenitra, Morocco
bahaaeddine.elbaghazaoui@uit.ac.ma

Abstract. Since the computational complexity is one among stations of interest of many interested researchers, numerous procedures are appeared for accelerating iterative methods and for reducing the memory bits required for computing machines. For solving Markov Decision Processes (MDPs), several tests are proposed in the literature and especially to improve the standard Value Iteration Algorithm (VIA). The Bellman optimality equation have played a central role for establish this dynamic programming tool.

In this work, we propose a new test based on the extension of some test for eliminating non-optimal decisions from the planning. In order to demonstrate the scientific interest of our contribution, we compare our result with those of Macqueen and Porteus by an illustrating example. Thus, we reduce the state and action spaces size in each stage as soon as it is possible.

Keywords: Markov processes · Computational complexity · Non-optimal decision · Stochastic optimization

1 Introduction

Markov Decision Processes (MDPs) continually model decision making in discrete, stochastic and sequential environments. Effectively, an agent controls an environment which changes state at random in response to the action choices made by the decision maker. The state of the environment influences the immediate reward obtained by the agent, as well as the probabilities of future state transitions. The agent's goal is to choose actions to maximize a utility function of the total discounted gain. For solving these Markov decision problems, efficient algorithms based on dynamic programming and linear programming have been

© Springer Nature Switzerland AG 2021
M. Fakir et al. (Eds.): CBI 2021, LNBIP 416, pp. 223–238, 2021.
https://doi.org/10.1007/978-3-030-76508-8_16

established. Sometimes, the computation of an optimal solution requires enormous efforts depending on the size of the state and action spaces. This implies the search for structured tools to reduce the complexity of calculations.

Numerous tests of sub-optimal actions were proposed in the literature after computing lower and upper bounds on the optimal objective function of discrete finite discounted MDPs over the infinite horizon. In this context, authors contributed by various results for solving these problems as accelerating techniques or action elimination tools. As soon as an action in a certain state is identified as non-optimal control in an MDP, it can be eliminated from the computation planning after this epoch. This idea was firstly used by MacQueen [1,2]. He proved how non-optimal decision can be discarded in discrete MDPs with finite state and action spaces. Furthermore, he showed that non-optimal actions can be tested if terminals of the optimal utility function are determined. Hence, these bounds can be exploited for reducing the action space size at each iteration as soon as it is possible. Next, Porteus [3] gave an extension of MacQueen test in the general framework of the MDPs under the discounted criterion. He established new bounds on the optimal discounted gain. His approach is based on to begin with a functional equation structure for a process rather than to show such a structure is induced by appropriate way from an alternative starting point. For reminding, we briefly present the Grinold [4] contribution which was inspired from MacQueen and Porteus results. He confirmed that lower and upper bounds on the optimum of objective function can be found via linear programming or policy iteration method. Since the action elimination techniques are the main focus of many interested authors in this background, we also appreciate the contributions of Puterman and Shin [5], White [6], Sladký [7] Semmouri and Jourhmane [8,9], Semmouri, Jourhmane and Elbaghazaoui [10] etc. by participating to this great challenge.

Fig. 1. Computational complexity

Complexity theory is a field of mathematics which is more precisely a part of theoretical computer science. It formally studies the quantity of resources (time, memory space, etc.) that an algorithm requires to solve an algorithmic problem. Therefore, it is a question of dealing with the intrinsic difficulty of the problems, of organizing them by classes of complexity and of studying the relations between the classes of complexity. Since the problem of complexity in space and time is present in many situations which call for enormous calculations. In fact, the large size of some problems is a challenge that can be faced. Thus, the theory of complexity aims to establish hierarchies of difficulty between algorithmic problems; it is based on a theoretical estimate of computation times and computer memory requirements. Besides, the complexity analysis is closely associated with a computational model which is one of the most widely used calculation models. This is the Turing machine proposed by Alan Turing in 1936 because it allows measuring the calculation time and the memory used. From our point of view, it is interesting to establish amazing results in the same literature and particularly in the context of action elimination procedures (AEPs). Hence, we will extend the test proposed by Sladký [7] for identifying non optimal action in discounted Markov decision problems with nonnegative costs in order to solve MDPs with discounted rewards. To demonstrate the effectiveness of this contribution on the theoretical scale, we give an illustrative example.

The structure of this manuscript, is organized as follows:

In Sect. 1 we have presented an overview on some works in the same context of our purpose. Next, Sect. 2 gives useful preliminaries related to Markov decision problems literature which will be used in this paper. Section 3 provides an efficient iterative algorithm as result of the standard value iteration algorithm (VIA) and a new action elimination procedure. Finally, we conclude by showing the strengths of our contribution in Sect. 4.

2   Preliminaries

2.1   Markov Decision Processes Model and Policies

In this subsection, we present the main concepts and definitions related to Markov decision problem model that are relevant to this manuscript. For more details, we refer the interested readers to Semmouri and Jourhmane [8,9], Semmouri et al. [10] and Semmouri et al. [11].

In infinite planning horizon, a descrete Markov decision problem can be described formally by the tuple

$$(S, A = \bigcup_{i \in S} A(i), Q = \{p(j/i, a)\}_{i,j,a}, r = \{r(i, a)\}_{i,a}, \gamma)$$

1. S is a finite set of states called state space in which the state process $(X_n)_{n \geq 0}$ takes place;
2. A is a set of available finite action sets $A(i)$ of all possible actions for the state $i \in S$ which control the state dynamics and in which the action process $(A_n)_{n \geq 0}$ takes place;
3. $p(j/i, a)$ denotes the state-transition probability distribution for moving from the current state i to the next state j by selecting an action $a \in A(i)$;

4. $r : S \times A \to \mathbb{R}$ is denoted to be the one-step reward function defined on state transitions and gives the rewards $r(i, a)$ of taking action $a \in A(i)$ in state i at stage n;

5. γ is a discount factor that determines the influence of future rewards. It is normally used to guarantee a bounded expected value.

At discrete time points $n = 1, 2, ...,$ a system is controlled by a decision maker to be in one of the states of a finite state space S under the state process $\{X_n, n = 1, ...\}$ which lives into S by taking values there. At each period n of time, the decision maker agent controls this system by the choice of an action a from the finite set of controls A. If the action a is selected in state i, the following happens independently of the history of the process at epoch n: an immediate reward $r(i, a)$ is earned and the state of the system at the next decision period of time is the state j with transition probability $p(j/i, a)$, such that $p(j/i, a) \geq 0$ and $\sum_{j \in S} p(j/i, a) = 1$.

Hence, the sample space is given by $\Omega = \{S \times A\}^\infty$, so that the typical realization ω can be represented as $\omega = (x_1, a_1, x_2, a_2, ...)$. The state and action random variables X_n, A_n for $n = 1, 2, ...$ are then defined as the coordinate functions $X_n(\omega) := x_n$ and $A_n(\omega) := a_n$. In the following, the sample space Ω will equipped with the σ-algebra $\mathcal{F}$ generated by the sequence of random variables $(X_n, A_n, n = 1, 2, ...)$.

According to the same horizon, a policy (or strategy) is a sequence of decision rules $\pi = (\pi^1, \pi^2, ...)$, where π^n denotes the decision rule at time point n, $n = 1, 2, ...$. The decision rule π^n at stage n is a mapping $\pi^n : S \longrightarrow A(i)$, which specify the action selection when the system is in state i at decision epoch n where $A(i)$ is the available action set for the state i. Furthermore, it may depend on all available information on the system until time point n.

For indicating the dependence of the probabilities on the policy π, the notation $\mathbb{P}_\pi(Z)$ will denote the probability of an event Z occurring when policy π is applied. From a standard application of the Kolmogorov consistency theorem, it is well known that there exists a unique probability measure $\mathbb{P}_\pi^i$, on $(\Omega, \mathcal{F})$ where $\mathbb{P}_\pi^i(Z) = \mathbb{P}_\pi(Z/X_1 = i)$ and $\mathbb{E}_\pi^i[X] = \mathbb{E}_\pi[X/X_1 = i]$ for each random variable X on Ω. Thus, for each policy π and starting state i, we have constructed a probability space $(\Omega, \mathcal{F}, \mathbb{P}_\pi^i)$.

Let Π the set of all policies. In terms of our model, we will be interesting in a particular class of policies that have very important stochastic properties. The decision rule that it depends on previous system states and actions only throught the current state of the system, it is said to be Markovian (memoryless). The set of all Markovian policies are denoted by Π^M. Among the policies of Π^M, each decision rule that specifies an action with certainty for each current state is called deterministic (or pure). Their set is denoted by Π^{MD}. In other hand, a policy is said to be stationary if the same decision rule is used in every epoch. The set of all stationary policy is denoted by Π^{MS} and the stationary policy is denoted by $\pi = (\pi^\infty, \pi^\infty, ...)$.

With this vocabulary we have

$$\Pi^{MD} \subset \Pi^M \subset \Pi \quad and \quad \Pi^{MS} \subset \Pi^M \subset \Pi$$

For $n \geq 1$ it can be easily seen that a decision rule π^n defines a probability transition matrix

$$\mathbf{P}(\pi^n) = [p(j/i, \pi^n)]_{i,j \in S}$$

with entries given by

$$p(j/j, \pi^n) = \sum_{a \in A(i)} p(j/i, a) \pi^n_{i,a}$$

Since $\sum_{j \in S} p(j/i, a) = 1$ for all $i \in S$ and $a \in A(i)$, $\mathbf{P}(\pi^n)$ is a stochastic matrix (all rows sum to 1) that uniquely defines a Markov chain.

The system evolves as follows. If the system is in state $X_n = i$ at epoch n and the control action $A_n = a \in A(i)$ is selected, then we earn a reward $r(i, a)$ and the system moves to a new state $X_{n+1} = j \in S$ according to the probability distribution $p(j/i, a)$ on the state space S.

Using the decision rule π^n at stage n, we define the immediate expected reward column vector by

$$r(\pi^n) := [r(i, \pi^n)]_{i \in S}$$

where, for each $i \in S$,

$$r(i, \pi^n) := \sum_{a \in A(i)} r_n(i, a) \pi^n_{i,a}$$

The main goal of the agent is to maximize certain objective function in order to find an optimal policy.

Definition 1. *For a policy $\pi \in \Pi$ and a starting state i, the total discounted expected reward under a discounting factor $\gamma \in [0, 1)$ is defined by*

$$\mathcal{V}_\gamma(i, \pi) := \sum_{n=1}^{\infty} \gamma^{n-1} \mathbb{E}^i_\pi [r(X_n, A_n)] \tag{1}$$

The vector expression of the value vector $\mathcal{V}_\gamma(\pi) = (\mathcal{V}_\gamma(i, \pi))_{i \in S}$ is given by the formula below:

$$\mathcal{V}_\gamma(\pi) = \sum_{n=1}^{\infty} \gamma^{n-1} \mathbf{P}(\pi^1) \mathbf{P}(\pi^2) ... \mathbf{P}(\pi^{n-1}) r(\pi^n), \quad for \, \pi = (\pi^1, \pi^2, ...) \in \Pi$$

where $\mathbf{P}$ is the transition probabilities matrix of the decision rule π^n, $\mathbf{P}(\pi^0) := I$ and I is the identity matrix of size n.

Definition 2. *Let $\epsilon \succ 0$ be a given nonnegative number and $\pi \in \Pi$.*

(i) *The optimal value function for the discounted criterion is defined by*

$$\mathcal{V}_\gamma(i) := \max_{\pi \in \Pi} \mathcal{V}_\gamma(i, \pi), \ i \in S$$

(ii) *A policy $\pi^* \in \Pi$ is said to be optimal if*

$$\mathcal{V}_\gamma(i, \pi^*) = \mathcal{V}_\gamma(i) \ for \ all \ i \in S \tag{2}$$

(iii) *A policy $\pi \in \Pi$ is called epsilon-optimal if*

$$\mathcal{V}_\gamma(i, \pi^*) \succ \mathcal{V}_\gamma(i) - \epsilon \ for \ all \ i \in S \tag{3}$$

Since the state and action spaces are finite, it is well known that it exists an optimal deterministic policy which optimize the performance of the system via the dynamic programming (see Howard [12], Bertsekas and Shreve [14], White [15], Puterman [16], Piunovskiy [17]) and the linear programming tool (see Derman [18], Bertsekas and Shreve [14], White [15], Puterman [16]).

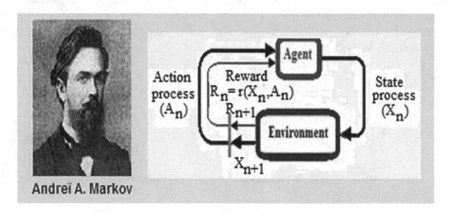

Fig. 2. Markov decision problem model

2.2   Overview on the Standard Value Iteration Algorithm

Markov decision processes have been known since the 1950^s. A great contribution comes from the work of R. A. Howard. They are employed in many disciplines, including robotics, automatic control, digital economics and manufacturing. For solving MDPs, the dynamic programming approach was originally proposed by Howard [12]. It is based on the direct resolution of the Bellman [13] optimality equation $V = \mathcal{L}_B V$ using an iterative fixed point method where the Bellman operator $\mathcal{L}_B$ is defined by

$$\mathcal{L}_B V(i) := \min_{a \in A(i)} \{c(i, a) + \gamma \sum_{j \in S} p(j/i, a) V_j\}, \ i \in S$$

By considering an MDP with discounted costs given by the tuple

$$(S, A = \bigcup_{i \in S} A(i), Q = \{p(j/i, a)\}_{i,j,a}, c = \{c(i, a)\}_{i,a}, \gamma)$$

the standard Value Iteration Algorithm (VIA) is presented as follows:

Algorithm 1: Standard Value Iteration Algorithm

Input: MDP=$(S, A = \bigcup_{i \in S} A(i), Q = \{p(j/i, a)\}_{i,j,a}, c = \{c(i, a)\}_{i,a}, \gamma)$

Output: The optimal value vector & an optimal policy

1 **Initialize** $W_i^0 \leftarrow 0$, $i \in S$, **specify** $\epsilon > 0$

2 **n $\leftarrow$ 0**

3 **repeat**

> **for** $i \in S$ **do**
>> $W_i^{n+1} \leftarrow \min_{a \in A(i)} \{c(i, a) + \gamma \sum_{j \in S} p(j/i, a) W_j^n\};$
>
> **n $\leftarrow$ n+1**
> **until** $\| W^{n+1} - W^n \|_\infty < \frac{1-\gamma}{2\gamma}\epsilon;$
> **for** $i \in S$ **do**
>> $\pi^*(i) \in \arg \min_{a \in A(i)} \{c(i, a) + \gamma \sum_{j \in S} p(j/i, a) W_j^n\};$

4 **return** W^n, π^*.

---

3 Results and Discussion

3.1 Extended Test for Identifying Suboptimal Policies

In this subsection, we transform the original MDP to new PDP which is described by the tuple

$$(S, A = \bigcup_{i \in S} A(i), Q = \{p(j/i, a)\}_{i,j,a}, \tilde{c} = \{c(i, a)\}_{i,a}, \gamma)$$

where $\tilde{c}(i, a) = R - r(i, a)$ for all $i \in S$, $a \in A(i)$ and $R := \max_{i \in S, a \in A(i)} |r(i, a)|$. The corresponding utility function is defined by

$$\forall \pi \in \Pi, \forall i \in S, \quad J_\gamma(i, \pi) := \sum_{n=1}^{\infty} \gamma^{n-1} \mathbb{E}_\pi^i [\tilde{c}(X_n, A_n)] \tag{4}$$

It is obvious to show that

$$\mathcal{V}_\gamma(i, \pi) = \frac{R}{1 - \gamma} - J_\gamma(i, \pi), \quad \pi \in \Pi, i \in S$$

For $\pi \in \Pi$ let $J_\gamma \in \mathbb{R}^{|S|}$ be the vector of components $J_\gamma(i) = \min_{\pi \in \Pi} J_\gamma(i, \pi)$, $i \in S$. By simple operations, we get following relation linking between the objective functions.

$$\mathcal{V}_\gamma(i) = \frac{R}{1 - \gamma} - J_\gamma(i), \quad i \in S \tag{5}$$

Let $\tilde{c}'(i) := \min_{a \in A(i)} \tilde{c}(i, a)$, $m' := \min_{i \in S} \tilde{c}'(i)$ and $M' := \max_{i \in S} \tilde{c}'(i)$ be respectively the minimum one-stage cost in state $i \in S$, the minimum and maximum value of all minimum one-stage costs.

Lemma 1. *Let $(W^n)_{n \geq 0}$ the sequence of vectors given by Algorithm 1 (VIA). Then, for all $n \in \mathbb{N}$ we have*

$$W_i^n + \frac{m' \gamma^n}{1 - \gamma} \leq J_\gamma(i) \leq W_i^n + \frac{M' \gamma^n}{1 - \gamma}, \quad i \in S, a \in A(i) \tag{6}$$

Proof. We use a reasoning by induction on n.

For $n = 0$, since $W_i^0 = 0, \forall i \in S$ and Section 5 in Sladký [7] guarantee the following assertion

$$\frac{m'}{1 - \gamma} \leq J_\gamma(i) \leq \frac{M'}{1 - \gamma}$$

Suppose the property is true for n. By assumption 6 we write

$$\tilde{c}(i, a) + \gamma \sum_{j \in S} p(j/i, a)[W_j^n + \frac{m' \gamma^n}{1 - \gamma}] \leq \tilde{c}(i, a) + \gamma \sum_{j \in S} p(j/i, a) J_\gamma(j)$$

$$\leq \tilde{c}(i, a) + \gamma \sum_{j \in S} p(j/i, a)[W_j^n + \frac{M' \gamma^n}{1 - \gamma}]$$

Identically equivalent to

$$\tilde{c}(i, a) + \gamma \sum_{j \in S} p(j/i, a)W_j^n + \frac{m' \gamma^{n+1}}{1 - \gamma} \leq J_\gamma(i) \leq \tilde{c}(i, a) + \gamma \sum_{j \in S} p(j/i, a)W_j^n + \frac{M' \gamma^{n+1}}{1 - \gamma}$$

Since

$$\tilde{c}(i, a) + \gamma \sum_{j \in S} p(j/i, a)W_j^n \leq J_\gamma(i) - \frac{m' \gamma^{n+1}}{1 - \gamma}$$

and

$$J_\gamma(i) - \frac{M' \gamma^{n+1}}{1 - \gamma} \leq \tilde{c}(i, a) + \gamma \sum_{j \in S} p(j/i, a)W_j^n$$

Then

$$\min_{a \in A(i)} \{\tilde{c}(i, a) + \gamma \sum_{j \in S} p(j/i, a)W_j^n\} \leq J_\gamma(i) - \frac{m' \gamma^{n+1}}{1 - \gamma} \tag{7}$$

and

$$J_\gamma(i) - \frac{M' \gamma^{n+1}}{1 - \gamma} \leq \min_{a \in A(i)} \{\tilde{c}(i, a) + \gamma \sum_{j \in S} p(j/i, a)W_j^n\} \tag{8}$$

Combining (7) and (8), we obtain

$$W_i^{n+1} + \frac{m'\gamma^{n+1}}{1-\gamma} \le J_\gamma(i) \le W_i^{n+1} + \frac{M'\gamma^{n+1}}{1-\gamma}, \quad i \in S, a \in A(i)$$

Which completes the proof. □

Theorem 1 (New Test of suboptimality). *Let $(W^n)_{n \ge 0}$ the sequence of vectors generated by Algorithm 1 (VIA). If*

$$\tilde{c}(i,a') + \gamma \sum_{j \in S} p(j/i,a')W_j^n + \frac{m'\gamma^{n+1}}{1-\gamma} \succ W_i^n + \frac{M'\gamma^n}{1-\gamma} \tag{9}$$

holds at some stage $n \in \mathbb{N}$, then any markovian policy which uses action a' in state i is nonoptimal.

Proof. Suppose that the action a' is optimal in state i and in epoch n. Then, we will get

$$J_\gamma(i) = \tilde{c}(i,a') + \gamma \sum_{j \in S} p(j/i,a')J_\gamma(i) \tag{10}$$

From (8), (9) and (10) we conclude that

$$J_\gamma(i) \ge \tilde{c}(i,a') + \gamma \sum_{j \in S} p(j/i,a')\{W_i^n + \frac{m'\gamma^n}{1-\gamma}\}$$

$$= \tilde{c}(i,a') + \gamma \sum_{j \in S} p(j/i,a')W_i^n + \frac{m'\gamma^{n+1}}{1-\gamma}$$

$$\succ W_i^n + \frac{M'\gamma^n}{1-\gamma} \ge J_\gamma(i)$$

Then, $J_\gamma(i) \succ J_\gamma(i)$. What is absurd. This ends the proof. □

3.2   Improved Value Iteration Algorithm

The value iteration algorithm is undergoing several improvements in the Markov decision problems literature. In this context, we will improve this algorithm by the addition of the non-optimality test (9) in order to establish a new algorithm which is very useful for dynamic programming.

We consider again the sequence of vectors $(W^n)_{n \ge 0}$ provided by algorithm 1 (VIA). Let Q^n and T^n be respectively the Q-learning function and the test function defined on the Cartesian product set $S \times A$ at stage n by

$$Q^n(i,a) := \tilde{c}(i,a) + \gamma \sum_{j \in S} p(j/i,a)W_j^n, \quad i \in S, a \in A(i)$$

and

$$T^n(i,a) := W_i^n - Q^n(i,a) - \frac{m'\gamma - M'}{1-\gamma}\gamma^n, \quad i \in S, a \in A(i)$$

If $T^n(i,a) \succ 0$, then we must eliminate the action $a \in A(i)$ in state i from the calculation at stage $n+1$.

Thus, we propose the following algorithm:

Algorithm 2: Improved Value Iteration Algorithm

Input: MDP=$(S, A = \bigcup_{i \in S} A(i), Q = \{p(j/i,a)\}_{i,j,a}, \tilde{c} = \{\tilde{c}(i,a)\}_{i,a}, \gamma)$

Output: The optimal value vector & an optimal policy

1 **Initialize** $W_i^0 \leftarrow 0$, $i \in S$, **specify** $\epsilon > 0$

2 **n** $\leftarrow$ **0**

  for $i \in S$ do

3     $\pi^*(i) \in \arg\min_{a \in A(i)} Q^n(i,a)$;

4 **repeat**

    for $i \in S$ do

      for $a \in A(i)$ do

5         $Q^n(i,a) \leftarrow \tilde{c}(i,a) + \gamma \sum_{j \in S} p(j/i,a) W_j^n$;

6         $T^n(i,a) \leftarrow W_i^n - Q^n(i,a) - \frac{m'\gamma - M'}{1-\gamma}\gamma^n$;

7   if $T^n(i,a) \prec 0$ then

8           $A(i) \leftarrow A(i) - \{a\}$

      $W_i^{n+1} \leftarrow \min_{a \in A(i)} Q^n(i,a)$;

  n $\leftarrow$ **n+1**

  until $\| W^{n+1} - W^n \|_\infty < \frac{1-\gamma}{2\gamma}\epsilon$;

  for $i \in S$ do

    $\pi^*(i) \in \arg\min_{a \in A(i)} Q^n(i,a)$;

9 **return** W^n, π^*.

For appreciating and illustrating our contribution, we present the following example.

Example 1. Consider an MDP with the following data:

$$S = \{1,2\}; A(1) = A(2) = \{a_1, a_2, a_3\}; \gamma = 0.6.$$

Transition probabilities:

$p(1/1,a_1) = 1$; $p(1/1,a_2) = 0.2$; $p(1/1,a_3) = 1$; $p(2/1,a_2) = 0.8$; $p(2/2,a_1) = 0.3$; $p(1/2,a_2) = 1$; $p(2/2,a_3) = 1$; $p(1/2,a_1) = 0.7$

Rewards:

$r(1,a_1) = -7$; $r(1,a_2) = 9$; $r(1,a_3) = -3$; $r(2,a_1) = 12$; $r(2,a_2) = -4$; $r(2, a_3) = 5$.

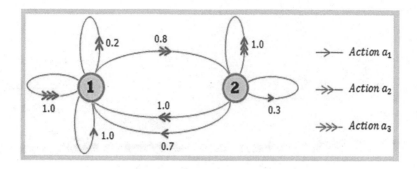

Fig. 3. Graph representation

By exploiting Algorithm 2 (IVIA), we get the following numerical results:

Table 1. Values of $T^n(.,.)$ by IFIA

Iteration n	$T^n(1, a_1)$	$T^n(1, a_2)$	$T^n(1, a_3)$	$T^n(2, a_1)$	$T^n(2, a_2)$	$T^n(2, a_3)$
1	-1.5	14.5	2.5	17.5	1.5	10.5
2	-7.3	10.14	-3.3	9.24	-7.3	3.5
3	-11.356	5.652	$--.356$	5.922	-10.456	-0.196
4	-13.6168	3.5208	-9.6168	3.4398	-12.9868	-2.5648
...	...		...		...	...
30	-17.107686	5.93E−06	-13.10769	5.93E−06	-16.415379	-6.030763
31	-17.107689	3.56E−06	-13.10769	3.56E−06	-16.415381	-6.030765
32	-17.10769	2.14E−06	-13.10769	2.11E−06	-16.415382	-6.030767

Each value of $T^n(i, a)$ colored in red indicates that the action a selected in state i is identified as a non-optimal action which must be eliminated. First, actions a_2 and a_3 must be eliminated in state 2 respectively at the 1st iteration and the 2nd iteration. Second, actions a_1 and a_3 must be eliminated in state 1 respectively after the 2^{sd} iteration and the 3rd iteration.

For the transformed problem, the optimal value vector J_γ is reached at iteration $n = 32$ for $\gamma = 0.6$ and is given by

$$J_{0.6} = (4.730768952, 2.423076645)$$

In addition, the new action spaces given by Algorithm 2 (IVIA) are: $A(1) = \{a_3\}$ after 2nd iteration and $A(2) = \{a_1\}$ after 3rd iteration.

By using (5), we conclude that the optimal value vector $\mathcal{V}_\gamma$ of the original problem is

$$\mathcal{V}_{0.6} = (25.26923105, 27.57692336)$$

Comparing MacQueen and Porteus tests with our test, we obtain

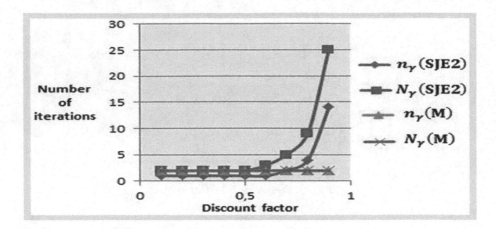

Fig. 4. MacQueen and Semmouri-Jourhmane tests

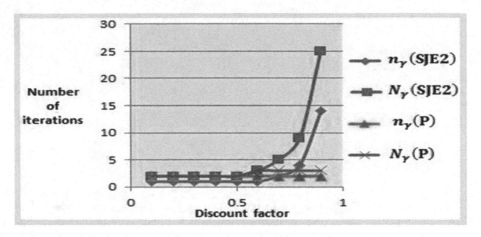

Fig. 5. Porteus and Semmouri-Jourhmane tests

where

$$n_\gamma(SJE2) = \inf\{n \in \mathbb{N} / \exists i \in S, \exists a \in A(i), \, T^n(i,a) \prec 0\}$$

and

$$N_\gamma(SJE2) = \sup\{n \in \mathbb{N} / \exists i \in S, \exists a \in A(i), \, T^n(i,a) \prec 0\}$$

The other concepts are given in appendix.

In the figures Fig. 4 and Fig. 5, for $\gamma \in (0, 0.7]$ our test is near from MacQueen and Porteus tests. This behavior is well convincing in the real life and practical applications.

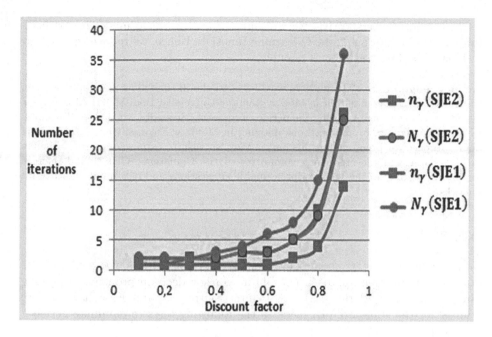

Fig. 6. Semmouri-Jourhmane-Elbaghazaoui and Semmouri-Jourhmane tests

From Fig. 6 below, it can be seen that the numbers $n_\gamma(SJE2)$ and $N_\gamma(SJE2)$ are respectively smaller than the numbers $n_\gamma(SJE1)$ and $N_\gamma(SJE1)$. This explains why the new test established in this manuscript is very efficient compared to the one given in [10].

4   Conclusion

The dynamic programming is an efficient framework which solves many stochastic situations related to Markov decision problems. In this context, the standard value iteration algorithm is a central gathering point of numerous authors. All ameliorations are aimed at speeding up this algorithm or reducing computational complexity in some way. In this regard, the algorithm improvements (VIA) are constantly increasing to provide more efficient results under dynamic programming approach.

We state that our work is distinguished from previous tests. In Macqueen and Porteus ones, we must store the vectors V^{n-1} and V^n for calculating V^{n+1}. However, our test requires only the storage of the vector V^n in the computer memory for calculation of V^{n+1}. Thus, the power and the efficiency of our result appear very important when the size of the state space is very large. Despite this challenge, most of sub-optimality tests that have been established up to now consider the MacQueen and Porteus tests as an essential reference.

In the future perspective, we will apply our theoretical model in Artificial Intelligence by improving the Q-learning function. Hence, we reduce the computational complexity as best as possible.

Acknowledgements. The authors would like to thank the following people. Firstly, Professor Dr. C. Daoui of Sultan Moulay Slimane University, Beni Mellal, Morocco for his help and encouraging during the period of research. Secondly, Mr. Lekbir Tansaoui, ELT teacher, co-author and textbook designer in Mokhtar Essoussi High School, Oued Zem, Morocco for proofreading this paper. We also wish to express our sincere thanks to all members of the organizing committee of the Conference CBI'21 and referees for careful reading of the manuscript, valuable suggestions and of a number of helpful remarks.

Appendix

Now, we give reminders about some famous tests that have played a crucial role in action elimination approach notably in Markov decision problems. Applying MacQueen's [1,2], Porteus [3], Semmouri, Jourhmane and Elbaghazaoui [10] bounds for the standard VIA leads to the following tests to eliminate action a in $A(s)$ permanently:

MacQueen test:

$$M^n(i,a) := Q^n(i,a) - V_i^n - \gamma \frac{a_n - b_n}{1 - \gamma} \prec 0$$

Porteus test:

$$P^n(i,a) := Q^n(i,a) - V_i^n - \gamma^2 \frac{a_{n-1} - b_{n-1}}{1 - \gamma} \prec 0$$

where

$$a_n = \min_{i \in S}(V_i^n - V_i^{n-1}), \quad b_n = \max_{i \in S}(V_i^n - V_i^{n-1})$$

and

$$Q^n = r(i,a) + \gamma \sum_{j \in S} p(j/, i, a) V_j^n$$

Let also

$$n_\gamma(M) = \inf\{n \in \mathbb{N}/\exists i \in S, \exists a \in A(i), M^n(i,a) \prec 0\}$$
$$N_\gamma(M) = \sup\{n \in \mathbb{N}/\exists i \in S, \exists a \in A(i), M^n(i,a) \prec 0\}$$

and

$$n_\gamma(P) = \inf\{n \in \mathbb{N}/\exists i \in S, \exists a \in A(i), P^n(i,a) \prec 0\}$$
$$N_\gamma(P) = \sup\{n \in \mathbb{N}/\exists i \in S, \exists a \in A(i), P^n(i,a) \prec 0\}$$

Semmouri, Jourhmane and Elbaghazaoui test:

$$SJE^n(i,a) := V_i^n - Q^n(i,a) + \frac{(1+\gamma)C}{1-\gamma}\gamma^n \prec 0$$

where

$$C = \max_{i,a}|c(i,a)|$$

and let too

$$n_\gamma(SJE1) = \inf\{n \in \mathbb{N}/\exists i \in S, \exists a \in A(i),\ SJE^n(i,a) \prec 0\}$$
$$N_\gamma(SJE1) = \sup\{n \in \mathbb{N}/\exists i \in S, \exists a \in A(i),\ SJE^n(i,a) \prec 0\}$$

References

1. MacQueen, J.B.: A modified dynamic programming method for Markovian decision problems. J. Math. Anal. Appl. **14**, 38–43 (1965). https://doi.org/10.1016/0022-247X(66)90060-6
2. MacQueen, J.B.: A test for suboptimal actions in Markovian decision problems. Oper. Res. **15**, 559–561 (1967). https://doi.org/10.1287/opre.15.3.559
3. Porteus, E.L.: Some bounds for discounted sequential decision processes. Manag. Sci. **18**, 7–11 (1971). https://doi.org/10.1287/mnsc.18.1.7
4. Grinold, R.C.: Elimination of suboptimal actions in Markov decision problems. Oper. Res. **21**, 848–851 (1973). https://doi.org/10.1287/opre.21.3.848
5. Puterman, M.L., Shin, M.C.: Modified policy iteration algorithms for discounted Markov decision problems. Manag. Sci. **24**, 1127–1137 (1978). https://doi.org/10.1287/mnsc.24.11.1127
6. White, D.J.: The determination of approximately optimal policies in Markov decision processes by the use of bounds. J. Oper. Res. Soc. **33**, 253–259 (1982). https://doi.org/10.1057/jors.1982.51
7. Sladký, K.: Identification of optimal policies in Markov decision processes. Kybernetika **46**, 558–570 (2010). MSC:60J10, 90C40, 93E20 — MR 2676091 — Zbl 1195.93148
8. Semmouri, A., Jourhmane, M.: Markov decision processes with discounted cost: the action elimination procedures. In: ICCSRE 2nd International Conference of Computer Science and Renewable Energies, pp. 1–6. IEEE Press, Agadir, Morocco (2019). https://doi.org/10.1109/ICCSRE.2019.8807578
9. Semmouri, A., Jourhmane, M.: Markov decision processes with discounted costs over a finite horizon: action elimination. In: Masrour, T., Cherrafi, A., El Hassani, I. (eds.) International Conference on Artificial Intelligence & Industrial Applications, pp. 199–213. Springer, Cham (2020). https://doi.org/10.1007/978-3-030-51186-9_14
10. Semmouri, A., Jourhmane, M., Elbaghazaoui, B.E.: Markov decision processes with discounted costs: new test of non-optimal actions. J. Adv. Res. Dyn. Control Syst. **12**(05-SPECIAL ISSUE), 608–616 (2020). https://doi.org/10.5373/JARDCS/V12SP5/20201796
11. Semmouri, A., Jourhmane, M., Belhallaj, Z.: Discounted Markov decision processes with fuzzy costs. Ann. Oper. Res. **295**(2), 769–786 (2020). https://doi.org/10.1007/s10479-020-03783-6

12. Howard, R.A.: Dynamic Programming and Markov Processes. Wiley, New York (1960)
13. Bellman, R.E.: Dynamic Programming. Princeton University Press (1957)
14. Bertsekas, D.P., Shreve, S.E.: Stochastic Optimal Control. Academic Press, New York (1978)
15. White, D.: Markov Decision Processes. Wiley, England (1993)
16. Puterman, M.L.: Markov Decision Processes: Discrete Stochastic Dynamic Programming. Wiley, New York (1994)
17. Piunovskiy, A.B.: Examples in Markov Decision Processes, vol. 2. World Scientific, London (2013)
18. Derman, C.: Finte State Markovian Decision Processes. Academic Press, New York (1970)

Learning Management System Comparison: New Approach Using Multi-Criteria Decision Making

Farouk Ouatik[1][(✉)] and Fahd Ouatik[2]

[1] Sultan Moulay Slimane University, Beni Mellal, Morocco
[2] Cadi Ayad University, Marrakech, Morocco

Abstract. Learning management system with its various means, enable us to track the level of learners based on the exploitation of the information that it provided to us, Especially with the closure of educational institutions due to the Corona virus. But these methods differ in their effectiveness and results according to the purpose for which they are used. This article presents a new method for comparing the means of a learning management system, as we relied on pointing features, but these features do not have the same distribution and the same effect as their importance varies according to the purpose of the study, in our case we consider all the weights equal 1, so we apply the Multi-Criteria Decision Making algorithm to solve this problem by using sum normalization for maximization of Adaptive learning, Technology collaboration, Security, mobile, Ratings, Learning Analysis reports, Easy use and inverse for the minimization of price. To experiment this method, we compared Moodle, Sakai, Claroline, TalentLM, Easy LMS and OpenedX for the purpose of tracking learners' performance. The result shows that Moodle is the best for tracking learners'.

Keywords: Learning management system · Moodle · Sakai · Claroline · TalentLM · Easy LMS and OpenedX

1 Introduction

Among of methods used in e-learning [1] we find the learning management systems, LMS [2], which also contains many e-learning platforms, for example Moodle, Sakai, edomodo,… etc. They allow the management of courses, content, students, etc. [3, 4]. But the problem that arises is how we can choose the right platform for our objective, given the diversity of these tools in terms of functions and performance.

Several comparative studies on LMS are available in the scientific literature. In [5] the author use user-experience (UX) to compare two Online Learning Systems: iQuality and Moodle. He finds that iQuality works better than Moodle. In another work [6] the authors compare the most popular LMS tools: EduBrite, Moodle, TalentLMS, Sakai, and Edomodo for implementing of adaptive e-learning model using these features: Target Customer Size, Platform, Mobile, E-commerce, Video conferencing, Blended Learning, Virtual Classroom, Synchronous Learning, Asynchronous Learning and Skill Tracking,

© Springer Nature Switzerland AG 2021
M. Fakir et al. (Eds.): CBI 2021, LNBIP 416, pp. 239–248, 2021.
https://doi.org/10.1007/978-3-030-76508-8_17

then they found that Moodle is the best. Also [7] analyse Moodle, Facebook, and Paper-based Assessment Tools and they found that the preferred learning tool is Moodle and it helps students to improve their performance. Similarly [8], evaluates the accessibility of Sakai, Moodle, and ABC platform using theses accessibility criteria: Login, Config-uration, Compatibility Testing, Personalization and Customization, Navigation, Forms, Help and Documentation, Common Student Facing Modules/Tools, Authoring Tools and Content Creation, Features Unique to LMS that Affect Accessibility, they found the accessibility of the open source platforms better than platforms developed in house. Furthermore, in [9] the authors present seven criteria: Basic, User Interface, Integration With Mobile Applications, Tools, Compatible Format File, Language Support and Basic Price to Compare three Learning Management System: Moodle, Edmodo and Jejak Bali. And the authors in [10] compare Moodle, ATutor and ILIAS based on two factors, Flex-ibility and Maintainability. The results indicate that Moodle is more flexible and easy to maintain. In order to choose the right Learning Management System, the authors in [11] compare Moodle, ATutor, Blackboard and Success Factors based on ease of use, flexi-bility, user friendliness and accessibility, but they don't make the decision, they leave it to readers. Another criteria used by [12] to compare Canvas, Blackboard, Moodle, and OpenedX using the user interface, market share, weaknesses and feature strengths. In addition [13] use tree features: administrator, the tutor's point of view, the student's point of view to compare ATutor, Moodle and Chamilo and also find that Moodle is better than Chamilo and at last ATutor. As well as. In the same way [14] evaluate + CMS, atutor, Claroline, Dokeos, Drupal, Ilias, Moodle, Mambo using these criteria: Security, Performances, Support, Interoperability, Flexibility, Easy of using, Management, Com-munication tools, Administration tools, Course delivery tools, Content development they take the best four LMS: Moodle, Claroline, Mambo, Atutor. All these studies consider the features used in the comparison have the same distribution and importance. to solve this problem we use Multi-Criteria Decision Making algorithm.this papper is structured as follows the Sect. 2 presents six learning management system: Moodle, Sakai, Claroline, TalentLM, Easy LMS and OpenedX, the Sect. 3 explain the features used to compare the previous LMS platforms, the Sect. 4 presents the Multi-Criteria Decision Making algorithm and the Sect. 5 presents the results and the conclusion in the Sect. 6.

2   E-learning

E-learning is a new form of learning, e-learning draws its appeal from being able to learn at your own pace, on your computer, educational content on a variety of subjects. Organized in sessions or modules, with evaluation tests, the training can be completely self-managed and monitored via a dashboard that lists each of the participant's progress.

2.1   Learning Management System (LMS)

The main objective of an LMS is to facilitate the distribution of your online training. Indeed, the learning platform is a tool that allows you:
  • To host your educational modules and content.
  • To distribute your modules.

• To facilitate the monitoring and management of the curriculum of the community of learners as well as of the individual curriculum. It is therefore used to support the learner in his learning process while ensuring the follow-up of his educational path.

In this paper, we compare four LMS platforms, Moodle, Claroline, Sakai and Openedx.

2.1.1 Modular Object-Oriented Dynamic Learning Environment (Moodle)

Moodle [15] is a free online learning platform used to create educating communities around educational content and activities. It is flexible [16] and it offers educators around the world an open source solution for online training that is scalable, customizable and secure, with the widest choice of activities available. It allows you to design questionnaires and exams, statistical monitoring, send reminders to people who have not completed a training or evaluation activity. Thus, it allows to know who has or has not completed a course and who has passed or failed.

2.1.2 OpenEdX

OpenedX is an Open Source platform implementing a Learning Management System [17] which supports the creation of courses and the provision of a large number of open courses online. Particularly easy to implement, it offers a clean interface and user experience, mainly based on the distribution of videos and the answer to quizzes.

2.1.3 Claroline

Claroline [18] is a distance learning and collaborative work platform. It makes it possible to create educational content that can be consulted remotely via an internet connection and at different rates by the users concerned. The course manager can at any time follow the progress of each user and interact with them through various tools such as the forum or the wiki, as well as users have the possibility of communicating with each other, either in groups, either within the whole class.

2.1.4 Sakai

Sakai [19] is a free software that can be used for teaching, research, and collaborative projects. It is written in the Java language [20], and is a service oriented application suite designed to support a strong increase in the number of users, reliable, interoperable and extensible. Sakai works without modification on Unix, Linux, FreeBSD, Windows, Mac OS X, NetWare, IBM AIX and other systems that support a Java EE servlet and JSP web container server such as Apache Tomcat and Database Management System (Mysql, MariaDB and Oracle Database).Sakai software has many features common to virtual learning environments and course management systems, including distribution of course material documents, grade book, discussion board, live chat room, authorization assignment of file repositories to render handwritten scanned or office work, and online tests.

2.1.5 TalentLMS

TalentLMS [21] is a very simple, flexible, adaptable and customizable learning management system and also mobile. It allows the reuse of already existing material, it is cloud-based. Rich in features, it allows the creation of resources and management of online training platforms, it allows content loading, certification, survey and blended learning.

2.1.6 Easy LMS

Easy LMS [22] is a customizable, easy and efficient learning management system, it allows to follow the progress and gives detailed statistics it is based on the cloud. It allows to structure and evaluate the trainings, To give the users the possibility to access training material in one place, Reduce training time thanks to courses adapted to the pace of users.

3 Criteria Used to Compare LMSs Platforms

We took the criteria from studies that have already been done [23–28] and we added other criteria according to the purpose of our study which concerns student performance and each criteria contains 3 sub features.

- Adaptive learning:
  Customization of the course interface, practical work by the learner as well as the adaptability of the content according to the learner's profile.
- Technology collaboration:
  This criterion takes into consideration the possibility of creating a group of learners, wiki, forum and also the possibility of integrating collaboration tools.
- Security:
  Security is a very important criteria, it includes: Validation of Input Feature, Data Encryption Feature, SSL Certificate Layer.
- Mobile:
  This criterion includes the existence of an android and an IPad version for these e-learning platforms and Multi-platform.
- Ratings:
  Ratings is the criteria that measures the level of learning of students, it includes: Online assessments, Online Grading Feature, QCM and quiz.
- Learning Analysis reports:
  It includes student progress statistics: assessment tracking, feedback tracking and student tracking.
- Easy use:
  This criteria signify the simplicity of use, it takes into account the installation time and the simplicity of maintenance and updating, as well as ergonomics and compatibility with different browsers.
- price:
  Presents platform prices in US dollars.

4   Research Methods

To find the platform that meets our needs, which is tracking learners' performance, we have to order these platforms: Moodle, OpenedX, Sakai, Claroline, TalentLMS and Easy LMS, using the Multi-Criteria Decision Making method [29], this method makes it possible to make the decision with the reduction of the impact biases and incidence.

The first step is to build the dataset and to do it, We compared the six learning management system platforms by 8 features: Adaptive learning, Technology collaboration, Security, mobile, Ratings, Learning Analysis reports, Easy use and price. And each feature contains 3 sub features that is already mentioned in Sect. 3. If the platform has the functionality of the sub feature we assign it 1 otherwise we assign 0. Then we sum the points of these sub features for each feature of each platform. The Table 1 presents a comparative study of Moodle, OpenedX, Sakai, Claroline, TalentLMS and Easy LMS after applying the first step.

Table 1. Comparative table of Moodle, Open edX, Sakai, Claroline, TalentLMS and Easy LMS after applying the first step

Platforms	Moodle	Openedx	Sakai	Claroline	TalentLMS	Easy LMS
Ratings	3	2	1	1	2	2
Adaptive learning	2	3	2	1	2	2
Technology collaboration	2	2	3	2	2	2
Security	3	3	2	3	3	2
mobile	2	2	1	3	3	2
Learning Analysis	3	2	2	1	1	2
Easy use	1	2	2	2	2	3
Price	0	0	0	0	429	250

The second step is to order these platforms, according to the data in Table 1 and take the first platform. But the problem is that the distribution of the criteria used are not the same, so we weighted the values of each criterion, and also we want a platform with high values in Adaptive learning, Technology collaboration, Security, mobile, Ratings, Learning Analysis carryovers, Easy use and low price values. Which is transformed into two, functions maximize and minimize. Therefore, maximize the values of the criteria: Adaptive learning, Technology collaboration, Security, mobile, Ratings, Learning Analysis reports, Easy use and minimize the values of the price. So we used Multi-Criteria Decision Making method to solve this problem.

Algorithm:

1- Define alternatives to be ranked

2- Define criteria used for evaluated

3- Define weights of every feature that represents the importance of this in our case all weights =1.

4- Define criteria to be maximize and criteria to be minimize.

5- Maximization function and minimization function

for the maximization function :

F_j:represents feature to be maximized, and $e_1,...,e_n$ ($1 \leq i \leq n$) its elements in the table 1. We have 3 methods: sum normalization, max normalization, max-min scaling.

sum normalization:

$$max(e_i, F_j) = \frac{e_i}{\sum_i^n e_i} \tag{1}$$

max normalization:

$$max(ei, F_j) = \frac{e_i}{max(F_j)} \tag{2}$$

max-min scaling:

$$max(ei, F_j) = \frac{e_i - min(Fj)}{max(Fj) - min(Fj)} \tag{3}$$

For the minimization function, we have 2 methods: inverse and subtract.

Inverse method:

$$min(ei, F_j) = \frac{1}{max(e_i, Fj)} \tag{4}$$

subtract methods:

$$min(ei, F_j) = 1 - max(ei, F_j). \tag{5}$$

6- Combine the scores by sum or product the values of platforms, then make decision by choosing the platform that has the highest score.

Table 2. Comparative table of Moodle, Open edX, Sakai, Claroline, TalentLMS and Easy LMS and the Sum

Platforms	Moodle	Openedx	Sakai	Claroline	TalentLMS	Easy LMS	Sum/max
Ratings	3	2	1	1	2	2	11
Adaptive learning	2	3	2	1	2	2	12
Technology collaboration	2	2	3	2	2	2	13
Security	3	3	2	3	3	2	16
Mobile	2	2	1	3	3	2	13
Learning Analysis	3	2	2	1	1	2	11
Easy use	1	2	2	2	2	3	12
Price	0	0	0	0	429	250	429

5   Results

This table presents the comparative study of Moodle, OpenedX, Sakai, Claroline, TalentLMS and Easy LMS and the sum of all values of every feature except the price in order to use it in maximization function and in minimization function we use the min function (5) of price that use the max (Table 2).

We applied the Multi-Criteria Decision Making method on the previous data by using sum normalization for the maximization function and subtract for the minimization function. Because we can't divide by 0 in inverse function for price. And this table presents the results of the Multi-Criteria Decision Making method.

After applying "subtract methods" function on the price and "sum normalization" function on the features, adaptive learning, technology collaboration, Security, mobile, Learning Analysis and Easy use. The scores are calculated by the sum of the results of each platform (Table 3). From Table 3 the platform that has the highest score is Moodle 2.29064685 then Openedx, Sakai, Claroline, Easy LMS and at last TalentLMS.

Table 3. Comparative table of Moodle, Open edX, Sakai, Claroline, TalentLMS and Easy LMS using scores

Platforms	Moodle	Openedx	Sakai	Claroline	TalentLMS	Easy LMS
Ratings	0,27,272,727	0,18,181,818	0,09,090,909	0,09,090,909	0,18,181,818	0,18,181,818
Adaptive learning	0,16,666,667	0,25	0,16,666,667	0,08,333,333	0,16,666,667	0,16,666,667
Technology collaboration	0,15,384,615	0,15,384,615	0,23,076,923	0,15,384,615	0,15,384,615	0,15,384,615
Security	0,1875	0,1875	0,125	0,1875	0,1875	0,125
mobile	0,15,384,615	0,15,384,615	0,07,692,308	0,23,076,923	0,23,076,923	0,15,384,615
Learning Analysis	0,27,272,727	0,18,181,818	0,18,181,818	0,09,090,909	0,09,090,909	0,18,181,818
Easy use	0,08,333,333	0,16,666,667	0,16,666,667	0,16,666,667	0,16,666,667	0,25
Price	1	1	1	1	0,36,818,851	0,63,181,149
scores	2,29,064,685	2,27,549,534	2,03,875,291	2,00,393,357	1,5,463,645	1,84,480,683

The figure presents the scores of Moodle, Openedx, Sakai, Claroline, Easy LMS and TalentLMS.

From Fig. 1 appears that moodle and Openedx have similar results and the highest Scores above 2 points, but TalentLMS has the lowest Score less than 1.5 points.

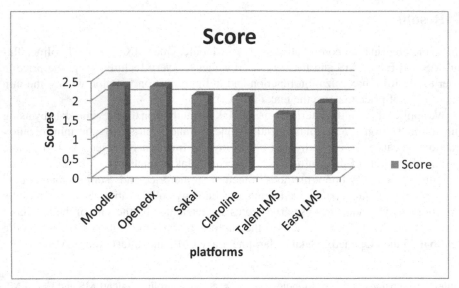

Fig. 1. Moodle, Openedx, Sakai, Claroline, Easy LMS and TalentLMS scores

6   Conclusion

In our study, and in order to help users to choose the LMS adequate to their purpose, we present a method to compare LMS platforms using 9 criteria Ratings, adaptive learning, technology collaboration, Security, price, mobile, Learning Analysis and Easy use. These criteria control the selection of the best learning management system according to its intended use. But the distribution of these features are not the same, so we use the Multi-Criteria Decision Making algorithm to solve this problem. To test this method, we compared six learning management systems: Moodle, Sakai, Claroline, TalentLMS, Easy LMS, and OpenedX, in order to study the performance of students. Where it became clear that Moodle is the best and the most suitable for studying the performance of students.

References

1. Alqudah, N.M., Jammal, H.M., Saleh, O., Khader, Y., Obeidat, N., Alqudah, J.: Perception and experience of academic Jordanian ophthalmologists with E-Learning for undergraduate course during the COVID-19 pandemic. Ann. Med. Surg. **59**, 44–47 (2020)
2. Aldiab, A., Chowdhury, H., Kootsookos, A., Alam, F., Allhibi, H.: Utilization of learning management systems (LMSs) in higher education system: a case review for Saudi Arabia. Energy Procedia **160**, 731–737 (2019)
3. Muhardi, M., Gunawan, S.I., Irawan, Y., Devis, Y.: Design of web based LMS (learning management system) in SMAN1 KAMPAR KIRI HILIR. J. Appl. Eng. Technol. Sci. **1**(2), 70–76 (2020)
4. Rabiman, R., Nurtanto, M., Kholifah, N.: Design and development e-learning system by learning management system (LMS) in vocational education. Int. J. Sci. Tech. Res. **9**(01) (2020)

5. Nichols, M.: A comparison of two online learning systems. J. Open Flexible Distance Learn. **20**(1), 19–32 (2016)
6. Alameen, A., Dhupia, B.: Implementing adaptive e-Learning conceptual model: a survey and comparison with open source LMS. iJET **14**(21) (2019)
7. Jeljeli, R., Alnaji, L., Khazam, K.: A comparison between moodle, Facebook, and paper-based assessment tools. iJET **13**(05) (2018)
8. Acosta, T., Sergio, L.-M.: Comparison from the levels of accessibility on LMS platforms that supports the online learning system. In: 8th International Conference on Education and New Learning Technologies (2016)
9. Dhika, H., Destiawati, F., Sonny, M., Jaya, M.: Comparison of learning management system Moodle, Edmodo and Jejak Bali. In: Advances in Social Science, Education and Humanities Research, vol. 422 (2020)
10. Anggrainingsih, R., Johannanda, B.O.P., Kuswara, A.P., Wahyuningsih, D., Rejekiningsih, T.: Comparison of maintainability and flexibility on open source LMS. In: 2016 International Seminar on Application for Technology of Information and Communication (2016)
11. Kasim, N.N.M., Khalid, F.: Choosing the right learning management system (LMS) for the higher education institution context: a systematic review. Int. J. Emerg. Technol. Learn. (iJET) (2016)
12. Öztürk, Y.E.: Evaluation of Moodle, Canvas, Blackboard, and Open EdX. Methods, and Programs in Tertiary Education, ICT-Based Assessment (2020)
13. Poulova, P., Klimova, B., Krizek, M.: Selected E-learning systems and their comparison. In: 2019 International Symposium on Educational Technology (ISET) (2019)
14. Saeed, F.A.: Comparing and evaluating open source E-learning platforms. Int. J. Soft Comput. Eng. (IJSCE) **3**(3) (2013)
15. Aikina, T.Y., Bolsunovskaya, L.M.: Moodle-based learning: motivating and demotivating factors. iJET **15**(2), 239–248 (2020)
16. Vaganova, O.I., Koldina, M.I., Lapshova, A.V., Khizhnyi, A.V.: Preparation of bachelors of professional training using MOODLE. In: International conference on Humans as an Object of Study by Modern Science: The Impact of Information on Modern Humans, pp. 406–411 (2017)
17. Liu, Z.-Y., Lomovtseva, N., Korobeynikova, E.: Online learning platforms: reconstructing modern higher education. iJET **15**(13), 4–21 (2020)
18. Vora, M., Barvaliya, H., Balar, P., Jagtap, N.: E-learning systems and MOOCs -a review. Int. J. Res. Appl. Sci. Eng. Technol. **8**(9) (2020)
19. Obeidallah, R., Shdaifat, A.: An evaluation and examination of quiz tool within open-source learning management systems. iJET **15**(10) (2020)
20. Wan, H., Yu, Q., Ding, J., Liu, K.: Students' behavior analysis under the Sakai LMS. In: 2017 IEEE 6th International Conference on Teaching, Assessment, and Learning for Engineering (TALE), Hong Kong, China, pp. 250–255 (2017)
21. Toma, I., Cotet, T.-M., Dascalu, M., Trausan-Matu, S.: ReadME–a system for automated writing evaluation and learning management. Int. J. User-Syst. Interact. **12**(2), 99–188 (2019)
22. Gaur, P.: Role of assessment in post COVID-19. Int. J. Multidisc. Educ. Res. **9**(3) (2020)
23. Blazheska-Tabakovska, N., Ristevski, B., Savoska, S., Bocevska, A.: Learning management systems as platforms for increasing the digital and health literacy. In: The 3rd International Conference on E-Education, E-Business and E-Technology (ICEBT) (2019)
24. Blagoev, I., Monov, V.: Criteria and methodology for the evaluation of e- Learning management systems based on the specific needs of the organization. Int. J. Educ. Inf. Technol. **12** (2018)
25. Yilmaz, S., Erol I.E.: Comparison of the online education platforms, and innovative solution proposals. In: Communication and Technology Congress – CTC 2019 (April 2019 – Turkey, İstanbul)

26. Karadimas, N.V.: Comparing learning management systems from popularity point of view. In: 5th International Conference on Mathematics and Computers in Sciences and Industry (MCSI) (2018)
27. Krouska, A., Troussas, C., Virvou, M.: Comparing LMS and CMS platforms supporting social e-Learning in higher education. In: 8th International Conference on Information, Intelligence, Systems & Applications (IISA) (2017)
28. Muhammad, M.N., Cavus, N.: Fuzzy DEMATEL method for identifying LMS evaluation criteria. In: 9th International Conference on Theory and Application of Soft Computing, Computing with Words and Perception, ICSCCW 2017, 22–23 August 2017, Budapest, Hungary (2017)
29. Nadkarni, R.R., Puthuvayi, B.: A comprehensive literature review of Multi-Criteria Decision Making methods in heritage buildings. J. Build. Eng. **32** (2020)

Finding Agreements: Study and Evaluation of Heuristic Approaches to Multilateral Negotiation

Mouna Azariz[1]([✉]) and Sihame Elhammani[2]

[1] Faculty of Sciences and Technology, University Sultan Moulay Slimane, Beni Mellal, Morocco
[2] Faculty of Sciences, University Mohammed V Agdal, Rabat, Morocco

Abstract. Negotiation is a process by which participants get an agreement that satisfies them all. Multilateral negotiation involves more than two agents. The complexity of negotiation intensifies when involving multiple agents with conflicting interests that need to reach a joint agreement. To model negotiation, we have three axes: object of negotiation, negotiation protocol, and negotiation model. There are Several negotiating approaches are available to study negotiation model. The heuristic is the most commonly used in multilateral negotiations, because it realistically simulates negotiating problems. Negotiation is made more difficult when there are multiple agents who need to reach an agreement on differing interests. An important facet of any successful negotiation is to lay out a clear and specific protocol guiding the interactions between the agents. In our research study, the negotiation protocol is based on the divide and conquer concept. In other words, agents are divided into several groups where they debate without a mediator. This article aims to evaluate and study two heuristic approaches. In the first approach, the agents negotiate on a single attribute; we will compare this approach to the case where the agents negotiate in a single group according to the results obtained. In the second approach, the agents negotiate on several attributes. This approach is compared with the first approach to determinate in which case of the second approach it is better to use the first approach to facilitate the search for agreement.

Keywords: Agreements · Negotiation · Multilateral approach · Heuristic approach

1 Introduction

Negotiation is a process by which participants get an agreement that satisfies them all. Negotiation is fundamental in multi-agent systems [1]. It is applied in various fields such as electronic commerce, companies. In the real world, agents can have divergent interests. Therefore, everyone can try to steer the negotiation towards their preferred solutions. The difficulty of finding agreement in negotiation depends on the complexity of the negotiating agent's organizations, the number of agents, the number of issues negotiating, and the nature of the information shared between these agents is what information is

© Springer Nature Switzerland AG 2021
M. Fakir et al. (Eds.): CBI 2021, LNBIP 416, pp. 249–266, 2021.
https://doi.org/10.1007/978-3-030-76508-8_18

complete, i.e., their utilities are shared between them, or it is incomplete information. Multiple bilateral models have received substantial attention in the literature, whereas little has been written about multilateral models because of their difficulties.

Most multilateral negotiation protocols are commonly based on bilateral negotiation protocols. For example, [2] proposes a generalization of the monotonic concession protocol in the case where there are more than two agents must reach an agreement (in the multilateral case) [3]. Another protocol for multilateral negotiation with mediation. In [4] the authors discuss multi-attribute multilateral negotiation involving a large number of agents and interdependent issues. This negotiation process uses an agenda management mechanism using a hierarchical model. The problem is seen as a system (S) or a process. The system is decomposed into subsystems $(S_1, S_2, S_3, \ldots)$, and if can be each subsystem is decomposed like this $(S_{i_1}, S_{i_2}, S_{i_3}, \ldots)$, so that each set of attributes is associated with a subsystem. There is another protocol. It is the alternating offers protocol, is a protocol in which the agents interact in turn. Each turn, an agent makes a proposal to the other agents. An agent begins negotiation with an offer and the other agents evaluate the given offer: agents can accept the given offer, then the negotiation ends with an agreement; agents can reject the given offer, in which case, agents can either end the negotiation or make a counter-offer. This process continues in turn while respecting the speaking position which is predefined, [5] presents the two variants of this protocol. Research in [6] and [7] proposes decentralized negotiation protocols, that negotiation is carried out in a decentralized manner. The authors propose a protocol inspired by the divide and conquer principle. The agents are divided into several groups. Both groups discuss the same issues.

In this article, we will study heuristic approaches for receiving in a real context. Thus, we focus on negotiating agents' organizational aspect to reduce the complexity of their interactions during the search for agreement. The agents' organization in the approaches studied is based on the divide-and-conquer principle [8]; agents are divided into several groups to negotiate without unity of control.

The contributions to this paper include the following: First, we will present the necessary ingredients within a negotiation process and define the different approaches to multilateral negotiation, and then we will present the heuristic approaches studied in detail. To study the performance of these two approaches studied, we compare the first approach to a multilateral negotiation model in which agents are gathered into a single ring, then we compare the first approach to the second.

2   Multilateral Negotiation

Multilateral negotiation involves more than two dimensions, i-e more than two participants. Life is a series of negotiations, and we negotiate either consciously or unconsciously with our parents, our children, within the creation of a project. Negotiation is like our shadow wherever we are, and it is used in economics, politics and other fields. Today, computers, phones, tablets, robots, and many machines are everywhere. We can say that the number of machines in the world is equal to or greater than the number of human beings. This is due to man's needs. The development of computing is, in particular, artificial intelligence, offers skill and a capacity to machines to meet the needs of man, i.e., the machines replace the human being with doing their work when they are busy, for example, negotiating with buyers. As a result of this development negotiation has become automatic. Automatic negotiation is studied in multi-agent systems to model machines as autonomous agents and resolve conflicts that occur during interactions between these agents. In most cases, these participants behave selfishly, and each defends their interests. To model negotiation in multi-agent systems, several researchers have put the necessary ingredients into the negotiation process.

2.1 Negotiation Ingredients

In [9], the authors identify three main axes to model the negotiation, which are:

- **Negotiation protocol**: is a set of rules shared by agents to manage interactions. For negotiation, its rules are called confrontation rules. It describes the steps in negotiating which messages can be sent and what actions agents can perform and when these actions should be performed.
- **Object of negotiation**: The object of negotiation answers the following question: "What are agents negotiating on?". It represents the subject that promotes interactions between agents during the negotiation, and this subject can contain a single attribute, for example, the agents negotiated on the price of a product, or several attributes, for example, the agents negotiated on the price, the quality and method of payment for a product.
- **Negotiation model**: It represents the agent reasoning model that provides the agent decision-making mechanism. We distinguish three approaches to studying negotiation models: game theory approach, heuristic approach, and argumentative approach.

2.2 Multilateral Negotiation Approaches

In the multi-agent literature, various models of automated negotiation have been proposed and studied. These include three approaches.

- **Game theory**: is a branch of mathematics introduced in 1994 by Oskar and John von Neumann [10]. It is used in economics and also applies to negotiation processes involving selfish agents. It allows the rational actions of negotiating agents to be studied in order to guarantee optimal solutions. In-game theory, agents are supposed to know the set of possible solutions, and their decisions depend only on the belief that they are actions that others will take. In a real context, all the solutions are not known by the agents. Each agent only knows his own space for agreements. Because of this, game theory can not be used in different areas of negotiation.
- **Heuristic:** The heuristic approach provides a more suitable basis for automated negotiation, as these models are based on realistic assumptions. Therefore, they can be used in various negotiation contexts. Models based on the heuristic approach do not provide optimal solutions but provide a sufficiently good solution. They aim at designing mechanisms that make agents freer in their interaction [1, 11]. This approach is a solution to the problems that pose game theory, and each agent knows only his propositions, which he exchanges with others during these interactions and makes their decision during the negotiation according to their reasoning and strategy.

- **Argumentation**: In game theory and heuristics, negotiations between agents are carried out based on an exchange of proposals. A proposal is a combination of values assigned to the attributes being negotiated. These two approaches do not allow officers to express the reason for their decision. The advantage of argumentation protocols is that they use persuasion techniques to lead the negotiation towards a result that best meets the agent's objectives and to facilitate the search for agreements.

3 Negotiation Principles

In our study, agents negotiate to reach an agreement i-e a solution that satisfies all these agents' aspirations. The principle of negotiation of the two approaches that we will evaluate and compare is the same. The mechanism is inspired by the divide and conquer principle. In this mechanism, the divide and conquer principle does not allow the problem to be divided into subproblems but divides the negotiating agents into several groups. In a group, each agent makes it possible to propose proposals and receive proposals from other group agents and evaluate them either by accepting them, by refusing them, by reinforcing them, or by attacking them. The agents allow listening to the exchanges made in other groups. When an agent acquires information from other groups, he can decide to migrate to another group in a strategic way to propose his proposals in order to increase the chance of acceptance of his proposals or to evaluate the proposals of other agents who are in this group. The negotiation protocol specifies the rules for agent transition between groups.

Negotiation takes place within a time limit d_l.

The two approaches studied have the same negotiation mechanism, which includes three phases (see Fig. 1).

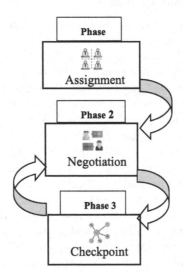

Fig. 1. The phases of the negotiation mechanism.

- **Assignment phase**: the agents are divided into several groups. The way of distributing agents in the two approaches is different. This phase will be explained in detail for each approach in the following sections.
- **Negotiation:** In this phase, the agents exchange their proposals according to the protocol rules, detailed in the following sections, and receive them and make these decisions. They migrate to other groups to defend their interests. The two approaches differ in part of this negotiation phase.
- **Checkpoint:** Checking the negotiation progress i-e is a proposal with an acceptance rate above a majority threshold. This proposal is said to be an acceptable solution. If an acceptable solution is found, the negotiation ends; if not, and the negotiation time has not expired, another negotiation phase begins.

4   Framework of Negotiation

Let $A = \{a_1, \ldots\ldots\ldots, a_n\}$ be the set of negotiating agents and $G = \{g_1, \ldots\ldots, g_m\}$ the set of groups in which the agents will negotiate. Each agent a_i is characterized by a tuple $(\omega_{a_i}, u_{a_i}, U_{min_{a_i}}, U_{max_{a_i}}, P_{a_i}, \chi_i)$ such that $\omega_{a_i} \in [0,1]$ representing its importance in decision-making P_{a_i} is the set of proposals of agent a_i, $u_{a_i}: P \rightarrow [0, 1]$ is the utility function of an a_i, Where P is the set of propositions submitted, this is used to establish an order of preference on the set of alternatives. i-e each agent gives a numerical value between 0 and 1 to each submitted proposal, which represents the happiness or satisfaction of the agent for that submitted proposal. The question that comes to our mind is how an agent will value a submitted proposition or how an agent will evaluate a proposition according to his aspirations?

Let $\chi = \{x_{(1)} \ldots x_{(r)}\}$ be a set of criteria. Each agent has a subset of the criteria on which an agent is based to give its utility function. The utility functions of agents are private.

The other elements of the tuple $U_{min_{a_i}}$ and $U_{max_{a_i}}$ represent the minimum utility on which the agent can accept a proposition, in other words, its reserve utility below which no proposition is acceptable, and its desired utility value, above which any proposal is acceptable. Before the negotiation, each agent defines his aspirations, which are represented by the interval $[U_{min_{a_i}}, U_{max_{a_i}}]$.

The set of speech acts that agents use to interact with each other are five acts, that is, $O = \{Propose, Accept, Refuse, Reinforce, Attack\}$. The set of these acts are formalized as follows:

- **Propose:** In each group, an agent submits a proposition with a set of positive arguments to any agent in his group.
- **Accept:** In each group, an agent accepts the proposal submitted by another agent in his group and sends a message to any agent.
- **Refuse:** In each group, an agent rejects the proposal submitted by another agent in his group and sends a message to any agent in his group.
- **Reinforce:** In each group, an agent reinforces the proposal submitted by another agent in his group and sends a message to any agent in his group with a set of positive arguments.

- **Attack:** In each group, an agent attacks the proposal submitted by another agent in their group and sends a message to any agent in their group with a set of negative arguments.

A tuple characterizes each proposition submitted by an agent $(T_{p_\alpha}^{ac}, T_{p_\alpha}^{re}, T_{p_\alpha}^{rf}, T_{p_\alpha}^{at}, v_{p_\alpha}, \omega_s(p_\alpha), \varepsilon_{p_\alpha})$ we will now explain each element of the tuple.

- $T_{p_\alpha}^{ac}$: the acceptance rate of the proposal p_α submitted by an agent, formally:

$$T_{p_\alpha}^{ac} = \frac{\left|a_{p_\alpha}^{ac}\right|}{n} \tag{1}$$

With $a_{p_\alpha}^{ac}$ the number of agents who accepted the proposal p_α and n the number of agents.

- $T_{p_\alpha}^{rf}$: the reinforcement rate of the proposal p_α submitted by an agent, formally:

$$T_{p_\alpha}^{rf} = \frac{\left|a_{p_\alpha}^{rf}\right|}{n} \tag{2}$$

With $a_{p_\alpha}^{re}$ the number of agents who reinforced the proposal p_α and n the number of agents.

- $T_{p_\alpha}^{re}$: the rate of refusal of the proposal p_α submitted by an agent, formally:

$$T_{p_\alpha}^{re} = \frac{\left|a_{p_\alpha}^{re}\right|}{n} \tag{3}$$

With $a_{p_\alpha}^{re}$ the number of agents who refused the proposal p_α.

- $T_{p_\alpha}^{at}$: the attack rate of the proposal p_α submitted by an agent, formally:

$$T_{p_\alpha}^{at} = \frac{\left|a_{p_\alpha}^{at}\right|}{n} \tag{4}$$

With $a_{p_\alpha}^{at}$ the number of agents who attacked the proposal p_α.
v_{p_α}: the support value of the proposition with:

$$v_{p_\alpha} = \frac{T_{p_\alpha}^{ac} + T_{p_\alpha}^{re}}{T_{p_\alpha}^{rf} + T_{p_\alpha}^{at} + 1} \tag{5}$$

- $\omega_s(p_\alpha)$: The social satisfaction value of the proposition p_α.

$$\omega_s(p_\alpha) = \frac{\sum_{i=0}^{q} \sigma_{s_i}}{n - n_{ap_\alpha}(s_q)} \tag{6}$$

- ε_{p_α}: A period of time between the dates of obtaining a majority between two proposals.

We will use these measurements in the checkpoint phase to verify the existence of a solution.

How does an agent make his decisions?

The decisions of the agents are made according to their utility u_{a_i} from the current proposals. There are three cases:

- If the utility u_{a_i} of the proposition is in the range $[U_{min_{a_i}}, U_{max_{a_i}}]$, the agent accepts this proposition.
- If the utility u_{a_i} of the proposition is strictly less than $U_{min_{a_i}}$, the agent attacks and rejects this proposition.
- If the utility u_{a_i} of the proposition is strictly greater than $U_{max_{a_i}}$, the agent reinforces this proposition.

For each case, we respectively associate the three levels of satisfaction: $\sigma_0, \sigma_1, \sigma_2$.

If two agents give the same utility value to a proposition, it does not imply that they have the same degree of satisfaction because its utility intervals are different. We will provide an example to understand.

Example 1: We have agent one and agent two, which have the utility intervals $[0.5, 0.8]$ and $[0.3, 0.5]$. These two agents give the same utility value to a proposition. This value is 0.4. For agent one, 0.4 is between 0.3 and 0.5, so it will accept this proposition. On the other hand, agent two will refuse it because its utility of 0.4 is strictly lower than its minimum utility (0.5). We conclude that the two agents do not have the same levels of satisfaction.

5 First Approach

5.1 Assignment Phase

In this section, we will explain how agents are partitioned into groups. This phase includes two steps: The first is the choice of the number of agents, and the second is the assignment of agents into groups.

- **The choice of the number of groups:** There are two ways to choose the number of groups for this step: either we have more than n/2 groups such that n is the number of agents in the negotiation. The second way, the number of groups of agents is calculated based on the disparity D_p of the weights of the agents as the difference between the largest and the smallest weight.

Example 2: Let a_1, a_2, a_3, a_4 four agents. Their weights are respectively: 0.708, 0.631, 0.162 et 0. 876. The largest weight is 0.876 and the smallest weight is 0.162, so it is calculated as follows: $D_p = 0.876–0.162$, so $D_p = 0.714$.

After calculating the disparity in weights, how will we calculate the number of groups from this measurement?

To calculate the number of groups, we will check the value of the weight disparity. If this value is between 0 and 1, then: m $= (D_p * \frac{n}{2})$ such that m is the number of groups, if not i-e $D_p = 0$ which implies that all the agents have the same weight and the number of groups is chosen randomly.

Example 3: From Example 2, we have .714, and we have $0 < 0.714 < 1$, where n $=$ 4. m $= (0.714 * 4/2)$, which implies m $= 1.42$, so we will take the integer part of this value, we conclude that the number of groups is a single group.

- **Assigning agents to groups**: In the two ways of calculating the number of groups, the assignment of agents to groups is carried out by respecting the following rule:

$$\bigcap_{k \geq 1}^{m} A_{gk} = \emptyset, \quad \bigcup_{k \geq 1}^{m} A_{gk} = A, \ |A_{gk}| \geq 2 \tag{7}$$

We will now explain each part of this rule: The first part implies that the intersection of the agents of all the negotiation groups is equal to the empty set, i.e., an agent can only belong to one group. The second part implies that the union of agents of all negotiation groups is equal to the numbers of negotiation agents, the third part implies that each group contains two or more agents.

5.2 Negotiation

Negotiation takes place in each group and outside each group according to inter-group interaction policies.

- **Negotiation inside a group**: Each agent submits these proposals incrementally according to his order of preference to the agents in his group, such that the utilities of these proposals are greater than his minimum utility $U_{min_{a_i}}$. Thus, an agent receives the proposals of the other agents of his group and evaluates them. He can accept, refuse, reinforce and attack these proposals. Agents negotiate according to the following protocol rules:

Rule 1: During the negotiation, each agent has the right to make a limited number of proposals δ_p and a limited number of refusals δ_r.
Rule 2: At any point in the negotiation, each agent can only submit their proposals to agents in their group.
Rule 3: At any time during the negotiation, an agent can accept a proposal from another agent.
Rule 4: An agent can not accept, refuse, reinforce or attack the same proposition several times.

These protocol rules allow agents to better coordinate their interactions and limit the message and processing rate.

- **Inter-group interaction policies groups**: An inter-group interaction policy must determine when agents can listen to exchanges made outside their group and thus when agents can express their opinions on the proposals they have listened to. Agents can express their views on other groups' proposals through inter-group communication. The inter-group interaction policy used is called (FIC) Free Inter-rings Communication. This policy allows agents to listen and respond to negotiations outside their group.

In the rules of negotiation, the protocol is more precisely in rule 2, agents of one group are not allowed to submit a proposal to agents of another group. Agents can decide to migrate to other groups at any time during the negotiation. According to the protocol, once an agent migrates to a new group, it immediately obtains all interaction rights within that group. Agent migrations are carried out according to two mobility rules, which are:

R_{m1}: an agent has the right to make a limited number of visits δ_v to the same group.
R_{m2}: an agent has the right to make a limited number of migrations δ_d throughout the negotiation.

- **Migration strategies:** An agent can listen to several groups simultaneously but can only join one group at a time. When an agent decides to migrate to another group, they need the policy to choose the best group to join. This choice is made according to three criteria: the number of agents in the group; the weight of the agents; The gap between the acceptance rates of proposals submitted in the ring and those of the agent who implements the strategy.

In the event that the agents listened to more than two groups, the choice of the best group is made as follows:
Let $G^O_{a_i} \subset G$ all the rings listened to by the agent $a_i \in g_k$. Each group $g_y \in G^O_{a_i}$ is represented by a utility vector $u(g_y) = u_1(g_y), u_2(g_y), u_3(g_y)$. The vector elements are calculated as follows:

- $u_1(g_y) = \dfrac{|A_{g_y}|}{|A_{g_k}|}$: the ratio between the number of agents belonging to the listened to group and the number of agents belonging to the agent's group.

- $u_2(g_y) = \dfrac{|\omega_{g_y}|}{|\omega_{g_k}|}$: Is the ratio between the average weight of agents belonging to the group listened to and others.

- $u_3(g_y) = \dfrac{|T^{ac}_{p\alpha}|}{|T^{ac}_{p\alpha}(i)|}$: is the ratio between the acceptance rate of the proposal with the highest acceptance rate in the listened group and the agent's current proposal acceptance rate.

These vectors are used in the Lorenz dominance [12] for the choice of the best group.

Definition 1: For each $x \in \mathbb{R}_+^n$, the generalized Lorenz vector associated with x is the vector:

$$L(x) = (x_{(1)}, x_{(1)} + x_{(2)}, \ldots, x_{(1)} + x_{(2)} + \ldots + x_{(n)})$$

Or $x_{(1)} \geq x_{(2)} \geq \ldots \geq x_{(n)}$ represent the sorted components in descending order. The k^{me} component of $L(x)$ is:

$$L_k(x) = \sum_{i=1}^{k} x_{(i)} \tag{8}$$

Definition 2: The relation of generalized Lorenz dominance over $\mathbb{R}_+^n$ is defined by:

$$\forall\, x, y \in \mathbb{R}^z, x \succsim_L y \;\Leftrightarrow\; L(x) \succsim_p L(y) \tag{9}$$

Example 4: Suppose we have 16 agents divided into 4 groups g_1, g_2, g_3, g_4. The agent a_1 belongs to the group g_1 and submits his proposal to the agents of the same group. At a certain point in the negotiation, a_1 decides to defend his proposal in the other groups to obtain a majority of acceptances. He takes note of the exchanges that took place in g_2 et g_3. The utility vectors of these are: u $(g_2) = (2, 0.3, 1.5)$ and u $(g_3) = (0.8, 0.5, 1.5)$. we have: u $(g_2) = (2, 0.3, 1.5)$ and u $(g_3) = (0.8, 0.5, 1.5)$ thus in the first definition, we have $x_{(1)} \geq x_{(2)} \geq \ldots \geq x_{(n)}$ So:

For u (g_2): $x_{(1)} = 2$ and $x_{(2)} = 1.5$ et $x_{(3)} = 0.3$
so: $L(u(g_2)) = (x_{(1)}, x_{(1)} + x_{(2)}, x_{(1)} + x_{(2)} + x_{(3)})$

$$L(u(g_2)) = (2, 3.5, 3.3)$$

In the same way we find $L(u(g_3)) = (1.5, 2.3, 2.8)$. We thus obtain the following dominance relation:

$$L\big(u(g_3)\big) \succsim_p L\big(u(g_2)\big) \;\Leftrightarrow\; u(g_3) \succsim_L u(g_2).$$

Therefore, the agent a_1 will join g_3.

5.3 Checkpoint

Negotiation takes place in several rounds separated by predefined checkpoints. Either the majority threshold $\emptyset_{maj}$, we say that a proposal is acceptable if this proposal's acceptance rate is higher than the majority threshold. Before the negotiation, we set the number of checkpoints; checkpoints.

Let $C_{pt} = \{C_{pt_1}, \ldots, C_{pt_q}\}$ the set of planned control points. Each checkpoint C_{pt_r} is represented by a tuple of elements (t_r, S_r), whose elements are respectively, the date, the set of acceptable proposals.

- If $|S_r| = \emptyset$ then there is no solution found. Negotiation ends with failure.
- If $|S_r| = 1$ Then there is only one acceptable proposal, and this proposal will be the solution of the negotiation. Negotiation ends successfully.

- If $|S_r| > 1$ Then there are several acceptable propositions. If we have two propositions with the same acceptance rate, we calculate for each the support value, and we choose the proposition with the highest support value as the solution of the negotiation. If these two propositions have the same support value, we calculate for each the social satisfaction value, and we take as the solution of the negotiation the proposition with the greatest social satisfaction value if these two propositions have the same satisfaction value. Social, these will be compared according to their stamps. There is always a lapse of time between two proposals that separates their dates of obtaining a majority.

6 Second Approach

In this approach, agents do not share their utility function. They are selfish, and each seeks to defend their interests. They negotiate on several attributes $E = (e_1, \ldots, e_k)$ with k the number of attributes and each agent is only interested in the part of the negotiated attributes, so these propositions only contain the values of the attributes that interest them (partial propositions). A proposal p_α is a tuple $(v(e_i), \ldots, v(e_k))$ whose elements are the values of the attributes. The agents form alliances to defend their proposals jointly. Agent attributes are public, but their values are private.

Now we will detail the phases of the negotiation mechanism for this approach. The phase of the checkpoint is the same as in the first approach.

6.1 Assignment Phase

The partitioning of agents is done in two steps: Each agent a_i classifies agents in descending order according to the number of attributes they share with them, and the formation of groups according to the order of classification of agents.

The formation of groups is done in turns. Each agent asks those who share a maximum number of attributes to form a group. The other agent accepts the request if and only if he shares a maximum number of attributes with him. If an agent accepts a request, he does not have the right to accept another request, but he can continue and ask others to join his group. This is repeated until all the agents join groups.

Let $\gamma_i = |E_{a_i}|$ the number of agent attributes a_i. We designate by $C_{O_{a_i}} = O_{\gamma_i}, O_{\gamma_{i-1}}, O_{\gamma_{i-2}}, \ldots \ldots, O_0$ the classification of a_i representing respectively all the agents with whom it shares γ_i attributes, the set of agents with whom he shares $\gamma_i - 1$ attributes, the set of agents with whom he shares $\gamma_i - 2$ attributes etc.

Example 5: Let $A = \{a_1, a_2, a_3, a_4\}$ four agents which aim to negotiate on the five attributes $(e_1, e_2, e_3, e_4, e_5)$ such that each agent is interested in a subset of attributes as follows: $E_{a_1} = (e_1, e_2, e_3)$, $E_{a_2} = (e_2, e_3)$, $E_{a_3} = (e_1, e_5)$, $E_{a_4} = (e_1, e_4, e_5)$. The agent a_2 have $\gamma_1 = 3$ attributes. Its classification is as follows: $C_{O_{a_i}} = \{O_3 = \emptyset, O_2 = \{a_2, a_4\}, O_1 = \{a_3, a_5\}, O_0 = \emptyset\}$ and the agent a_2 a $\gamma_2 = 2$ attributes. Its classification is as follows: $C_{O_{a_i}} = \{O_2 = \{a_1\}, O_1 = \{a_4\}, O_0 = \{a_3\}\}$. So, the agent a_1 send a request to the agent a_2 to make a group because he shares a maximum number of attributes. The agent accepts his request because he also shares with him a maximum number of attributes. This procedure is repeated until all agents joining groups.

6.2 Negotiation

The phase of negotiation is the same as the first approach. The difference is that in the protocol rules, two rules are added, which are:

Rule 5: Each agent a_i can only listen to groups with which share attributes.
Rule 6: Having read the exchanges carried out (proposals) in another group, each agent a_i has the right to either accept or refuse or attack these proposals.

The second difference is the formation of an alliance during the negotiation. A tuple $(A, U_{min_{a_i}}, U_{max_{a_i}}, p_\alpha)$ represents each alliance c_x. An alliance is formed if at least two agents accept the same proposal. The agents define a common decision-making strategy i-e an alliance utility interval $[U_{min_c}, U_{max_c}[$, this is calculated according to the utilities expected from agents such as $U_{min_c} = \{U_{min_{a_i}}\}$ et $U_{max_c} = \{U_{max_{a_i}}\}$. Agents make these decisions based on the alliance utility interval. During the negotiation, each alliance aims to make with other alliance groups to join forces to have a maximum number of agents with a proposal that addresses all the attributes, either by merging their proposals or by negotiating on conflicting attributes to refine a joint proposition. The structure of alliance groups is shown in Fig. 2 below:

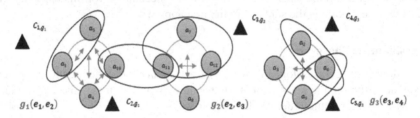

Fig. 2. Alliance groups.

Alliance group formation is decentralized i-e, each alliance chooses the alliances with which it wishes to associate. Alliances that are formed in the same group can not combine.

How does an alliance choose which alliance a group will make with it?

Each alliance chooses the most beneficial alliances based on two criteria: the number of agents in the other alliance and the number of attributes it shares with that alliance.

Each alliance agrees to associate with alliances, which maximizes its utility. The expected utility of an alliance c_x if it forms a group with another alliance c_y which maximizes its utility is formal as follows:

$$u(c_x, c_y) = f_{gain}(c_x, c_y) - f_{cost}(c_x, c_y) \qquad (10)$$

Such as:

$$f_{gain}(c_x, c_y) = \frac{|E_{c_x} \cup E_{c_y}|}{k} \times \frac{|A_{c_x} \cup A_{c_y}|}{n} \qquad (11)$$

$$f_{cost}(c_x, c_y) = \frac{|E_{c_x} \cap E_{c_y}|}{|E_{c_x} \cup E_{c_y}|} \times \frac{|A_{c_x} \cup A_{c_y}| - |A_{c_x} \cap A_{c_y}|}{|A_{c_x} \cup A_{c_y}|} \tag{12}$$

n, k represent respectively the number of agents in the system and the number of attributes. (c_x, c_y) are two different alliances, but the cost and the gain can be generalized to more than two alliances.

When alliances form groups, it remains to combine the proposals for alliances in each group. There are two processes for this operation: a simple alliance merger process, and the proposal refinement process.

- **Merging process:** This process is performed when we have two propositions that do not share any attribute between them. For example, if we have two alliances that are going to support respectively, the proposals $p_1 = (v(e_1), v(e_2))$ et $p_2 = (v(e_3), v(e_4))$, where $v(e_i)$ is the value e_i of being accepted by an alliance. The resulting proposition of the merger is as follows.: $p_1 = f_m(p_1, p_2) = (v(e_1), v(e_2), v(e_3), v(e_4))$ with f_m the merge function.
- **Refinement process:** This process is carried out when alliances have to negotiate attributes in common whose values are different. Suppose that two alliances were respectively supporting the propositions. $p_1 = (v(e_1), v(e_3), v(e_4))$ et $p_2 = (v(e_2), v(e_3), v(e_5))$ form a group to renegotiate their proposals. A new proposal submitted in group is:

$p_\alpha = (v(e_1), v(e_2), set(v(e_3)), v(e_4), v(e_5))$, with $set(v(e_3))$, a new value of the attribute e_3.

7 Experimental Results

7.1 First Approach

The first approach will be evaluated according to three performance criteria, which are:

- The rate of negotiation convergence.
- The average negotiation time.
- The average quality of the solution, i.e., the agents' satisfaction compared to the final solution found: The quality of each agent's solution is the ratio between the utility of the solution and its maximum aspiration. The average quality Q_s^m is the ratio of the sum of individual qualities to the total number of agents. Formally:

$$Q_s^m = \frac{1}{n} \times \sum \frac{u_{a_i}}{U_{max}} \tag{13}$$

To illustrate the first negotiation mechanism, we consider a negotiation scenario based on a project to set up a new tram line. This project involves several agents. Their purpose is to determine the new tram line route, that is to say, the coordinates of its stops. These participants may have divergent interests.

In the first approach, agents negotiate in several groups. We will compare the performance measures with the case in which the agents negotiate in a single group. i-e we have the same negotiation phases of the first approach, except we will create a single group, and migration does not exist in this case.

We varied the number of n agents from 4 to 32 and the number of groups $m \in$ {2, 4, 8, 12, 16}. For n agents, we created most [n/2] groups. We performed 20 tests for each case. We added 20 tests in the case of a large number of agents to have the impact of the number of agents on the negotiation of a single group, and in several groups, we chose n = 100 and m of 1 to 2. For each negotiation test, each agent randomly selects a subset of criteria and 30 random proposals. We have created two files called Proposals and Criteria; the first contains 20 proposals. From these 20 propositions, 30 propositions are randomly generated for each. Each proposition is generated as follows: "stop_1, street_num_1, street_name_1, postal_code_1", such that each address of a stop i is made up of three attributes: street_num_i, street_name_i, postal_code_i representing respectively the street number, the name of the street and zip code. The criteria file includes criterion 1, criterion 2, and criterion 3.

Table 1 shows the convergence rate results for each case.

Table 1. The convergence rates.

	n = 4	n = 8	n = 16	n = 24	n = 32	n = 100
m = 1	1	1	11	1	1	0.6
m = 2	0.05	0.7	0.8	1	0.95	0.9
m = 4		0	0.65	0.75	0.95	--
m = 8			0	0.05	0.35	--
m = 12				0	0	--
m = 16					0	--

Table 1 shows that the negotiation convergence rate obtained when the number of agents between 4 and 32 forms a single group equal to 1 is higher than those obtained when the agents are divided into several groups, but for 100 agents, the rate of convergence when they form a single group is low than when they are divided into several groups. This result shows that the negotiation mechanism studied facilitates the search for agreements when the number of agents is large.

We also observe that the formation of [n/2] leads to a very low rate of convergence. For example, in the case of n = 4 and m = 2 the rate of convergence is 0.05, or zero for example in the cases: n = 8 and m = 4, n = 16 and m = 8. However, the rates are not low when the number of rings is smaller than [n/2].

We will now know the average negotiation time. In the negotiation parameters, we have a negotiation time limit of 2 min, so the number of checkpoints is 4. We will know for each case at which checkpoint we find a solution.

Table 2 shows the average negotiation time for each case to reach an agreement.

Table 2. Average negotiation time.

Number of agents	One group	Two groups	Four groups	Eight groups
4	30 s	30 s	-	-
8	30 s	30 s	-	-
16	30 s	30 s	30 s	30 s
24	30 s	30 s	30 s	30 s
32	30 s	30 s	30 s	30 s
100	60 s	30 s	30 s	30 s

Table 2 shows that the negotiation time to find an agreement when the number of agents is between 4 and 32 negotiate in a single group or are divided into several groups converges faster in these cases, but when the number of agents becomes important (100 agents in this case) negotiate in a single group does not converge faster (the solution is found in the second checkpoint).

This result shows that the studied mechanism limits the negotiation time when the number of agents is large.

To evaluate the performance criterion and the quality of the negotiation solution, the number of groups to be formed is calculated according to the agents' weights.

Figure 3 shows the quality of the negotiation solution.

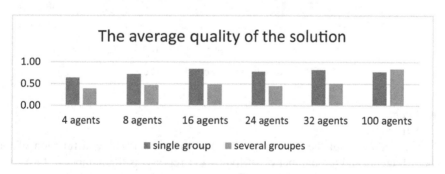

Fig. 3. The average quality of solution.

Figure 3 shows the comparison between the studied distributed negotiation model and the centralized negotiation model (when a single group is formed) according to the solution's average quality.

When agents between 4 and 32 negotiate in a single group, the negotiation solution's quality is better compared to when they are distributed in several groups. However, from 100 agents, the negotiation solution's quality is better when the agents are divided into several groups and weak when a single group is formed. This result shows that the studied model gives a better negotiation solution when the number of agents is large.

The experimental results show that the first approach facilitates agreements among the agents and reduces negotiation runtime compared to the single ring model.

7.2 Second Approach

In this approach, we first evaluated the number of turns agents take to form the initial groups. Next, we compared the negotiation time as the number of attributes increases for the same number of agents.

For each test, each agent randomly chooses a subset of attributes from a predefined set of attributes. We created these attributes in a text file. Each agent seeks to form a group with those with whom he shares the maximum number of attributes.

Now we will see the impact of the number of attributes k and the number of agents n on the formation of groups and the negotiation time to find a solution, i.e., at what point of control the solution is found. For this, we have varied the number of agents from 4 to 32 and the number of attributes from 2, 5, and 20.

Figure 4 shows the result of the second approach.

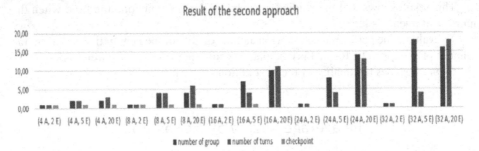

Fig. 4. Result of the second approach.

A: Agents.
E: Attributes.

Figure 4 shows that the number m of initial groups formed as a function of the number of agents and the number of attributes k. r represents the number of turns that the agents made to form the m groups. The results obtained show that the number of rounds of negotiation r depends on the number of attributes to be negotiated and the number of agents. The number of groups formed has an impact on the negotiation time and the solution's result. In the case where the number of groups is close to the number of agents, a negotiated solution can not be found, for example in the following case: 10 groups and 11 and 16 agents. In this case, the negotiation becomes complex because each agent has the right to migrate only two times to another group, so in this case, we find groups with two agents, are undesirable because they negatively affect performance. For example, the cost of migrating between the rings and the negotiation time is expected to increase; blockages related to migration will be more frequent, so the choice of a random attribute impacts the outcome of the negotiation.

In this approach, we find three situations:

- A situation where each agent is interested in all the attributes. In this case, according to dividing the agents into groups, a single negotiating group will be formed, and the agents will exchange their proposals. The first negotiation approach studied can be used because each proposition of an agent is complete; that is to say, it concerns all the attributes and can become the negotiation solution if the majority accepts it.
- In a situation where each agent does not share any attributes with the others, the final negotiation solution is a combination of the values that each agent gives to the attributes that interest him. In this case, there is no negotiation.
- In a situation where not, all agents are interested in all attributes, their attribute subsets may have common attributes, and different negotiation groups will be formed.

From the results obtained from the two approaches, we conclude that the first approach studied facilitates the search for agreements when there is a large number of agents, on the other hand, in the case where the agents negotiate in a single group. For the second approach studied, we concluded that the number of agents and attributes have an impact on the formation of the initial groups and that the random choice of attributes has an impact on the result of the negotiation. It is better to choose the attribute according to the weight of the agents to solve this problem, so when we have the situation where each agent is interested in all the attributes, it is better to use the first approach to facilitate the search for agreements.

8 Conclusion

This article focuses on studying two approaches to multilateral negotiations to facilitate the search for agreements. These approaches are heuristics. The goal of this choice is to take into account the different aspects inherent in a real negotiation. In these two approaches, the solution space is not known by all agents, unlike work based on game theory. The negotiation approaches studied concern selfish agents who do not share any information concerning their utility functions. Negotiating in such a context requires the design of sophisticated protocols to facilitate the search for agreements. The protocols used in both mechanisms are based on the "divide and conquer" principle. The agents are divided into several groups in which they negotiate to find partial solutions. In our experimental study, for the first approach that we compare with the centralized approach (a single group formed), we have proved that when the number of agents is large, the studied approach facilitates the search for agreements and reduces the number of agents. Negotiation time is due to the distribution of agents into groups, which reduces the agents' reasoning complexity. For the second approach, we have proved that the number of agents and attributes have an impact on the formation of the initial groups and that the random choice of the attribute has an impact on the result of the negotiation. It is better to choose the attribute according to the weight of the agents to facilitate the search for agreements.

References

1. Michael, W.: An Introduction to Multiagent Systems. Wiley, Hoboken (2009)

2. Endriss, U.: Monotonic concession protocols for multilateral negotiation. In: AAMAS 06 Proceedings of the Fifth International Joint Conference on Autonomous Agents and Multiagent Systems, pp. 392–399 (2006)
3. Rosenschein, J.S., Zlotkin, G.: Rules of Encounter. MIT Press, Cambridge (1994)
4. Zhang, X.S., Klein, M., Midsemester, I.: Scalable complex contract negotiation with structured search and agenda management. In: AAAI, pp. 1507–1514 (2014)
5. Aydoğan, R., Festen, D., Hindriks, K., Jonker, C.: Alternating offers protocols for multilateral negotiation. In: Fujita, K., et al. (eds.) Modern Approaches to Agent-based Complex Automated Negotiation. SCI, vol. 674, pp. 153–167. Springer, Cham (2017). https://doi.org/10. 1007/978-3-319-51563-2_10
6. Diago, N.A., Aknine, S., Ramchurn, S., Shehory, O., Sène, M.: Distributed negotiation for collective decision-making. In: ICTAI (2017)
7. Diago, N.A., Aknine, S., Shehory, O., Arib, S., Cailliere, R., Sene, M.: Decentralized and fair multilateral negotiation. In: 2016 IEEE 28th International Conference on Tools with Artificial Intelligence (ICTAI), pp. 149–156. IEEE (2016). Parsons, S., Maudet, N., Moraitis, P., Rahwan, I.: Argumentation in multi-agent systems. In: Second International Workshop, ArgMAS. Springer, Heidelberg (2005)
8. Posner, E., Spier, K., Vermeule, A.: Divide and Conquer, Harvard Law School John M. Olin Center for Law, Economics and Business Discussion Paper Series. Paper 636 (2009)
9. Jennings, N.R., Faratin, P., Lomuscio, A.R., Parsons, S., Wooldridge, M.J., Sierra, C.: Automated negotiation: prospects, methods and challenges. Group Decision Negotiation **10**(2), 199–215 (2001)
10. Von Neumann, J., Morgenstern, O.: Theory of games and economic behavior (commemorative edition). Princeton University Press (2007)
11. Zheng, R., Chakraborty, N., Dai, T., Sycara, K.: Multiagent negotiation on multiple issues with incomplete information. In: Proceedings of the 2013 international Conference on Autonomous Agents and Multi-agent Systems, pp. 1279–1280. International Foundation for Autonomous Agents and Multiagent Systems (2013)
12. Golden, B., Perny, P.: Infinite order lorenz dominance for fair multiagent optimization. International Foundation for Autonomous Agents and Multiagent Systems (2010).Author, F.: Contribution title. In: 9th International Proceedings on Proceedings, pp. 1–2. Publisher, Location (2010).

Signal, Image and Vision Computing
(Full Papers)

A New Approach Based on Steganography to Face Facial Recognition Vulnerabilities Against Fake Identities

Saadia Nemmaoui[1] (iD) and Sihame Elhammani[2](✉)

[1] Information Processing and Decision Support Laboratory, Sultan Moulay Slimane University TIAD, Beni Mellal, Morocco
[2] Faculty of Sciences, University Mohammed V Agdal, Rabat, Morocco

Abstract. By the huge use of biometric identification and authentication systems, securing user's images is one of the major recent researches topics. In this paper we aim to present a new approach which is based on detecting face in any picture (even with low quality) a detection that can goes to 93%. As a second work we aim to test the ability of this algorithm to identify people while wearing medical masks. In fact, with the spread of Covid-19, people are now obliged to wear medical masks. These masks cover almost 60% of persons faces, this lack of information can prevent the identification of the person or can create some confusions. And To secure images and prevent identity theft we propose an approach that consists of hiding a generated key in each person's image. This unique key is based on user's personal information, the key will be verified by Luhn algorithm, which is considered as a widely used algorithm to verify generated IDs. As a third work, we aim to hide the person's ID in the image using a steganographic algorithm. Our work main objective is to protect pictures and prevent any attempt of creating a fake model of the owners. Therefore, the ID hidden in the picture will be destructed in every attempt of creating a fake image or model.

Keywords: Face recognition · Steganography · Luhn algorithm · Secure pictures · Covid-19

1 Introduction

Facial recognition is the process of identifying a person based on his face, nowadays most facial recognition algorithms are based on deep learning and Artificial Intelligence (AI) technologies. It can detect and identify persons identities based on their facial details. Facial recognition consists of three principal processes: detection, capture and match. The first process represents the process of detecting person's, or human in general, face in an image or a video. The second process (i.e., capture) transforms analogue information, detected faces, into digital information. In addition, the last process consists of comparing the detected face with faces that are already stored in a database. Facial recognition is now used in biometric systems to identify and authenticate a person, it is considered as the most secure method to verify the identity of a user. One, because it is easy to deploy,

© Springer Nature Switzerland AG 2021
M. Fakir et al. (Eds.): CBI 2021, LNBIP 416, pp. 269–283, 2021.
https://doi.org/10.1007/978-3-030-76508-8_19

two there is no need for physical justifications (legal papers or passwords) and three, and the most important, the face recognition process can be done in a very short time.

With the huge use of biometric systems, the security of biometric data is considered now more important than the security of passwords. Because, passwords can be changed in case the user has forgot it or in case the password has been stolen, but biometric data are unique for each person and should not be leaked to wrong persons. Ultimately, every company is responsible for the security of its user's data.

Although, biometric systems suffer from fake faces attacks. In fact, an attacker can easily build a fake 2D or 3D fake picture based on a picture stolen from the persons profile on the web, or on a social media website. Moreover, an attacker can build a fake video of the person using multiple pictures.

In this paper, Our main goal is to propose a secure method to highly increase security of web users pictures, in the first section we propose an algorithm of facial recognition based on a the GaussianFace algorithm developed in 2014 by researchers at the Chinese university of Hong Kong which has achieved and identification score of 98.52% [4]. In the second section we propose an ID calculation method, the ID will be based on the user personal information and will be verified by Luhn algorithm. In the next section we aim to use a basic steganographic algorithm to hide the ID in the picture and keep the picture identifiable with the facial recognition algorithm.

2   Facial Recognition Algorithm

Human face is the feature that allows to identify a person, it represents lot of details, and due to its complexity, analyzing and identifying persons faces is considered as a challenging task in all previous researches. Nowadays, many researchers has successfully implemented algorithms [1–3] able to detect and identify persons. And also there has been many improvements attempts [5–9].

Python including CMake library develop the algorithm used in this paper for facial recognition, it is based on the study made by researchers at the Chinese university of Hong Kong. The model, based on deep learning, has an accuracy of 99.38%.

The first function is used to translate all the pictures stored in the database and encode them into machine language. The images in the database should imperatively be stored with the name or the ID of its owner as shown in Fig. 1.

All the encoded images are stored in a table, which will be used to compare given image with the stored images in order to find a match.

The function classify is used to rename the output file and shows a box around the face with the name (the image label) of its owner.

Fig. 1. Data example

Algorithm 1: Main face recognition algorithm functions

```
Function get_encoded_faces():
  Face:=null;
  For f_name in faces loop:
    For f in f_names loop:
      If f.endswith (".jpg") or f.endswith (".png")
        Face:=f.loadimagename;
      End if;
    End for;
  End for;
Return face;
End get_encoded_faces;

Function classify_face(im):
    Faces:= get_encoded_faces();
    Known_face_name=faces.keys();
    If known_face_name == Null
      Name:= "Unknown";
    Else
      Name:= known_face_name;
    End if;
  Return Name;
End classify_face;
```

2.1 Pictures Without Medical Mask

In this section, we tested the previous algorithm on 686 pictures, including free copyright images, a free database and a free celebrity database images. Figure 2 represents one of the tested images and Fig. 3 represents the output image.

Fig. 2. Tested image

Fig. 3. The output of face_recognition algorithm

To test the identification score we used PIL library to make modification on brightness, sharpness, and colors of the image. We choose to focus on sharpness and brightness evaluation because these two parameters have an essential role to show facial features in images. Results are shown in Index 1.

As results, the facial recognition algorithm was able to identify persons till a value of 50 or brightness and 4 for sharpness.

2.2 Picture with Medical Mask

In the previous section, the algorithm has shown its ability to identify people even in pictures with low quality. Nowadays, and due to the spread of Covid-19, people are now obliged to wear medical masks. These masks hide almost 60% of people faces which make it difficult to identify a person – and if he/she was wearing sunglasses-. In this section we aim to test the facial recognition algorithm on pictures of people wearing medical masks.

2.2.1 Step 1

According to the given database we defined 3 main categories (Table 1):

- Category 1: contains pictures in which the mask hides 50% of face details.
- Category 2: contains pictures in which the mask hides 60% of face details.
- Category 3: contains pictures in which the mask hides 75% of face details.

Examples:

Table 1. Example of pictures.

Category 1	Category 2	Category 3

2.2.2 Step 2

We tested the facial recognition algorithm on faces from the 3 categories. Table 2 contains examples and the output of the algorithm.

2.2.3 Step3: Analyzing Results

As shown in Table 2, the algorithm is unable to detect people faces while using masks, and if it detects the face it is unable to identify the person.

We tested the algorithm on 690 picture of different people wearing medical mask, and results are shown in Fig. 4.

Figure 4 represents the average of the identification rate for the 3 categories, and as shown, the algorithm is able to detect people faces in pictures of categories 1 and 2, however the more the percentage of hidden details increases the more the identification rate decreases.

Table 2. Tested pictures.

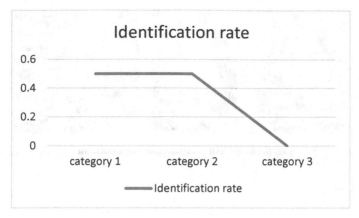

Fig. 4. Identification rate

Unlike human eyes which can identify people even with 80% of hidden details. In general, and as University of California, Santa Barbara (UCSB) researchers have found, humans tend in general to look below the eyes and above the nose to identify a person. And can easily identify real pictures and fake pictures.

For example, we took the picture shown in the Fig. 5(a) and we build a 3D model using blender, shown in Fig. 5(b), and we tested the face recognition algorithm on it, the output is represented in Fig. 6.

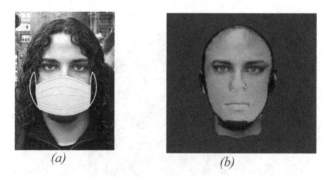

Fig. 5. Tested picture (a) and its corresponding 3D model (b)

The 3D generated model was identified as a real person with the face recognition algorithm, even if the model has a very low quality. The results represent an important vulnerability in face recognition algorithm, which is the fact of identifying a fake 2D and 3D models as real persons.

Identity theft and spoofing are the main attacks on biometric identification systems, in the next section we propose an approach based on steganography algorithm. In this method we aim to generate a unique ID and hide it in the picture, this ID will not be identified if the picture has been used to create a fake 3D model.

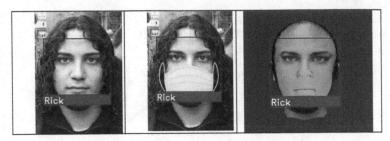

Fig. 6. Results of face recognition algorithm

3 Calculate ID

IDs or UIDs, Unique Identifiers, are defined as a combination of numbers and letters that allow the identification of an object. Many previous researchers have provided different approaches in all domains (medical, education, technology, etc.) of ID generators, among them the NHash algorithm used in medical domain and which can generate patients identifier based on demographic information [14], and generate IDs for epidemiological or clinical studies [15].

In this paragraph we define a mathematical function to generate a unique ID. This ID will be based on facial details of the picture's owner. Facial features represent (Eyes position, lips position, nose position, etc.) each feature is defined by separated points as shown in Fig. 7 (Table 3).

Fig. 7. Identify face features

We assume that ID is the function that defines the person's ID starting from Face details (N represents the number of face details), for each that equals to No it is considered

Table 3. Face features parameters

Face details	Inputs
Age between 1 and 15	No
Age between 18 and 30	Yes
Above 30	No
Eyebrows	Yes
Bald	No
Beard	No
Big lips	No
Big nose	No
Black hair	No
Blond hair	No
Brown hair	Yes
Wavy hair	No
Straight Hair	Yes
Glasses	No
Female	Yes
Male	No
White skin	Yes
Young	Yes
Wearing lipstick	Yes
Wearing hat	No
Heavy makeup	Yes
Blurry	No
Bushy eyebrows	No
...	

as null, and for each input where the value equals to Yes, the value is represented with a specific Code. Assuming that g is the function responsible for generating code for face details.

$$\forall x \epsilon N; \, g(x) = c \, where \, c \, \epsilon \, Facial\_Codes \qquad (1)$$

The resulting code will be the sum of small codes generated with the function g

$$ID = \sum g(x) \qquad (2)$$

The g(x) function expression is based on Linear Congruential Generator algorithm (LCG), which is one of the oldest and best-known pseudorandom number generator algorithms. Recurrence relation defines the generator:

$$X_{n+1} = (aX_n + c) \bmod m \tag{3}$$

Where:

$$m > 0 \text{ the modulus}$$

$$0 < a < m \text{ the multiplier}$$

$$0 \leq c < m \text{ the increment}$$

$$0 \leq X_0 < m \text{ the start value}$$

Note: when $c = 0$, the generator is called multiplicative congruential generator (MCG).

Algorithm 2: Linear Congruential Generator algorithm

```
Function lcg (modulus, a, c, seed):
While True loop:
  seed := (a * seed + c) % modulus;
End while;
Return seed
End lcg;
```

Before proceeding to generate an ID this function verifies informat/parameters given, if these information has already been used it returns the same ID (a previously generated value), the ID will be attached to time/the moment that this function has been used to generate the ID.

Example: ID = AAAAA20/04/2020.

To verify the validity of the generated Code we use Luhn algorithm. This algorithm, also called Luhn formula or mod 10 algorithm, has been created by IBM scientist Hans Peter Luhn. And it is widely used to validate identification numbers such as credit card numbers, IMEI numbers and National Provider Identifier Numbers in the united states.

To assume that the generated ID is valid, it should verify following points:

1- Starting from the rightmost digit, we double the value of every second digit. And then check if the digit is doubled.
2- The first digit doubled is the digit located at left of the check digit. If the result of this doubling operation is greater than 9 then we will add the digits of the result or, by subtracting the digits.
3- We calculate the sum of all the digits.
4- If the total modulo 10 is equal to 0 then the number is valid.

For example:
We assume that ID of the tested image is: 4599832674

N	4	5	9	9	8	3	2	6	7	4	X
D	4	10	9	18	8	6	2	12	7	8	X
S	4	1	9	9	8	6	2	3	7	8	X

(N: Number, D: double or the second digit, S: Sum of digits greater than 9).
The sum of the digits is:
$4 + 1 + 9 + 9 + 8 + 6 + 2 + 3 + 7 + 8 = 57$.
The check digit is: 3
60 mod 10 is equal to 0.
Then the number 4599832674 is considered a valid ID.

4 Steganography Algorithm

Steganography represents the process of hiding information in images, the information hidden remains secret as far as the receiver is not aware of it. There many techniques of Steganography [10], in [11, 12] the techniques is based on LSB, while in [13] the author proposed a technique based on 3D.

In our paper we propose a basic steganographic model, developed by python, the main objective is to encode the ID and to use almost all the surface of the image and hide the given information randomly in its pixels.

Billow is the algorithm of the encode and decode functions.

Algorithm 3: Encode function

```
function encode():
    img := input("Enter image name:");
    image := Image.open(img, 'r');
    data := input("Enter data: ");
    if (len(data) == 0):
        raise ValueError('Empty data');
End if:
    newimg := image.copy();
    encode_enc(newimg, data);

    new_img_name := input("new name ");
end encode;
```

Figure 8 represents the execution of the algorithm and Fig. 9 represents the output image.

To extract the hidden information from the picture we use the decode function (Fig. 10).

```
:: Welcome to Steganography ::
1. Encode
 2. Decode
1
Enter image name(with extension): Anna.png
Enter data to be encoded : Anna
Enter the name of new image(with extension): Anna_Encoded.png
```

Fig. 8. Execution of Steganography algorithm – Encoding

Fig. 9. Image + ID

Algorithm 4: Decode function

```
function decode():
    img := input("Enter name:");
    image := Image.open(img, 'r');
    data := '';
      date := get_dataFromImage(image);
      return data
End decode()
```

```
:: Welcome to Steganography ::
1. Encode
 2. Decode
2
Enter image name(with extension) :Anna_Encoded.png
Decoded word- Anna
```

Fig. 10. Execution of Steganography algorithm – Decoding

Adding the generated ID should not affect the facial recognition process. Therefor we executed the facial recognition algorithm on the resulting picture. (Fig. 9: Image + ID) (Fig. 11).

Fig. 11. Identifying the owner

Hiding information on random pixels, do not allow the reuse of the picture to create a fake images, in fact creating a fake image or a fake video consists of taking various pictures of the target and from different angles to get a complete and almost perfect image of the target. Hiding the ID in an image do not allow hackers to use the image, because when the image is reused the ID will be destructed.

We took Fig. 9 that contains the ID = "Anna' and we use it to build a fake 3D picture (Fig. 12).

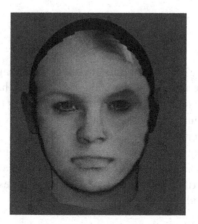

Fig. 12. Fake 3D model

When executing the encoding function on Fig. 12 the Key generated do not match the key hidden which is Anna (Fig. 13).

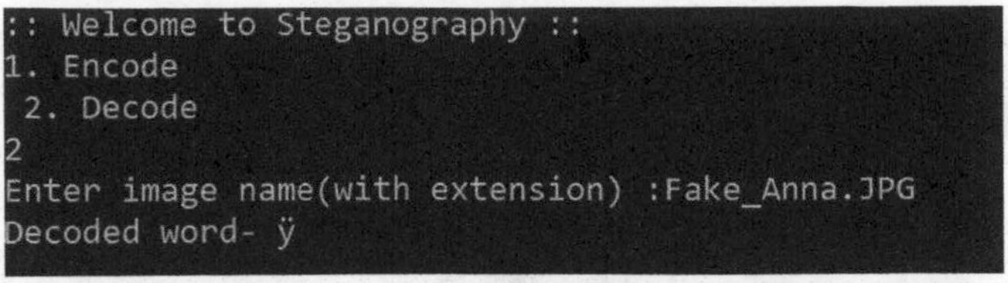

```
:: Welcome to Steganography ::
1. Encode
 2. Decode
2
Enter image name(with extension) :Fake_Anna.JPG
Decoded word- ÿ
```

Fig. 13. Execution of decoding function on Fake picture

Even if Fig. 14 represents a fake model it is identified as a real person by the facial recognition algorithm.

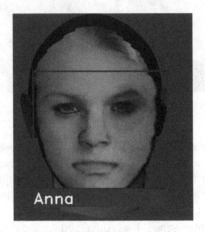

Fig. 14. Execution of the facial identification algorithm on a fake 3D model

5   Conclusion

In this paper we used a face detection algorithm which matching result can goes to 98%. This algorithm is not able to detect people on pictures that has brightness more than (50.0) and sharpness more than (40.0), and almost all pictures that has these parameters are hardly identified by humans. The algorithm was able to identify people wearing medical masks which covers 50% of face details.

The main issue was that this algorithm can identify fake pictures as real persons. As a solution we proposed a basic steganographic algorithm that allows to hide a generated ID (verified by Luhn algorithm) in the picture, this ID gets destructed in every attempt of using the picture to build a fake video or a fake 3D face.

References

1. Jadhav, A., Jadhav, A., Ladhe, T., Yeolekar, K.: Automated attendance system using face recognition. Int. Res. J. Eng. Technol. (IRJET) **04**(01), 1467–1471 (2017). ISSN 2395-0072

2. Punjani, A.A., Obaid, C., Yasir, C.: Automated attendance management system using face recognition. Int. J. Adv. Res. Comput. Eng. Technol. (IJARCET), **6**(8), 1210–1212 (2017)
3. Chaudhari, C., Raj, R., Shirnath, S., Sali, T.: Automatic attendance monitoring system using face recognition techniques. Int. J. Innov. Eng. Technol. (IJIET), **10**(1), 103–106 (2018)
4. https://github.com/ageitgey/face_recognition
5. Kowsalya, P., Pavithra, J., Sowmiya, G., Shankar, C.K.: Attendance monitoring system using face detection & face recognition. Int. Res. J. Eng. Technol. (IRJET) **6**(3), 6629–6632 (2019). ISSN 2395-0072
6. Mallikarjuna Reddy, A., Venkata Krishna, V., Sumalatha, L.: Face recognition based on cross diagonal complete motif matrix. Int, J, Image Graph Signal Process 359–66 (2018)
7. Kamencay, P., Benco, M., Mizdos, T., Radil, R.: A new method for face recognition using convolutional neural network. Digit. Image Process. Comput. Graph. **15**(4), 663–672 (2017)
8. Beli, I., Guo, C.: Enhancing face identification using local binary patterns and K-nearest neighbors. J. Imaging **3**(37), 1–12 (2017)
9. Gaikwad, A.T.: LBP and PCA based on face recognition system, pp. 368–373, November 2018
10. Sumathi, C.P., Santanam, T., Umamaheswari, G.: A study of various steganographic techniques used for information hiding. Int. J. Comput. Sci. Eng. Surv. (IJCSES) **4**(6), 9–25 (2013)
11. Juneja, M., Sandhu, P.S.: A new approach for information security using an improved steganography technique. J. Inf. Pro. Syst. **9**(3), 405–424 (2013)
12. Shanmuga Priya, S., Mahesh, K., Kuppusamy, K.: Efficient steganography method to implement selected least significant bits in spatial domain. Int. J. Eng. Res. Appl. **2**(3), 2632–2637 (2012)
13. Thiyagarajan, P., Natarajan, V., Aghila, G., Pranna Venkatesan, V., Anitha, R.: Pattern based 3D image steganography. In: 3D Research center, Kwangwoon University and Springer 2013, 3DR Express, pp.1–8 (2013)
14. Zhang, G.Q., et al.: Randomized N-Gram hashing for distributed generation of validatable unique study identifiers in multicenter research. JMIR Med. Inform. **3**(4), e4959 (2015)
15. Olden, M., Holle, R., Heid, I.M. and Stark, K.: IDGenerator: unique identifier generator for epidemiologic or clinical studies. BMC Med. Res. Methodol. 16, 1–10 (2016). https://doi.org/10.1186/s12874-016-0222-3

Mean Square Convergence of Reproducing Kernel for Channel Identification: Application to Bran D Channel Impulse Response

Rachid Fateh$^{(\boxtimes)}$ ⓘ and Anouar Darif ⓘ

Mathematics and Informatics Department, LIMATI Laboratory,
Sultan Moulay Slimane University, Beni Mellal, Morocco

Abstract. Nowadays, in the field of nonlinear system identification, the function approximation builds on the theory of reproducing kernel Hilbert spaces (RKHS) is of high importance in kernel-based regression methods. In this paper, we are focused on the finite impulse response identification problem for single-input single-output (SISO) nonlinear systems, whose outputs are detected by binary value sensors. In the one hand, we have used kernel adaptive filtering methods, such as, kernel least mean square (KLMS) and kernel normalized least mean square (KNLMS) to identify the practical frequency selective fading channel called Broadband Radio Access Network (BRAN). In the other hand, the mean square convergence is also investigated to indicate the robustness of the kernel normalized LMS algorithm. Monte Carlo simulation results in noisy environment and for various data length, shows that the kernel-normalized LMS algorithm can provide superior accuracy as opposed to the kernel-based LMS algorithm.

Keywords: Identification · Single-input single-output · Kernel normalized least mean square · Kernel least mean square · Binary sensors

1 Introduction

Over the last decade, multiple adaptive kernel filtering algorithms based on the Hilbert space theory with reproducing kernels have been developed [1,2]. Nowadays, kernel methods are increasingly developed, they are a significant source of advances, not only in terms of computational cost but also in terms of the obtained efficiencies in solving complex tasks. As these methods largely determine the efficiency of the treatments, by their ability to reveal existing similarities in the processed data. They are founded on a fundamental idea known as "kernel trick", which was first exploited with the Support Vector Machine (SVM) [3,4], then also used to transform numerous linear dimensionality reduction algorithms into non-linear algorithms [5]. The kernel's trick allows to give

© Springer Nature Switzerland AG 2021
M. Fakir et al. (Eds.): CBI 2021, LNBIP 416, pp. 284–293, 2021.
https://doi.org/10.1007/978-3-030-76508-8_20

a non-linear character to several originally linear methods and, without restriction, they can be expressed only in terms of inner products of the observations. The several kernel adaptive filtering algorithms, including the kernel least mean square (KLMS) [6] and kernel affine projection algorithms (KAPA) [7], have recently been noted for nonlinear signal processing [8,9]. Furthermore, to optimize the quality of the basic kernel adaptive filtering algorithms [5], subtypes of these algorithms have also been mentioned [10–13]. In this paper, we will focus on the nonlinear system identification with binary-valued output, where the goal is to identify a practical, i.e. measured, frequency-selective fading channel, called Broadband Radio Access Network (BRAN D), representing the outdoor propagation. This channel of the model is normalized throughout the European Telecommunications Standards Institute (ETSI) [14,15].

The present paper is organized as follows: the identification problem is formulated in Sect. 2. In Sect. 3, we review some basic concepts of nonlinear regression in reproducing kernel Hilbert spaces. In Sect. 4, we presented the derivation of the kernel least mean square algorithm, and kernel nomalized least mean square algorithm is presented in Sect. 5. Section 6 shows some simulations to measure the efficiency of the kernel normalized LMS and, finally, Sect. 7 presents the conclusion and future research.

2   Problem Statement

Let us consider a non-linear model as represented in Fig. 1; it consists of a Linear Time Invariant block followed by a static memoryless nonlinearity.

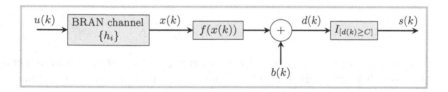

Fig. 1. Block of the nonlinear system with binary outputs and noises.

The following equations are used to model the system (Fig. 1):

$$x(k) = \sum_{i=0}^{P-1} h(i)u(k-i), \quad d(k) = f(x(k)) + b(k) \tag{1}$$

Where $u(k)$, $b(k)$ and $f(.)$ represent the input signal, the measurement noise and the nonlinearity, respectively. $\{h_i\}_{i=0}^{P-1}$ is the impulse response coefficients of channel and P is the order of FIR system.

The measurement of the system output $d(k)$ is carried out by binary-valued sensor $I_{[\cdot]}$, with a finite threshold $C \in \mathbb{R}$, such that:

$$s(k) = I_{[d(k) \geq C]} = \begin{cases} 1 & \text{if } d(k) \leqslant C, \\ -1 & \text{otherwise.} \end{cases} \tag{2}$$

During this paper, we will assume the following system propositions:

- $b(k)$ is a gaussian additive noise that is independent of $u(k)$ and $d(k)$.
- For any finite u, the non-linearity $f(.)$ is continuous and invertible.
- The system has no delay, i.e. $h_0 \neq 0$.
- The threshold C is known.

All these assumptions are employed to simplify the system analysis and to find a good comparison between the kernel methods (KNLMS and KLMS).

3   Nonlinear Regression Principles in RKHS

Nonlinear adaptive filtering problems can be formulated as linear finite order problems, where the original input data $\mathcal{U}$ has been mapped to a nonlinear Hilbert space $\mathcal{H}$ (infinite-dimensional) with a Mercer kernel [1]:

$$\kappa(u_i, u_j) = \Phi(u_i)\Phi^\top(u_j), \; for \; \forall u_i, u_j \in \mathcal{U} \tag{3}$$

Where Φ uses an inner product to identify $\mathcal{U}$ into a greater space $\mathcal{H}$.

A feature map Φ is used to reconstruct the sample sequence:

$$\begin{aligned} \Phi \; : \; \mathcal{U} &\longrightarrow \mathcal{H} \\ u_i &\longrightarrow \kappa(u_i, .), \; 0 \leq i \leq N \end{aligned} \tag{4}$$

The Gaussian kernel is used in the reproducing kernel Hilbert spaces model construction, which is generally a default choice due to its universal approximation capability:

$$\kappa(u_i, u_j) = \exp\left(-\frac{\|u_i - u_j\|^2}{2\sigma^2}\right), \quad \forall (u_i, u_j) \in \mathcal{U}^2. \tag{5}$$

Where, $\sigma > 0$ denotes the kernel bandwidth.

So as to define the attributes of Hilbert space from their exact location, we can give a Hilbert space $\mathcal{H}$ an orthonormal base. The identified kernel (κ) should be a positive definite, uninterrupted, symmetric and normalized function with the following properties: $\kappa : \mathcal{U} \times \mathcal{U} \to \mathbb{R}$, where the subset $\mathcal{U} \subset \mathbb{R}^N$ is compact.

The mapping of the data space ($\mathcal{U}$) and the Hilbert space ($\mathcal{H}$) inferred by the kernel reproducing κ is presented in Fig. 2. This correspondence mapping depicts every input point u by its commonality $\kappa(u, .)$ with all the remaining points in the $\mathcal{U}$ space.

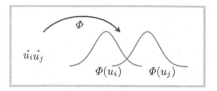

Fig. 2. From the input space $\mathcal{U}$ to the feature space $\mathcal{H}$ is a mapping.

4   Kernel Least Mean Square Algorithm

The kernel least mean square algorithm [6,16] is classified as a stochastic gradient algorithms category. The basic idea is to run the linear LMS algorithm in the kernel feature space, the function map $\Phi(.)$ described in Eq. (4) is often used to create the data input, which is identified with the kernel reproducing κ. The setting equations of the LMS algorithm are as follow [17]:

$$e(n) = s(n) - \theta(n-1)^{\top} u(n) \tag{6}$$

$$\theta(n) = \theta(n-1) + \mu e(n)u(n) \tag{7}$$

Where the term $e(n)$ represents the prediction error, $\theta(n)$ is the estimate of the weight vector and μ is called: the algorithm step-size parameter.

Now assume that the sample sequence is transformed using the Φ function map, then apply the LMS logic to the following transformed data sequence:

$$\{(\Phi(u(1)), s(1)), (\Phi(u(2)), s(2)), ...(\Phi(u(n)), s(n)), ...\}. \tag{8}$$

We have the kernel LMS:

$$e(n) = s(n) - \theta(n-1)^{\top} \Phi(u(n)) \tag{9}$$

$$\theta(n) = \theta(n-1) + \mu e(n)\Phi(u(n)) \tag{10}$$

The main difference with LMS is that in Eq. (11) it is in a space of possibly infinite dimensional characteristics and directly updating it would be virtually impossible. Instead, we will use each $\theta(n)$ of them to link to their initialization $\theta(0)$:

$$\theta(n) = \theta(0) + \mu \sum_{i=1}^{n} e(i)\Phi(u(i)). \tag{11}$$

$$\theta(n) = \mu \sum_{i=1}^{n} e(i)\Phi(u(i)). \quad \text{(supposing that } \theta(0) \text{ equals 0)} \tag{12}$$

We can obtain the prediction function by employing the kernel trick:

$$\langle \theta(n), \Phi(u(n)) \rangle_{\mathcal{H}} = \mu \sum_{i=1}^{n} e(i) \langle \Phi(u(i)), \Phi(u(n)) \rangle \qquad (13)$$

$$= \mu \sum_{i=1}^{n} e(i) \kappa(u(i), u(n)),$$

Where $\kappa(u(i), u(n))$ is a Mercer kernel and n is the number of training samples.

5   Kernel Normalized Least Mean Square Algorithm

The kernel normalized least squares (KNLMS) algorithm generally shows increased performance compared to the KLMS. The update rule of KNLMS is written as follows [18]:

$$\theta(n) = \theta(n-1) + \frac{\mu \Phi(u(n))}{\varepsilon + \kappa(u(n), u(n))} \left[s(n) - \Phi(u(n))^{\top} \theta(n-1) \right] \qquad (14)$$

Where, ε is the control parameter.

6   Simulation and Results

The goal of this section was to test the efficiency of the kernel NLMS algorithm with that of the kernel LMS algorithm. Let's define the MSE as follows:

$$MSE(n, snr) = \frac{1}{MC} E \left[\| s(n) - \theta^{\top}(n-1)\Phi(u(n)) \|^2 \right] \qquad (15)$$

Where MC being the total amount of Monte Carlo iteration and $E[.]$ is the expectation operator. We used a BRAN D channel impulse response parameters for the linear part and a hyperbolic function ($\tanh(\mathbf{x})$) for the nonlinear part to simulate the system. We apply the two kernel algorithms in the following situation: the threshold is $C = 0.5$, the step-size parameter $\mu = 0.05$, the kernel width $\sigma = 0.2$ and 50 Monte Carlo iterations.

6.1   BRAN D Radio Channel

In this paragraph, we present the values associated with the model BRAN D data (see Table 1). The impulse response of the BRAN D radio channel is described as follows:

$$h(n) = \sum_{i=0}^{L-1} A_i \delta(n - \tau_i) \qquad (16)$$

Where A_i, $\delta(n)$ and τ_i represent the path i magnitude, the Dirac function and paths i time delay, respectively. $A_i \in N(0, 1)$ and $L = 18$ is the path number.

Table 1. Delay and magnitudes of 18 targets of BRAN D radio channel.

Delay $\tau_i(ns)$	Magnitude $A_i(dB)$	Delay $\tau_i(ns)$	Magnitude $A_i(dB)$
0	0	230	−9.4
10	−10	280	−10.8
20	−10.3	330	−12.3
30	−10.6	400	−11.7
50	−6.4	490	−14.3
80	−7.2	600	−15.8
110	−8.1	730	−19.6
140	−9.0	880	−22.7
180	−7.9	1050	−27.6

6.2   BRAN D Channel Identification

The estimation of the amplitude and phase of the ETSI BRAN D radio channel, using the two algorithms is presented in Fig. 3. It is observed from Fig. 3 that both the amplitude and phase curves estimated by means of the kernel normalized least mean square algorithm are similar in their forms to those of the BRAN data. As opposed to the kernel least mean square algorithm, we notice a slight difference amid our estimated and measured data (amplitude and phase).

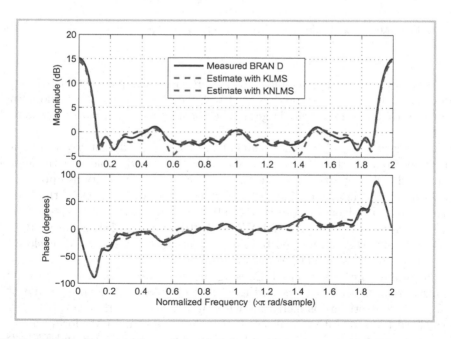

Fig. 3. Estimated magnitude and phase of the BRAN D channel, for an $SNR = 15\,\mathrm{dB}$ and a numbers of samples $N = 2048$.

We have plotted the estimated parameters of the model BRAN D radio channel impulse response as a function of targets in the time domain as well, for a numbers of samples $N = 1024$ and an $SNR = 15\,\mathrm{dB}$ (Fig. 4).

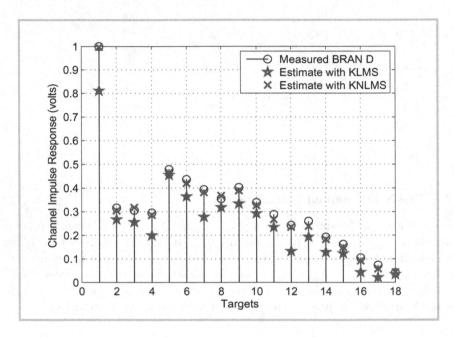

Fig. 4. Estimated magnitude of the BRAN D channel, depending on the targets in the case $N = 1024$ and $SNR = 15\,\mathrm{dB}$.

As summarized in Figs. 3 and 4, which presented the parameters estimated respectively, for $N = 2048$ and $N = 1024$. We can conclude that the kernel algorithm therefore requires a large number of samples to get closer to the real values.

The plot of Fig. 5 shows the results of the MSE, for different SNR obtained using the two algorithms (KNLMS and KLMS). It is noticeable that the kernel normalized LMS algorithm has the quickest convergence velocity as opposed to the kernel LMS algorithm, due to the fact that the MSE values of the kernel normalized LMS algorithm are very small as compared to those of the kernel LMS algorithm. For example, if SNR is 15 dB, we have MSE equal to $10^{-0.4}$ in the case of kernel normalized LMS algorithm, however, we get MSE close to $10^{-0.2}$ through using kernel LMS algorithm.

The evolution curve of the MSE, using the two algorithms in the case where μ ranges from 0.1 to 1 is shown in Fig. 6. From this result, we observe that the MSE of both algorithms is increasing with an increase of μ. However, the kernel normalized LMS algorithm is always better than the kernel LMS for all μ.

In conclusion, the convergence of both kernel methods (KLMS and KNLMS) is ensured by selecting the step-size parameter (μ) for updating the weight vector (θ) in order to control the generalization ability of the kernel algorithm. When

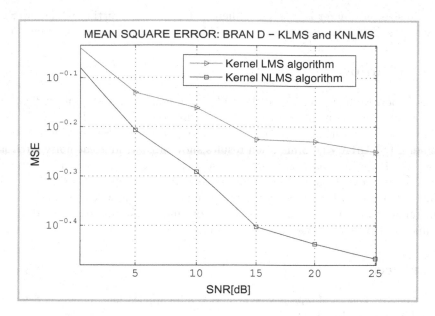

Fig. 5. Performance comparison of both algorithms with various values of SNR, and for a numbers of samples $N = 1024$.

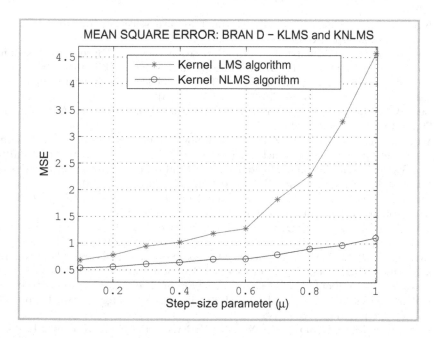

Fig. 6. Performance comparison of both algorithms with various values of μ, for a numbers of samples $N = 1024$ and $SNR = 15\,\text{dB}$.

the step size parameter is increased, there is a risk of overfitting while lowering the step-size helps the mean square error (MSE) to converge to its optimal value.

7   Conclusion

In this paper, we have considered the problem of nonlinear system identification, with binary measurements on the output, using kernel methods (KNLMS and KLMS). Both algorithms are used to estimate the parameters of the measured channel ETSI BRAN D. Simulation results showed that, in great noisy environments, the kernel-normalized least mean square approach is efficient and robust way to estimate the parameters (amplitude and phase) of the BRAN D radio channel. Future work focuses on the implementation of new kernel methods for identifying communication channels using non-linear systems constructed by the Hammerstein-Wiener model from input binary measurements.

References

1. Saitoh, S.: Theory of reproducing kernels and its applications. Longman Scientific and Technical (1988)
2. Muller, K.R., Mika, S., Ratsch, G., Tsuda, K., Scholkopf, B.: An introduction to kernel-based learning algorithms. IEEE Trans. Neural Netw. **12**(2), 181–201 (2001)
3. Cortes, C., Vapnik, V.: Support-vector networks. Mach. Learn. **20**(3), 273–297 (1995)
4. Nguyen, M.H., De la Torre, F.: Optimal feature selection for support vector machines. Pattern Recogn. **43**(3), 584–591 (2010)
5. Liu, W., Park, I., Principe, J.C.: An information theoretic approach of designing sparse kernel adaptive filters. IEEE Trans. Neural Netw. **20**(12), 1950–1961 (2009)
6. Liu, W., Pokharel, P., Principe, J.C.: The kernel least-mean-square algorithm. IEEE Trans. Signal Process. **56**(2), 543–554 (2008)
7. Liu, W., Príncipe, J.C.: Kernel affine projection algorithms. EURASIP J. Adv. Signal Process. **2008**(1), 1–12 (2008)
8. Wu, Q., Li, Y., Jiang, Z., Zhang, Y.: A novel hybrid kernel adaptive filtering algorithm for nonlinear channel equalization. IEEE Access **7**, 62107–62114 (2019)
9. Van Vaerenbergh, S., Via, J., Santamaria, I.: Nonlinear system identification using a new sliding-window kernel RLS algorithm. JCM **2**(3), 1–8 (2007)
10. Zhao, J., Liao, X., Wang, S., Chi, K.T.: Kernel least mean square with single feedback. IEEE Signal Process. Lett. **22**(7), 953–957 (2014)
11. Wang, S., Zheng, Y., Ling, C.: Regularized kernel least mean square algorithm with multiple-delay feedback. IEEE Signal Process. Lett. **23**(1), 98–101 (2015)
12. Liu, W., Park, I., Wang, Y., Principe, J.C.: Extended kernel recursive least squares algorithm. IEEE Trans. Signal Process. **57**(10), 3801–3814 (2009)
13. Slavakis, K., Bouboulis, P., Theodoridis, S.: Online learning in reproducing kernel Hilbert spaces. In: Academic Press Library in Signal Processing, vol. 1, pp. 883–987. Elsevier (2014)
14. ETSI. Broadband Radio Access Network (BRAN); HYPERLAN Type 2 Physical layer (PHY). Technical report, December 2001
15. ETSI. Broadband Radio Access Network (BRAN); HYPERLAN Type 2 Requirements and architectures for wireless broadband access, January 1999

16. Li, K., Principe, J.C.: No-trick (treat) kernel adaptive filtering using deterministic features. arXiv preprint arXiv:1912.04530 (2019)
17. Kwong, R.H., Johnston, E.W.: A variable step size LMS algorithm. IEEE Trans. Signal Process. **40**(7), 1633–1642 (1992)
18. Saide, C., Lengelle, R., Honeine, P., Richard, C., Achkar, R.: Nonlinear adaptive filtering using kernel-based algorithms with dictionary adaptation. Int. J. Adapt. Control Signal Process. **29**(11), 1391–1410 (2015)

Deep Learning for Medical Image Segmentation

Bouchra El Akraoui[✉] and Cherki Daoui

Laboratory of Information Processing and Decision Support, Faculty of Science and Technique,
Sultan Moulay Slimane University, Beni-Mellal, Morocco

Abstract. Alzheimer's disease is recognized as a progressive loss of memory and often leads to a total loss of autonomy, which renders it difficult to tolerate. Hippocampus is among the first cerebral regions to be infected in AD. An accurate diagnosis at an early stage of AD is crucial for the intervention process. The low contrast of supporting tissues and organs makes segmentation of the CC from brain MRI very difficult. In this paper we introduced both automatic brain image segmentation methods to extract the Corpus Callosum (CC), the first based on SLIC using parallel implementation which gives accelerated results over classical SLIC, the second method motivated by using Deep Learning approach. Finally, we compare the outcome given by the previous approaches.

Keywords: Magnetic resonance imaging · Segmentation · Corpus callosum · Alzheimer's disease · Simple linear iterative clustering · Deep learning

1 Introduction

Alzheimer's Disease (AD) is a neurological brain cells disorder. Memory loss is the one of the most frequent signs of AD, which generally progresses in three stages: early, middle, and late. Patients suffer from an inability to continue daily activities, confusion, and eventually, they may forget their life and cannot identify their family members.

AD is the highest frequency kind of dementia that often appears in old people. Accurate early-stage diagnosis of AD is primordial for the intervention process of this neurodegenerative disease progression. Many scientific research projects, such as neuroimaging, clinicians and researchers focused to help diagnosis of AD. One of the first affected brain regions in AD is Hippocampus. Structural Magnetic Resonance Imaging (MRI) scans allows the observation of the brain anatomical structures information, which can help to detect and measure brain atrophy of Corpus Callosum (CC) patterns in AD (see Fig. 1). Furthermore, precise, and dependable segmentation of the CC in brain MRI can predict informative biomarkers in advance of clinical symptoms. Moreover, the segmentation methods of CC from brain MRI in clinical studies of chronic diseases is still a challenge with regards to medical images having very same gray level and texture of the concerned objects, also, their structure has relatively low contrast. Another difficulty may arise due to undergo significant changes of brain structures over time making significant segmentation error may provide. Furthermore, the MRI scans to be segmented often adapted to the changes in image quality.

© Springer Nature Switzerland AG 2021
M. Fakir et al. (Eds.): CBI 2021, LNBIP 416, pp. 294–303, 2021.
https://doi.org/10.1007/978-3-030-76508-8_21

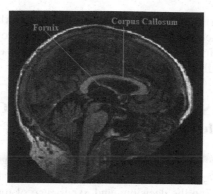

Fig. 1. Example midsagittal brain Magnetic Resonance (MR) image.

Traditionally, clinical analysis of brain MRI almost is done manually, which is be very tedious, time-consuming, especially, when in large and complex MRI datasets cases, it can be challenging for clinicians to extract important information manually [1]. To address these issues, many machine learning (ML) models have shown exceptional performance over the last decade using mid sagittal magnetic resonance imaging for CC segmentation [2, 3]. Various recently techniques presented for obtaining CC segmentation in literature based on Region or sub-region Of Interest (ROI) [4–6]. While is easier to implement, but maybe depend on the selection of the scheme to define the sub-regions. [7] also used adaptive region growing with level set framework to segment hippocampus automatically. [8] suggested a fully automated hybrid model based on Adaptive Mean Shift (AMS) clustering technique of image. They combine region analysis and template mapping with shape and feature analysis to detect the CC region, and the authors using the Geometric Active Contour (GAC) model to obtain the resulting segmentation outcome. [9] have further developed a technique to extract the CC using (ACM) as a combined hybrid method which merges the geometric method with the statistical active shape model (ASM) method, published by [10] based on prior knowledge. Other methods, like the multi-atlas voting scheme and the patch-based method, are also frequently used to segment CCs. The author [11] blends both methods by using a Bayesian inference setting to retrieve information of a sparse performance and a multi-atlas voting. Many researchers have focused on AD algorithms. In this paper, the main goal is to develop an efficient framework for the analysis of brain MRI images that will be able to extract the CC, the first contribution is to propose a fully automatic method based on the split/merge approach. The pixels of the image were grouped into a smaller number of super pixels by a Simple Linear Iterative Clustering (SLIC) algorithm that results in compact and almost uniform high quality super pixels. Some properties of this algorithm such as flexibility and simplicity of its method.

In recent years, Deep Learning (DL) schemes are the new trends in Artificial Intelligence (AI) in the scientific research community, in which various studies have shown outstanding performance. In this work, we discuss another success of DL in CC segmentation.

The reminder of the paper is organized as follows: The second section a briefly reviews some studies related to CC segmentation approaches; it describes the proposed method and a brief overview of the convolutional neural network architecture (in short, ConvNet or CNN), segmentation (CC) using deep learning and our CNN configurations are detailed in third section. Finally, we close with a summary description and discuss the prospects for brain MRI segmentation.

2   Briefly Reviews Some Relative Studies About CC Segmentation and Proposed Methods

The analysis of segmentation is an extensive area in medical diagnostic images. Reliable and accurate disease diagnostics vary depending on the acquisition and evaluation of images. Yet, it is limited to the decisions and experience of the specialist. In the last few decades, there have been enormous advances in image acquisition equipment that must run large data sets. In fact, the manual interpretation of medical images can be very complex, time consuming and a medical expert can be led to errors. This has led to great success in machine learning (ML) and artificial intelligence (AI) algorithms in medical fields such as medical image processing, computer-aided diagnosis, image segmentation... For some diseases like Alzheimer's disease, to date, there is no effective treatment to cure AD. Moreover, detection of AD at its early stage is crucial for prevention and intervention of its progression. Segmentation of CC is a vital step in the analysis of CC on medical brain images. The low contrast of MRI, the detailed shape is unstable, and the fornix has similar intensity with the CC making CC segmentation a challenging and intensive process. Various methods have been proposed in the literature regarding conventional of technologies related to CC segmentation, although any have used model-based methods that incorporate prior domain knowledge are commonly used to iteratively match a contour or landmark model to the expected final shape of the edges or approximate shapes, [12] have reported 3D level-set method that relies on a similarity score as a speed parameter. Also, [13] used a level-set approach to regularize the shape of the target. Other techniques have focused on region-based methods, that segments the image pixels into sets by assessing the similarity criteria. The segmentation is often initiated with starting points, this technique has been extensively applied in the area inclusive the Watershed Transform segmentation [3, 14], the Mean Shift algorithm (MS) [15], and graph-based methods [16]. There have been more studies based on Machine Learning (ML) which employs computational algorithms for learning complicated relationships from the empirical data. Some of the studied ML algorithms are regression [17], clustering techniques [11], Artificial Neural Networks (ANNs) [18], and probabilistic classifiers [17]. Other search uses hybrid methods, that combine different techniques from categories described previously [19, 20].

There are two principal main contributions of this study: The first is to advocate a non-geometric and fully automatic SLIC (Simple Linear Iterative Clustering) segmentation technique, to obtain the initial seed regions (super pixels) quickly, we implement our version parallelized on the GPU. SLIC method efficiently produces super pixels by K-means clustering pixels according on their color similarity and proximity in the image. This approach is commonly used to reduce computational cost in diverse computer vision tasks, and it outperforms and more memory efficient than existing methods [12,

13]. Furthermore, the estimated size of every super pixel of the image with N pixels is approximately N/K pixels. In our work, we implement with N = 256 and k = 400.

To efficiently generate super pixels, the SLIC algorithm selects cluster centers Ck = $[l_k; a_k; b_k; x_k; y_k]^T$ with k = [1; K] from the sampled evenly spaced grid S given by following equation:

$$S = \sqrt{\frac{N}{K}}. \tag{1}$$

The color of a pixel in the CIELAB color space is noted by $[l_k; a_k; b_k]^T$ and $[x_k; y_k]^T$ designates the position of pixel. For each center, we calculate the distance of the pixels within a radius of 2S to the center. If the distance is smaller than the memory distance for the pixel, it is changed in memory and the pixel is now part of the super pixel. SLIC associates the spatial distances with the color proximity in a distance measure D defined in Eq. 2:

$$Ds = d_{lab} + \frac{m}{S} \times d_{xy} \tag{2}$$

where d_{lab} indicates the color distance; m designates the proportion of the color value and spatial information in the similarity calculation; S represents the seed point distance and d_{xy} refers the space distance.

The computation of the super pixels of the offered method as illustrated in Fig. 2. In particular, the Fig. 2-a shows the input 2D image, whereas the Fig. 2-b represents the outcome of applying SLIC segmentation to generate over segmented regions (400 super pixels) where each super pixel region has a yellow edge.

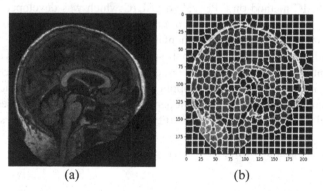

(a)   (b)

Fig. 2. Applying SLIC method: (a) The input MRI; (b) Result of the SLIC method

After grouping the image into a small number of super pixels in the previous step using SLIC method, which generates an over-segmentation. Efficient and faster-merging algorithms are necessary to merge adjacent and similar regions like as Region Adjacency Graph (RAG) which can simplify the problem of merging and thus built a spatial view of the image as shown in Fig. 3, whose vertices correspond to super pixel regions and edges represent connections between adjacent regions. The edges between the super pixels have been colored belonging to their weight. The dark color designed the edges between similar regions, while the purple color refers to the edges between dissimilar regions.

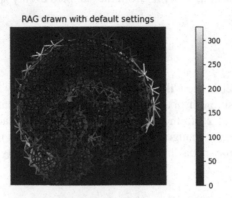

Fig. 3. Spatial overviews of the relationships between areas: Region Adjacency Graph (RAG).

In the next step, we achieve the outcome of the segmentation by performing a hierarchical region merging. The algorithm continuously iteratively merges the small regions until some threshold is satisfied. The hierarchical fusion scheme continues the bottom-up method. Figure 4 presents an illustration of the hierarchical clustering task. We can see that the over-segmented region is merged into areas with the similar properties.

This paper is motivated by the simplicity of the SLIC method and the speed of the parallel implementation as well as the online availability of the source code of the modified SLIC method on GPU called gSLIC which was developed by [21] and showed that the speed of this method is almost 10 times the sequential method, so we are interested in applying it for CC segmentation.

With the advent of Deep Learning approaches, several new techniques and architectures are invented, and various images segmentation problems have overcome. DL has successfully solved more complicated applications with growing efficiency over time. More recent studies have shown excellent performance, and outstanding results of deep learning-based model, such as image classification [22, 23], and semantic segmentation [24, 25].

The following sections will give a brief overview of DL approach.

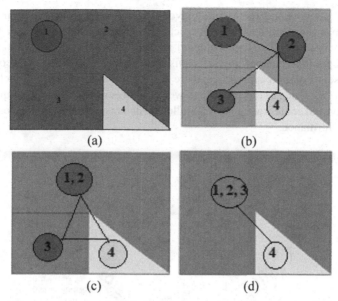

(a)   (b)

(c)   (d)

Fig. 4. Hierarchical merging algorithm including two steps: (a) over segmented area containing four regions (1), (2), (3), (4); (b) Associated RAG graph; (c) the first step of merging method (1, 2), (3), (4); (d) the second step of merging method (1, 2, 3), (4).

3 Deep Learning for CC Segmentation

In this section, we briefly describe Deep Neural Networks (DNN), and recent improvements and breakthroughs of them. The major advantage of DL is that in contrast with traditional machine learning the computer learns itself to recognize patterns. The performance of many Deep Leaning models is very promising. In this paper, we also pointed out the benefice of major advances in DL, this is the second contribution.

3.1 Convolutional Neural Networks

The DL approach uses CNN models, which is an extended standard Neural Network (NN) with several hidden layers, the typical structure of which is shown in Fig. 5. The CNN architecture typically consists of three kinds of layers, as convolution, pooling, and fully connected layers. Convolutional layers are composed of filters and image maps, and purpose to extract features from the input image, and pooling layers aggregate similar features into one [26]. One of the greatest successful state-of-the-art DL methods is built on the Fully Convolutional Networks (FCN) [27]. Another variant of FCN was also proposed which is known as U-Net neural network was initially introduced by Ronneberger et al. [28] for better segmentation on biomedical images.

3.2 Overview of the U-Net Architecture

The main idea behind U-Net framework is built on a contraction and an expansion path (left side and right side) as illustrated in Fig. 5. The contracting path is a traditional convolutional network and max-pooling layers that allows getting context information, while the expanding path leads to constructs the segmentation map using transposed convolutions. Therefore, U-Net considered as an FCN with only convolutional layers which it accepts the image of any size, this is a major advantage when compared to patch-based segmentation approaches, there are several advantages of U-Net model for segmentation tasks such as this model can works with very few training samples and the global location and context are done in parallel, which can provide quicker segmentation and better performance.

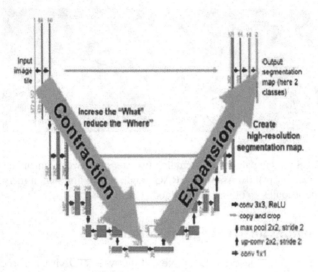

Fig. 5. U-Net architecture

In the next section, we presented the various datasets employed, the architecture of our framework, and the outcome of the proposed methods.

4 Experimental Results

To evaluated the proposed CC segmentation scheme, we used the available Open Access Series of Imaging Studies (OASIS) [29] at www.oasis-brain.org and Autism Brain Imaging Data Exchange (ABIDE). which is consisted of two wide collections that are ABIDE-I and ABIDE-II. Which gives structural and functional MRI datasets gathered from various MRI centers worldwide [30].

Building useful and beneficial deep learning models requires sufficient training data. However, large training sets are often unavailable, collecting new data can therefore be expensive, particularly in the medical imaging domain. One way to address this challenge is to use efficient techniques such as data augmentation, which can consist of simple transformations such as rotations, distortions, random cropping, zooms, color space increases, horizontal and vertical flips, etc. For other types of data, simple mathematical functions can be applied which resulted in a core sample set of 9600 sagittal images with their masks.

We used an Intel Core i7-6500 (2.60 GHz) with an NVIDIA graphics card to perform the gSLIC segmentation since it is faster than the sequential implementation, the method performance reaches 89%. the outcome of SLIC parallel implementation for CC segmentation is presented in Fig. 6.

The Deep Learning solved more and more complex applications with growing efficiency over time in our case to still show its success with an accuracy of 94%. The Fig. 7 illustrated the CC segmentation using DL approach.

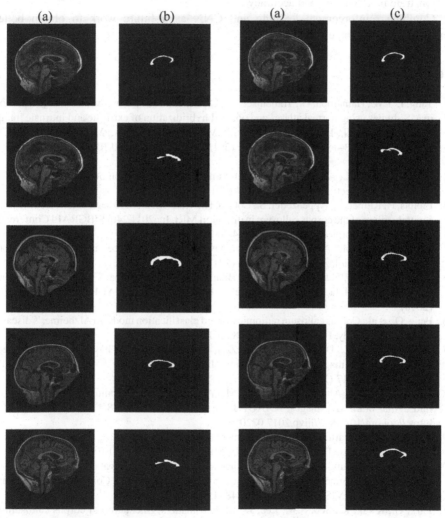

Fig. 6. The gSLIC implementation results of CC segmentation: *(a) Original image, (b) CC segmentation*

Fig. 7. The DL results of Corpus Callosum segmentation: *(a) Original image, (c) CC segmentation*

5   Conclusion

The Alzheimer's disease is progressive loss is memory of the patient, it is recommended to diagnose it earlier. The most affected part of the brain is the Hippocampus. The CC segmentation is the main process for medical image analysis, but it has major challenges due to the low contrast of tissues and supporting organs. In this research, we've used the parallel SLIC method for faster results compared to sequential approach, the deep learning succeeds in many fields including health and medicine that is why we decided to demonstrate its efficiency and accuracy.

Combine reinforcement learning and CNN as a future work to obtain better performing CC region extraction.

References

1. Tang, L.Y.W., Hamarneh, G., Traboulsee, A., Li, D., Tam, R.: Corpus callosum segmentation in MS studies using normal atlases and optimal hybridization of extrinsic and intrinsic image cues. In: Navab, N., Hornegger, J., Wells, W.M., Frangi, A.F. (eds.) MICCAI 2015. LNCS, vol. 9351, pp. 123–131. Springer, Cham (2015). https://doi.org/10.1007/978-3-319-24574-4_15
2. Patel, J., Doshi, K.: A study of segmentation methods for detection of tumor in brain MRI. Adv. Electron. Electr. Eng. 4(3), 279–284 (2014)
3. Freitas, P., Rittner, L., Appenzeller, S., Lotufo, R.: Watershed-based segmentation of the mid-sagittal section of the corpus callosum in diffusion MRI. In: 2011 24th SIBGRAPI Conference on Graphics, Patterns and Images, pp. 274–280 (2011)
4. Bennett, I.J., et al.: Age-related white matter integrity differences in oldest-old without dementia. Neurobiol. Aging 56, 108–114 (2017)
5. Taifi, K., Taifi, N., Safi, S., Malaoui, A., Bita, H.: Segmentation par Croissance de Région: Application à la Maladie d'Alzheimer. Rev. L'ENTREPRENEURIAT L'INNOVATION 2(6) (2018)
6. Feng, Q., et al.: Corpus callosum radiomics-based classification model in Alzheimer's disease: a case-control study. Front. Neurol. 9, 618 (2018)
7. Jiang, X., Zhou, Z., Ding, X., Deng, X., Zou, L., Li, B.: Level set based hippocampus segmentation in MR images with improved initialization using region growing. Comput. Math. Methods Med. 2017 (2017)
8. Li, Y., Wang, H., Ahmed, N., Mandal, M.: Automated corpus callosum segmentation in midsagittal brain MR images. ICTACT J. Image Video Process. 8(1), 1554–1565 (2017). https://doi.org/10.21917/ijivp.2017.0220
9. Rabeh, A.B., Benzarti, F., Amiri, H.: Segmentation of brain MRI using active contour model. Int. J. Imaging Syst. Technol. 27(1), 3–11 (2017). https://doi.org/10.1002/ima.22205
10. Ettaïeb, S., Hamrouni, K., Ruan, S.: Statistical models of shape and spatial relation-application to hippocampus segmentation. In: 2014 International Conference on Computer Vision Theory and Applications (VISAPP), vol. 1, pp. 448–455 (2014)
11. Park, G., Kwak, K., Seo, S.W., Lee, J.-M.: Automatic segmentation of corpus callosum in midsagittal based on Bayesian inference consisting of sparse representation error and multi-atlas voting. Front. Neurosci. 12 (2018). https://doi.org/10.3389/fnins.2018.00629
12. Nazem-Zadeh, M.-R., et al.: Segmentation of corpus callosum using diffusion tensor imaging: validation in patients with glioblastoma. BMC Med. Imaging 12(1), 1–16 (2012)
13. Liu, W., Ruan, D.: A unified variational segmentation framework with a level-set based sparse composite shape prior. Phys. Med. Biol. 60(5), 1865 (2015)

14. Rittner, L., Campbell, J.S., Freitas, P.F., Appenzeller, S., Pike, G.B., Lotufo, R.A.: Analysis of scalar maps for the segmentation of the corpus callosum in diffusion tensor fields. J. Math. Imaging Vis. **45**(3), 214–226 (2013)
15. Ju, Z., Zhou, J., Wang, X., Shu, Q.: Image segmentation based on adaptive threshold edge detection and mean shift. In: 2013 IEEE 4th International Conference on Software Engineering and Service Science, pp. 385–388 (2013)
16. Felzenszwalb, P.F., Huttenlocher, D.P.: Efficient graph-based image segmentation. Int. J. Comput. Vis. **59**(2), 167–181 (2004)
17. Zhang, J., Zhou, S.K., Comaniciu, D., McMillan, L.: Conditional density learning via regression with application to deformable shape segmentation. In: 2008 IEEE Conference on Computer Vision and Pattern Recognition, pp. 1–8 (2008)
18. Elnakib, A., Casanova, M.F., Gimel'farb, G., Switala, A.E., El-Baz, A.: Autism diagnostics by centerline-based shape analysis of the corpus callosum. In: 2011 IEEE International Symposium on Biomedical Imaging: From Nano to Macro, pp. 1843–1846 (2011)
19. Gass, T., Szekely, G., Goksel, O.: Simultaneous segmentation and multiresolution nonrigid atlas registration. IEEE Trans. Image Process. **23**(7), 2931–2943 (2014)
20. Hunt, B.R.: Super-resolution of images: Algorithms, principles, performance. Int. J. Imaging Syst. Technol. **6**(4), 297–304 (1995)
21. Ren, C.Y., Reid, I.: gSLIC: a real-time implementation of SLIC superpixel segmentation. University Oxford, Department of Engineering. Technical report, pp. 1–6 (2011)
22. Perez, L., Wang, J.: The effectiveness of data augmentation in image classification using deep learning. arXiv Preprint arXiv:1712.04621 (2017)
23. Li, S., Song, W., Fang, L., Chen, Y., Ghamisi, P., Benediktsson, J.A.: Deep learning for hyperspectral image classification: an overview. IEEE Trans. Geosci. Remote Sens. **57**(9), 6690–6709 (2019)
24. Ghosh, S., Das, N., Das, I., Maulik, U.: Understanding deep learning techniques for image segmentation. ACM Comput. Surv. **52**(4), 73:1–73:35 (2019). https://doi.org/10.1145/332 9784
25. Esteva, A., et al.: Deep learning-enabled medical computer vision. Npj Digit. Med. **4**(1), 1–9 (2021)
26. LeCun, Y., Bengio, Y., Hinton, G.: Deep learning. Nature **521**(7553), 436–444 (2015)
27. Long, J., Shelhamer, E., Darrell, T.: Fully convolutional networks for semantic segmentation. In: Proceedings of the IEEE Conference on Computer Vision and Pattern Recognition, pp. 3431–3440 (2015)
28. Ronneberger, O., Fischer, P., Brox, T.: U-net: convolutional networks for biomedical image segmentation. In: Navab, N., Hornegger, J., Wells, W.M., Frangi, A.F. (eds.) MICCAI 2015. LNCS, vol. 9351, pp. 234–241. Springer, Cham (2015). https://doi.org/10.1007/978-3-319-24574-4_28
29. Marcus, D.S., Wang, T.H., Parker, J., Csernansky, J.G., Morris, J.C., Buckner, R.L.: Open access series of imaging studies (OASIS): cross-sectional MRI data in young, middle aged, nondemented, and demented older adults. J. Cogn. Neurosci. **19**(9), 1498–1507 (2007)
30. Craddock, C., et al.: The neuro bureau preprocessing initiative: open sharing of preprocessed neuroimaging data and derivatives. Neuroinformatics **4** (2013)

Networking, Cloud Computing and Networking Architectures in Cloud (Full Papers)

Optimal Virtual Machine Provisioning in Cloud Computing Using Game Theory

Abdelkarim Ait Temghart[1]([⊠]), Driss Ait Omar[1], and Mbarek Marwan[2]

[1] TIAD Laboratory, FST, Sultan Moulay Slimane University, Beni-Mellal, Morocco
[2] Smart Systems Laboratory, ENSIAS, Mohammed V, Rabat, Morocco

Abstract. Virtualization of resources on the cloud computing allow fast and easy scaling of infrastructure to achieve a specified QoS. This will undoubtedly improve the allocation efficiency of physical machines in large-scale data centers. In fact, virtualization technologies not only have a direct effect on the performance and energy consumption but also can reduce the operating costs considerably. As there is fierce competition among cloud providers, there is an urgent need to adopt an elastic resource management system and an optimal pricing strategy as well. Such a strategy is designed principally to satisfy the requirements of a given service and simultaneously increase its associated profit margins. This study highlights why game theory is a very useful tool to analyze the impact of both energy efficiency and response time on virtual machine (VM) provisioning. To this aim, we present a formal policy that ensures fairness in the resource sharing and also takes into account the reasonable expectations market participants. More precisely, we rely on a non-cooperative game model in choosing the right provisioning scheme, and then discuss its equilibrium and stability. In a duopoly market, we conduct some numerical experiments to illustrate the efficiency and performance of the proposed approach and its proof of convergence with a certain number of iterations.

Keywords: Game theory · Noncooperative game · Cloud computing · System performance · Energy saving · Healthy competition

1 Introduction

As it supports flexible usage-based billing models, cloud computing is a new attractive alternative to on-premise solutions in terms of efficiency and price [1]. Formally, cloud stack, including platform-as-a-service (PaaS), infrastructure-as-a-service (IaaS) and software-as-a-service (SaaS), are built on top of one another to meet customers' needs. As one would expect, the demand for online services grew quickly and steadily, proving the solid interest for outsourced IT services. In response to the high demand, it is recommended to adopt a strategy that promotes healthy competition among available virtual machine (VM) for the benefit of both consumers and providers [2]. Technically, game theory models are used to analyze the major features of a market structure so as to select the appropriate policy, particularly in terms of price and service quality (QoS) [3, 4]. Undoubtedly, this would raise new questions about the best way to accurately

© Springer Nature Switzerland AG 2021
M. Fakir et al. (Eds.): CBI 2021, LNBIP 416, pp. 307–321, 2021.
https://doi.org/10.1007/978-3-030-76508-8_22

capture many aspects linked to the interactions between cloud providers and clients [5]. In strategic interaction situations, it is particularly important to remove barriers that prevent new firms from entering into cloud market [6]. In this case, the Third Party Auditor (TPA) may take any measures deemed necessary to avoid the fierce price competition and establish a long-term and win-win business relationship between providers and clients.

This study aims to analyze factors that can influence resource allocation decisions in cloud computing, especially in the context of virtual machine (VM). More specifically, we investigate the impact of performance and energy consumption on the profit margin ratio, which in turn will result in better and more revenue for cloud providers. In light of this fact, we need to determine the optimal strategy based on the energy saving and performance for supporting on-demand provisioning of VM and healthy competition among cloud providers as well.

The remainder of the paper is organized as the following: Sect. 2 provides the related works concerning VM allocation issue in cloud computing. Section 3 and Sect. 4 present the proposed game-based approach to f analyze the impact of energy saving and performance in reducing operating costs and improving profit margins. Section 5 provides the results of simulation and statistical analysis. And finally, Sect. 7 provides conclusion and suggests on economic performance and competitiveness of resource allocation in cloud computing through game theory.

2 Related Work

In most cases, various cloud providers are participating in a market for creating and maintaining a competitive environment that supplies online services. This section provides some works that analyze this competition based on game theory models, especially those that deal with performance and energy. Ye and Chen rely on noncooperative games models to improve the load balancing and virtual machine placement [7]. They prove the existence of an optimal solution for resources allocation strategy. Cardosa et al. in [8] propose an approach to improve energy consumption and performance in Hadoop MapReduce framework. To this aim, they suggest a spatiotemporal tradeoff method to scale MapReduce clusters dynamically. In [9], the authors rely on game theory to reduce resource fragmentation during VM allocation in cloud computing. The proposal is designed to ensure fairness among users as well as minimizing resource wastage in cloud computing. In [10], an effective load-balancing mechanism in used to improve performance in distributed systems. The proposed game-theoretical model ensures fairness and also provides a good response time to all end-users. Authors in [11] use game-theoretic model to dynamically control the computation offloading for mobile devices. Concretely, they seek to optimize the overall energy consumption in mobile cloud computing. In [12], authors rely on game theory to develop an energy efficient job scheduling method for mobile cloud computing environment. The main objective is to reduce energy cost and enhance computational performance during the transfer of resource intensive computational tasks to cloud computing.

Over past decades, various work that have been carried out for an efficient trade-off between the system performance and energy consumption of cloud computing. Many attempts have been made to do this, but there are still improvements to be made in order to satisfy the evolving needs of cloud services. In this respect, we will propose a game theatrical model to effectively address the need to ensure the maximum performance and energy saving during VM provisioning.

3 Resource Allocation in Cloud Computing

Cloud computing is typically designed to balance the demand and supply of IT resources within any organization. It usually involves a pool of shared resources that can be allocated and released based on user's request. Thus, resource allocation is a new challenge that requires prompt action to be taken. To best support their clients', cloud providers are continuously primarily engaged in guaranteeing a low VM provisioning time with the providing the resources needed by each job.

This study deals with a mathematical model of a cloud cluster composed of T physical servers. In this case, each server has $'I$ kinds of resources with different capacities. Hence, the capacity vector Ψ of each physical server is expressed as follows:

$$\Psi^{\varphi} = \Psi_1^{\varphi}, \Psi_2^{\varphi}, ..., \Psi_1^{\varphi}, \text{ where } 1 \leq \varphi \leq T \tag{1}$$

The goal of game theory in resource estimating is to assign the required VM to each user's request. To meet the users' job requirements with available VM, cloud provider relies on $'I$ kinds of resources to create a variety of VM types. We assume that the cloud will offer different VMs, which represented as vector V, when a user i submit allocation request for the job J_i.

$$V_i = V_{i1}, V_{i2}, ..., V_{ij}, V_{i1} \tag{2}$$

Obviously, the performance is improved by allocating more VMs to a specific job. More importantly, this strategy would significantly reduce the number of task migrations between the VMs and provisioning operations that can occur on each cloud data center. Despite its many advantages, this usually implies increased number of deployed VMs. Moreover, it has a negative effect on operating expenses and profit margin as well. In light of this fact, the cloud provider should ensure optimum utilization of available resources and the number of VMs assigned to each job. More specifically, it is necessary to consider the required resources, such as CPU, memory, storage, to be mapped to each VM to meet job's requirements.

In addition to handling the increased traffic loads, we need to choose the resources required for VM provisioning that maintain a balance between keeping costs down and satisfying the service-level agreement (SLA).

Under these circumstances, the proposed framework for VM provisioning is composed of five main components, Hypervisor, as shown in Fig. 1.

Functionally, each module has a specific role to support dynamic VM management.

Virtual Machine Manager (VMM). This parameter refers to the ability to create or release a VM in cloud computing.

Server Manager Console (SMC). This parameter refers to the possibility of adding or removing the storage devices in physical server.

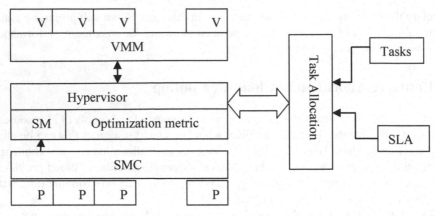

Fig. 1. VM management architecture in the cloud computing

Server Monitoring (SM). This tool provides the ability to check the state of physical server in terms of monitoring memory and storage usage.

Hypervisor. It is the core element of this architecture, used mainly to perform VM provisioning by using two parameters (1) optimization metric and (2) SM module.

Task Allocation. This module allows associating each job with the required SLA level to perform this job.

For an effective performance management, an integrated approach involving both resources allocation and decision-making strategies are of utmost importance for effective VM provisioning and migration techniques.

Definition (VM provisioning). To meet the job requirements of N cloud users, provisioning $'I$ type of VMs with different capacity can formally represented by as a matrix $Nx'I$.

$$VM = \begin{pmatrix} V_i \\ V_2 \\ \vdots \\ V_N \end{pmatrix} = \begin{pmatrix} V_{i1} & \cdots & V_{i1} \\ \vdots & \ddots & \vdots \\ r_{N1} & \cdots & r_{N1} \end{pmatrix} \qquad (3)$$

Another key concept of resources allocation is the ability to predict the resource requirement matrix and the capacity sets of all physical servers. Accordingly, cloud providers can determine a reasonable mapping scheme to assign virtual resources to each activity and job. This goal will be achieved when different kinds of physical resources are effectively distributed among all cloud users so as to create their required VMs. In this respect, both the efficiency and fairness of resource allocation policy have a considerable influence on the global cloud computing market.

Given the fierce competition in the cloud market, it is crucial for a cloud provider to find an efficient allocation decision during VM provisioning. More precisely, this would help cloud provider determine the best way for mapping VMs to physical servers based on the actual needs of each job and user's requests.

4   Game Theoretic Virtual Machine Provisioning

To minimize the number of active physical machines (AMPs), virtualized data centers are serving the ever-growing demand for IT resources. Besides, dynamic virtual machine provisioning in cloud computing is an efficient approach to reduce the resource fragmentation when mapping virtual machines to physical server [13]. In virtualized environments, there is fierce competition among the various existing consolidated virtual machines (VMs). This competition among available VMs can be seen as a non-cooperative game where each VM possesses individual characteristics like storage capacity, security, costs, energy consumption, etc. Notice that a VM's utility depends not only on its characteristics but also on the characteristics of other VMs of the same provider. On one hand, service providers can increase their revenue by using VMs with lower-cost. On the other hand, this strategy has serious negative impact on energy efficiency and the level of service quality. To this aim, the problem of optimal VM provisioning under fairness constraints can be formulated as the state in which no VM is allocated much better than others. In this scenario, fair allocation aims to equalize the resource allocation of total availability VMs to meet SLA requirements and reducing energy consumption of physical servers [14].

Motivated by these goals, the VMs allocation is modeled as a non cooperative game model that takes into account both response time and energy consumption. Synthesis of parameters required for this study is summarized in Table 1.

Table 1. Mathematical notation

Notation	Description
μ_i	Average service rate
λ_i	Arrival rate of each job to V_i
s_i	Total requests sent by the users i to V_i
Tw_i	Expected waiting time at V_i
Tex_i	Expected execution time at V_i
e_i	Power consumption of V_i
β_i	Factor of combined competitiveness
n_i	Number of hypervisor modules
δ_i	Efficiency of power usage

4.1   Performance Model

In general, SLA is considered a critical metric in evaluating performance and also in making decisions about what to expect from each V_i. It outlines the duties and responsibilities of the cloud provider and end-user during VM provisioning. To ensure a minimum level of quality of service (QoS), we use several parameters to evaluate the QoS of each

V_i; among them response time, availability, throughput, security and fail-over. In this context, Table 2 shows the most metrics used in previous works for an effective and dynamic utilization of V_i in cloud computing.

Table 2. Metrics for VM's performance

Reference	Metric	Objective
[15, 16]	Execution time	Minimize the total completion time for delivering cloud services
[17, 18]	Waiting time	Reduce the average amount of time it takes to find the required VMs
[19–21]	Resource utilization	Maximize resource utilization over a cloud platform
[22]	Migration rate	Minimize the number of migration during the operation of VM provisioning
[23]	Throughput	Maximize the number of tasks that are allocated to a V_i

In case of large data centers, we assume that the hypervisor module selects only the right VMs to execute a specific user's jobs. Thus, we suppose that VM provisioning process is modelled as an M/M/1 queue. Based on this consideration, the flowchart of the VM provisioning in cloud can be summarized in Fig. 2.

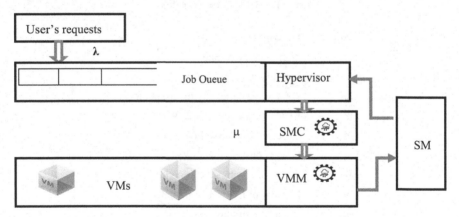

Fig. 2. Virtual machine provisioning modelled as M/M/1 Queue model

Let s_i denoted the sum of requests sent by users to the V_i, μ_i refers to the average service rate and λ_i the arrival service rate to hypervisor module. Then we define Tw_i as the expected waiting time and Tex_i as required time by a job for it's execution.

$$Tw_i = \frac{s_i}{\mu_i(\mu_i - \lambda_i)} \tag{4}$$

$$Tex_i = \frac{\lambda_i}{\mu_i} \tag{5}$$

Let T_i be denoted the finishing time include waiting and execution time. The finishing time can be formulated as following:

$$T_i = Tw_i + Tex_i \tag{6}$$

Hence, the performance q_i of a specific VM_i is expressed as follows:

$$q_i = Tmax_i - T_i \tag{7}$$

Where $Tmax_i$ represent the time response maximum that can a job request attends.

As illustrated in Fig. 3, an increase in the arrival service rate λ_i typically leads to a higher response time for cloud computing providers during VM provisioning.

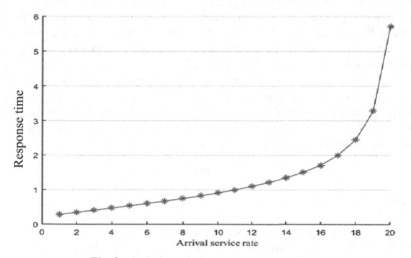

Fig. 3. Arrival service rate w.r.t. response time

Based on the response time, hypervisor module decides the allocation of new VMs for next incoming requests to support VM performance. However, this strategy would lead to the low efficiency of energy consumption for cloud data center in the long-run. The future of data centers will rely also on the energy-efficiency techniques used in the cloud computing environment.

4.2   Energy Efficiency Model

In most cases, each cloud provider relies on various large data center to manage clients' data in terms of storage and processing. Obviously, maintaining these services in optimum operation has a negative effect on operating costs. Despite significant growth in cloud services demand, the energy-efficiency in cloud computing can be successfully applied to reduce the amount of costs required to provide remote services. Here are some of the most common metrics to evaluate the energy- efficiency in cloud computing, as shown Table 3.

Table 3. Metrics for energy consumption.

Reference	Metric	Objective
[19, 24–26]	Resource utilization	Maximize the amount of resource used during services execution
[27, 28]	Energy overhead	Use the exact amount of the needed VM
[29]	Number of migration	Minimize the required migration operations
[22, 24, 30, 31]	Energy consumption	Reduce the amount of energy required in a cloud data center

For simplicity purposes, we focus only on the energy consumed by servers while ignoring the energy used by the other facilities such as cooling devices, fire-detection equipment, fire detection, alarm system and access control system. Accordingly, the proposed model aims to gives an accurate approximation of the power consumption for various types of servers in a cloud data center.

The energy model is conceptualized on the basis that energy consumption has a linear relationship with processor utilization [32, 33].This means, for a particular user's job, the processing time of a job and the processor utilization rate are the two parameters that determines the energy consumption for each job. In the same line, the expect power consumption by a server can be formulated as shown in Eq. (8):

$$P_i = \left(P_i^{peak} - P_i^{idle} \right) U_i + P_i^{idle} \tag{8}$$

Where

P^{peak} is the power when server's capacity is completely utilized.
P^{idle} is the amount of power consumption of the server in the idle state.
U_i is the fraction of CPU capacity being used.

In general, the total energy consumption of the data center can essentially be attributed to the amount of VM used to process clients' requests. In this case study, we aim to determine the energy consumed by a data center that consists of η_i different VMs. According to a recent study [34], we estimate the expected energy e_i consumed by any V_i in cloud data.

$$e_i = \frac{\left(P_i^{peak} - P_i^{idle} \right) \delta_i Y_i}{\eta_i} \tag{9}$$

Where δ_i is the power usage efficiency (PUE) $= \frac{Power\ used\ by\ data\ center\ facility}{Power\ consumption\ of\ servers}$.
Y_i is the number of active servers in the data center.

5   Utility of Non-cooperative VM Provisioning

In a data center, the hypervisor module selects a virtual machine that maximizes its revenue and minimizes the consumption of energy. Formally, we modelled the utility of

VM provisioning in cloud computing as a multinomial logit (MNL) model in the process of determining the appropriate V_i for each client's request. Note that cloud provider has a revenue policy that mainly depends on the performance and energy consumption of each V_i. In light of this fact, we can compute the expectation of the outcome U_i associated with cloud investment in V_i is follows:

$$U_i = \beta_i\, q_i - e_i \tag{10}$$

Where β_i is the factor of combined competitiveness that reflects the degree to which a cloud provider chooses a specific V_i with performance q_i.

Hence, the utility function measures the preferences of cloud provider to a given VM. This parameter determines the optimal payoff related to each V_i based on the performance level and energy consumption. Thus, the expected profit π_i of VM provisioning for each V_i is calculated using the formula (11).

$$\pi_i(e_i, q_i) = \frac{n e^{U_i}}{1 + \sum_j^N e^{U_j}} (\alpha T_i e_i \varphi - (1 - \alpha) q_i) \tag{11}$$

Where

n is number of hypervisor module in a data center ($n = 1$ in this case study).
e_i is the expected consumption of energy calculated using (9).
φ is the cost per unit of energy consumption.
α is weight factor, $\alpha \in [0, 1]$.

6 Game Formulation

In this paper, we consider a non-cooperative game model that involves two essential parameters, i.e. energy consumption and performance. Accordingly, the proposed model is denoted as:

$$G = (\mathbb{I}, \{q_i, e_i\}, \pi_i) \tag{12}$$

Note that $\mathbb{I} = (1, \ldots, \mathbb{I})$ identifies different available virtual machines indexed by i. For simplicity purposes, we assume that there are only two VMs in the data center. Moreover, each V_i chooses a level of performance q_i ($q_i^{min} < q_i < q_i^{max}$). As outlined above, e_i represents the level of power consumption in a data center. In general, e_i should lie within the interval of feasible power consumption of each V_i, which is represented by $e_i^{min} < e_i < e_i^{max}$.

Consequently, to determine the optimal VM provisioning decision involves looking at strategies that improves the expected profit that derives from the utility function, denoted by π_i. Moreover, VM provisioning policy should be adopted to ensure fair and equitable competition among VMs available in a cloud cluster. Under given constraints, the utility function can be expressed as:

$$\pi_i(q^*, e^*) = max_{q_i \in Q_i, e_i \in E_i} \pi(q_i, e_i) \tag{13}$$

Theorem (energy game). In cloud computing, Nash equilibrium for energy based game exists and is unique in VM provisioning.

Proof. Note that $k = T_i\varphi$ and $\theta = \frac{e^{U_i}}{1+\sum_j^N e^{U_j}}$.

Firstly, The condition to prove the existence of Nash equilibrium is: $\frac{\partial^2 \pi_i}{\partial q_i^2} \leq 0$.
After calculus, we have:

$$\frac{\partial^2 \pi_i}{\partial e_i^2} = -n\theta(1-\theta)(\alpha ke - (1-\alpha)q_i)(2\theta - 1) - 2n\theta(1-\theta)\alpha k \qquad (14)$$

if $\frac{\partial^2 \pi_i}{\partial q_i^2} \leq 0$ then $e_i \geq (1-\alpha)q_i - \frac{2}{(2\theta-1)}$ which is the condition to be satisfied for the game to admits Nash equilibrium.

Secondly, Nash equilibrium is unique if: $-\frac{\partial^2 \pi_i}{\partial e^2} - \sum_{j,j\neq i} \frac{\partial^2 \pi_i}{\partial e_i \partial e_j} \geq 0$.
We have:

$$-\frac{\partial^2 \pi_i}{\partial e^2} - \sum_{j,j\neq i} \frac{\partial^2 \pi_i}{\partial e_i \partial e_j} = n\theta(1-\theta)\alpha k \qquad (15)$$

While, $n\theta(1-\theta)\alpha k \geq 0$ Then we conclude that Nash equilibrium is unique.

Theorem (performance game). In cloud computing, Nash equilibrium for performance based game is exists and unique in VM provisioning.

Proof. Note that $k = T_i\varphi$ and $\theta = \frac{e^{U_i}}{1+\sum_j^N e^{U_j}}$.

The 1st derivative of function (11) is follows:

$$\frac{\partial \pi_i}{\partial q_i} = n\beta_i\theta(1-\theta)(\alpha ke - (1-\alpha)q_i) - n\theta(1-\alpha) \qquad (16)$$

Hence, the 2nd derivative of utility function is:

$$\frac{\partial^2 \pi_i}{\partial q_i^2} = n\beta_i^2\theta(1-\theta)(\alpha ke - (1-\alpha)q_i)(1-2\theta) - 2n\beta_i\theta(1-\alpha)(1-\theta) \qquad (17)$$

Then the performance game admits Nash equilibrium $\frac{\partial^2 \pi_i}{\partial q_i^2} \leq 0$, if the condition follow is satisfied: $q_i \geq \frac{\alpha ke}{1-\alpha} - \frac{2}{\beta_i(1-2\theta)}$.

Secondly, to prove the uniqueness of Nash equilibrium, we have:

$$\frac{\partial \pi_i}{\partial q_i \partial q_j} = -n\beta_i^2\theta(1-\theta)(1-2\theta)(\alpha ke - (1-\alpha)q_i) + n\beta_i\theta(1-\theta)(1-\alpha) \qquad (18)$$

Then

$$-\frac{\partial^2 \pi_i}{\partial e^2} - \sum_{j,j\neq i} \frac{\partial^2 \pi_i}{\partial e_i \partial e_j} = n\theta(1-\theta)(1-\alpha) \qquad (19)$$

While, $n\theta(1-\theta)(1-\alpha) \geq 0$ Then we conclude that Nash equilibrium is unique.
Note that the proof of the above mentioned theorems are presented in [3].

Theorem (energy-performance game). In cloud computing, there exists a Nash equilibrium $\pi_i(q_i, s_i)$ in VM provisioning, which satisfies the following system of equations:

$$
\begin{cases}
\frac{\partial \pi_i(q_i, e_i,)}{\partial q_i} = 0 \\
\frac{\partial \pi_i(q_i, e_i)}{\partial e_i} = 0
\end{cases}
\tag{20}
$$

6.1 Numerical Simulation and Results

The following section aims at providing a test case demonstrating the ability of the proposed game-theoretical model to enhance resource allocation in cloud computing. In particular, we consider a data center with only two virtual machines to accurately simulate the situation of duopoly market. Based on this consideration, the cloud providers rely on hypervisor console to select the appropriate VMs that maximize their payoff, while simultaneously promoting fair competition among VMs. As outlined above, the selection is based mainly on two parameters, i.e. performance and energy saving. To this aim, we analyze the impact of these two parameters on the issue of VM provisioning in cloud computing using best-response algorithm.

Algorithm. Find best-response function

   1: Step 1
       Each cloud provider chooses the level of performance and energy consumption for each VM.
   2: Step 2
       For each V_i /iϵ I at iteration t;
$$BRq_i^{t+1} = \text{argmax}_{q_i \epsilon Q} (\pi_i (q_i, e_i))$$
$$BRe_i^{t+1} = \text{argmax}_{e_i \epsilon E} (\pi_i (q_i, e_i))$$

As illustrated in Fig. 4 and Fig. 5 below, the competition among virtual machines in a cloud data center converges to Nash equilibrium. For the performance based competition, the chosen strategy converges to unique Nash equilibrium that guarantees the optimal outcome of available VM types in cloud data center, as shown in Fig. 4.

To achieve energy fairness, the proposed strategy converge to an equilibrium point that maximizes the benefit gain by VM provisioning in cloud computing, as shown in Fig. 5. Hence, the proposed game theory model can help cloud provider reach optimal decision-making when allocating and deallocating virtual machines.

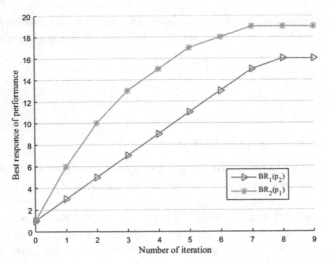

Fig. 4. Convergence to NE in performance based game

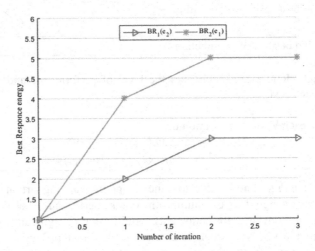

Fig. 5. Convergence to NE in energy consumption based game

7 Conclusion

The central idea behind the concept of virtual machine provisioning is to accurately determine the required VM to process each client's requests. Despite its great advantages, VM provisioning is becoming a major issue for a large-scale cloud environment. The cloud providers should ensure that there is a good balance between delivering high-quality services and remaining competition in the market. Under this perspective, we propose a non-cooperative game-theoretic approach to maximize the payoff (profits) of available VM in a data center and promote fair competition among them as well. More specifically, we use both the response time (RT) and the power consumption (PC) for each

VM to choose the optimal mapping strategy. Such a measure would undoubtedly help cloud providers generate more profitability by reducing operating costs and increasing their revenue. This study provides an efficient strategy for maintaining and enhancing the trade-off decisions encountered under the VM provisioning in cloud computing. The simulation results prove the efficiency and performance of the proposed solution in terms of convergence iterations. In the future work, we plan to investigate how to extend this game theoretical model to enhance allocation in data center with several virtual machines. Another direction involves considering the VM provisioning under the job priority case.

References

1. Buyya, R., Yeo, C.S., Venugopal, S., Broberg, J., Brandic, I.: Cloud computing and emerging IT platforms: vision, hype, and reality for delivering computing as the 5th utility. Future Gener. Comput. Syst. **25**, 599–616 (2009)
2. Hassan, M., Hossain, M., Sarkar, A., Huh, E.-N.: Cooperative game-based distributed resource allocation in horizontal dynamic cloud federation platform. Inf. Syst. Front. **16**(4), 523–542 (2012). https://doi.org/10.1007/s10796-012-9357-x
3. Ait Omar, D., Garmani, H., El Amrani, M., Baslam, M., Fakir, M.: A customer confusion environment in telecommunication networks: analysis and policy impact. Int. J. Coop. Info. Syst. **28**, 1930002 (2019)
4. Baslam, M., El-Azouzi, R., Sabir, E., Bouyakhf, E.-H.: New insights from a bounded rationality analysis for strategic Price-QoS war. In: Proceedings of the 6th International Conference on Performance Evaluation Methodologies and Tools, Cargèse, France. IEEE (2012)
5. Anglano, C., Canonico, M., Castagno, P., Guazzone, M., Sereno, M.: Profit-aware coalition formation in fog computing providers: a game-theoretic approach. Concurr. Comput. Pract. Exper. e5220 (2019)
6. Daoud, A., Agarwal, S., Alpcan, T.: Brief announcement: cloud computing games: pricing services of large data centers. In: Keidar, I. (ed.) DISC 2009. LNCS, vol. 5805, pp. 309–310. Springer, Heidelberg (2009). https://doi.org/10.1007/978-3-642-04355-0_32
7. Ye, D., Chen, J.: Non-cooperative games on multidimensional resource allocation. Future Gener. Comput. Syst. **29**, 1345–1352 (2013)
8. Cardosa, M., Singh, A., Pucha, H., Chandra, A.: Exploiting spatio-temporal tradeoffs for energy-aware MapReduce in the cloud. In: 2011 IEEE 4th International Conference on Cloud Computing, Washington, DC, USA, pp. 251–258. IEEE (2011)
9. Xu, X., Yu, H.: A game theory approach to fair and efficient resource allocation in cloud computing. Math. Prob. Eng. **2014**, 1–14 (2014)
10. Kishor, A., Niyogi, R., Veeravalli, B.: A game-theoretic approach for cost-aware load balancing in distributed systems. Future Gener. Comput. Syst. **109**, 29–44 (2020)
11. Ge, Y., Zhang, Y., Qiu, Q., Lu, Y.-H.: A game theoretic resource allocation for overall energy minimization in mobile cloud computing system. In: Proceedings of the 2012 ACM/IEEE International Symposium on Low Power Electronics and Design - ISLPED 2012, Redondo Beach, California, USA, p. 279. ACM Press (2012)
12. Ahn, S., Lee, J., Park, S., Newaz, S.H.S., Choi, J.K.: Competitive partial computation offloading for maximizing energy efficiency in mobile cloud computing. IEEE Access **6**, 899–912 (2018)
13. Son, S., Jung, G., Jun, S.C.: An SLA-based cloud computing that facilitates resource allocation in the distributed data centers of a cloud provider. J. Supercomput. **64**, 606–637 (2013)

14. Ghodsi, A., Zaharia, M., Hindman, B., Konwinski, A., Shenker, S., Stoica, I.: Dominant resource fairness: fair allocation of multiple resource types. In: Proceedings of the 8th USENIX Conference on Networked Systems Design and Implementation, USA, pp. 323–336. USENIX Association (2011)

15. Wang, S., Urgaonkar, R., Zafer, M., He, T., Chan, K., Leung, K.K.: Dynamic service migration in mobile edge-clouds. In: 2015 IFIP Networking Conference (IFIP Networking), Toulouse, France, pp. 1–9. IEEE (2015)

16. Souza, V.B., et al.: Towards a proper service placement in combined Fog-to-Cloud (F2C) architectures. Future Gener. Comput. Syst. **87**, 1–15 (2018)

17. Rodrigues, T.G., Suto, K., Nishiyama, H., Kato, N.: Hybrid method for minimizing service delay in edge cloud computing through VM migration and transmission power control. IEEE Trans. Comput. **66**, 810–819 (2017)

18. Velasquez, K., Abreu, D., Curado, M., Monteiro, E.: Service placement for latency reduction in the internet of things. Ann. Telecommun. **72**(1–2), 105–115 (2016). https://doi.org/10.1007/s12243-016-0524-9

19. Dashti, S.E., Rahmani, A.M.: Dynamic VMs placement for energy efficiency by PSO in cloud computing. J. Exp. Theor. Artif. Intell. **28**, 97–112 (2016)

20. Liaqat, M., Ninoriya, S., Shuja, J., Ahmad, R.W., Gani, A.: Virtual Machine Migration Enabled Cloud Resource Management: A Challenging Task. arXiv:1601.03854 [cs] (2016)

21. Jaumard, B., Pouya, H.: Migration plan with minimum overall migration time or cost. J. Opt. Commun. Netw. **10**, 1 (2018)

22. Ahmed, A., Ibrahim, M.: Analysis of energy saving approaches in cloud computing using ant colony and first fit algorithms. IJACSA **8** (2017)

23. Svärd, P., Hudzia, B., Walsh, S., Tordsson, J., Elmroth, E.: Principles and performance characteristics of algorithms for live VM migration. SIGOPS Oper. Syst. Rev. **49**, 142–155 (2015)

24. Lucas-Simarro, J.L., Moreno-Vozmediano, R., Montero, R.S., Llorente, I.M.: Cost optimization of virtual infrastructures in dynamic multi-cloud scenarios: cost optimization of virtual infrastructures in dynamic multi-cloud scenarios. Concurr. Comput.: Pract. Exp. **27**, 2260–2277 (2015)

25. Ding, Y., Qin, X., Liu, L., Wang, T.: Energy efficient scheduling of virtual machines in cloud with deadline constraint. Future Gener. Comput. Syst. **50**, 62–74 (2015)

26. A-Shehri, H.A., Hamdi, K.: Multi-objective VM placement algorithms for green cloud data centers: an overview. In: 2018 21st Saudi Computer Society National Computer Conference (NCC), Riyadh, pp. 1–8. IEEE (2018)

27. Nathan, S., Bellur, U., Kulkarni, P.: Towards a comprehensive performance model of virtual machine live migration. In: Proceedings of the Sixth ACM Symposium on Cloud Computing, Kohala Coast Hawaii, pp. 288–301. ACM (2015)

28. Chen, Y.-H., Chen, C.-Y.: Service oriented cloud VM placement strategy for Internet of Things. IEEE Access **5**, 25396–25407 (2017)

29. Li, H., Zhu, G., Cui, C., Tang, H., Dou, Y., He, C.: Energy-efficient migration and consolidation algorithm of virtual machines in data centers for cloud computing. Computing **98**(3), 303–317 (2015). https://doi.org/10.1007/s00607-015-0467-4

30. Zhou, Z., et al.: Minimizing SLA violation and power consumption in Cloud data centers using adaptive energy-aware algorithms. Future Gener. Comput. Syst. **86**, 836–850 (2018)

31. Khosravi, A., Andrew, L.L.H., Buyya, R.: Dynamic VM placement method for minimizing energy and carbon cost in geographically distributed cloud data centers. IEEE Trans. Sustain. Comput. **2**, 183–196 (2017)

32. Hsu, C.-H., et al.: Energy-aware task consolidation technique for cloud computing. In: 2011 IEEE Third International Conference on Cloud Computing Technology and Science, Athens, Greece, pp. 115–121. IEEE (2011)

33. Bojanova, I., Samba, A.: Analysis of cloud computing delivery architecture models. In: 2011 IEEE Workshops of International Conference on Advanced Information Networking and Applications, Biopolis, Singapore, pp. 453–458. IEEE (2011)
34. Borgetto, D., Casanova, H., Da Costa, G., Pierson, J.-M.: Energy-aware service allocation. Future Gener. Comput. Syst. **28**, 769–779 (2012)

Comparative Study Between RFID Readers Anti-collision Protocols in Dense Environments

El hassania Rouan[1](✉), Said Safi[2](✉), and Ahmed Boumezzough[2](✉)

[1] Faculty of Sciences and Technics, University Sultan Moulay Slimane,
Beni-Mellal, Morocco
[2] Faculty Poly-disciplinary, University Sultan Moulay Slimane, Beni-Mellal, Morocco

Abstract. Radio Frequency IDentification (RFID) is a wireless communication technology, that suffers from a recurring issue and one of the most important challenges of RFID networks: Collision between readers that occurs when a several readers are placed very closely to each other, thus decreasing efficiency of RFID systems. In literature, many anti-collision protocols have been proposed to reduce reader collisions. In this paper we present a review of the best performing anti-collision protocols for RFID collision avoidance which are Geometry Distribution Reader Anti-collision (GDRA), Distance based RFID reader Collision Avoidance protocol (DRCA), Beacon Analysis based Collision Prevention (BACP), Distributed Efficient & Fair Anti-collision for RFID (DEFAR). This study is on the performance of these protocols in a dense RFID environment, advantages and disadvantages of each protocols in a static deployment where readers are immobile using MATLAB.

Keywords: Anticollision protocols · BACP · DRCA · DEFAR · Dense RFID environment · GDRA · Interference

1 Introduction

RFID is a non-contact automatic identification technology using wireless radio frequency signals. As shown in Fig. 1, an RFID system mainly consists of two components, which are: Reader(s) and tag(s).

An RFID reader identifies the tags affixed to an item, an animal, or a person within a given radio frequency range through radio waves without the need of human intervention or data entry for tracing purposes. The principle is to emit a radio frequency signal towards tags that in return use the energy issued from the transmitted reader signal to reflect their response [1]. Due to the low power of the backscattered signal, the response of a passive tag can be decoded only within a certain distance, which is called interrogation range.

RFID is a promising technology for ubiquitous computing. RFID has numerous applications in many fields due to its various benefits and features such

© Springer Nature Switzerland AG 2021
M. Fakir et al. (Eds.): CBI 2021, LNBIP 416, pp. 322–334, 2021.
https://doi.org/10.1007/978-3-030-76508-8_23

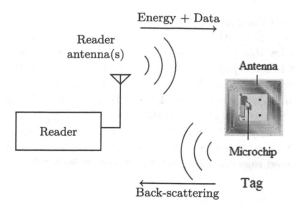

Fig. 1. The main components of an RFID system

as supply chain management, production and delivery of products, transportation and logistics management [2], object tracking [3], smart hospitals [4], smart parking systems [5].

1.1 Problem Statement

With the development of applications based on RFID, the deployment of numerous readers simultaneously in a small geographic area will be necessary to ensure a large coverage and retrieve more information. For example, in the case of a warehouse where every good is attached to a tag, it should be possible to track the different positions or movement of the goods inside the warehouse: entrance/exit of goods, checkpoints, delivery and many other operation flows [6]. In order to overcome the full coverage of the deployment area, the dense RFID systems are deployed. This kind of environment is called Dense Reader Environments. It leads to collisions between readers, which decrease the performance of the RFID systems. A reader collision is classified into two categories [1,6–12]:

Reader-to-Reader Collisions (RRC) occur when multiple readers in the interference range of each other and use simultaneously the same frequency. They will disrupt each reader's read operations. As shown in Fig. 2, $R1$ is within the interference range of another reader $R2$. Consequently, $R2$ interference range alters $R1$ interrogation range and $R1$ will not be able to read replies of $T1$. To avoid RRC effects, the distance between each other must be higher than d_{rr} which can reach until 1000 meters with output power equals $3, 2$ watts of effective isotropic radiated power (EIRP) or reader within the interference range of each others' readers has to operate at different frequencies and/or at different times [7]. In addition, this reader's output power limits the reader-to-tag read range d_{rt} to a maximum distance of $10\,$m.

Reader-to-Tag Collisions (RTC) occur when more than one readers concurrently interrogate a tag within their active interrogation range. Consequently, this tag will not be able to reply to any of those readers even if their frequencies are different because it cannot correctly listen to the interrogations sent by readers. As represented in Fig. 3, $T4$ is within the reading range of both readers, which makes it unable to decode all of these requests and therefore will not be able to respond. To avoid RTC, those readers should not operate simultaneously.

Readers affected by RTC are certainly also suffers from RRC due to $d_{rr} \gg 2 \times d_{rt}$. Both of these negatively affect system performance by reducing the number of identified tags.

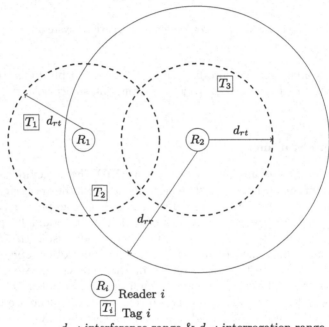

R_i Reader i

T_i Tag i

d_{rr}: interference range & d_{rt}: interrogation range

Fig. 2. Reader-to-Reader Collision

1.2   Categorization of Anticollision Protocols

The existing anti-collision protocols are broadly classified based on the methodologies used for scheduling of the resources. In order to coordinate the communication between readers, some schemes of communication are built. These scheme depending on how readers communicate [8]. Generally, there are two paradigms to follow:

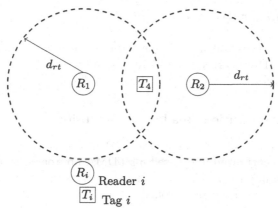

R_i Reader i

$\boxed{T_i}$ Tag i

d_{rr}: interference range & d_{rt}: interrogation range

Fig. 3. Reader to Tag Collision

Centralized Scheme: the readers communicate with a superior entity called central server through wired or a wireless link. The central server manages, allocates resources to readers and organizes reading rounds.

Distributed Scheme: the readers are not dependent on a central server for obtaining resources and they communicate with themselves directly in a peer-to-peer fashion and agree on their reading behaviors.

To resolve collisions problem, there many multiple access forms have been used in this anti-collision protocols which are:

Time Division Multiple Access (TDMA): the available time is divided into time slots. In TDMA scheme, more than one of readers share the same medium by allocating transmission of each reader into different time slots. Each reader can use one time slot to access the medium and read tags in their reading range.

Frequency Division Multiple Access (FDMA): the available frequencies is divided into several channels, each channel is dedicated to one reader. ETSI standards specify the use of four channels [9].

Carrier Sense Multiple Access (CSMA): in this mechanism, the readers listen to the channel to check of its inactivity beforehand with the aim of entering the competition mode and gain access to the channel if this latter was free [9].

In literature, many protocols have been proposed to solve Reader collisions. In this paper, we focus on the most known anti-collision protocols that provide higher performance for dense RFID environment. The remainder of this paper is organized as follows: Sect. 2 provides a literature review of the main anti-collision protocols. Section 3 identifies the main parameters that allow us to evaluate these protocols. In Sect. 4, a discussion is done regarding the different protocols with a comparative study. Conclusions are finally drawn in Sect. 5.

2   Anticollision Protocols

The most known protocols which are suggested to coordinate readers in a dense RFID environment (DRE) are discussed in this part, with emphasis on their requirements and problems.

2.1   Geometry Distribution Reader Anti-collision

Geometry Distribution Reader Anti-collision (GDRA) [7] is an anti-collision protocol multichannel, centralized and based on TDMA approach. It aims to minimize RRC by managing frequencies and time slots as available resources.

In this protocol, readers are controlled by a central server (CS) that indicate the beginning of each identification round by sending an AC (Arrangement Command) packet which contains number of slots.

In every identification round, readers randomly select one of the four frequencies recommended in the EPC and ETSI standards while slots are selected based on the Sift distribution function that favors the choice of a unique slot by a reader in contention. After choosing the time slot k, the reader listens to the channel during the $(k-1)^{th}$ slot. If the channel is busy, the reader will leave it else, the reader will wait until its slot arrives and send a beacon message to its neighbors. In this case, there are two possibilities:

- the reader will earn the contention if it is the only one who sends a beacon among its neighbors, then it will start the R-T communication phase in its own channel and keep the same frequency for the next identification round.
- Otherwise, the reader detects a collision, leaves the contention with a new frequency, and waits a new AC packet.

Although this protocol has a high performance, it is still suffering from a large number of disabled readers in DRE.

In GDRA protocol, each slot is picked using the sift function. The probability of selecting a slot among N_{slot} is defined in Eq. (1):

$$p_i(k) = \frac{(1-\alpha_i)\alpha_i^{N_{Slot}}}{1-\alpha_i^{N_{Slot}}}\alpha_i^{-k} \tag{1}$$

Where: N_{Slot} is the number of slots and $N_{Neighbor}$ is the maximum number of contenders.

$$1 \leq k \leq N_{Slot}, \quad 0 < \alpha_i < 1, \qquad \alpha_i = N_{Neighbor}(i)^{\frac{-1}{(N_{Slot}-1)}}$$

Note that when $N_{Neighbor}(i) = 1$ then $\alpha_i = 1$, the Eq. (1) corresponds to a uniform probability distribution function: $\lim_{\alpha_i \to 1} p_i(k) = \frac{1}{N_{Slot}}$.

2.2   Distance Based RFID Reader Collision Avoidance Protocol

Distance based RFID reader Collision Avoidance protocol(DRCA) [10] is a developed algorithm for dense reader environments based on GDRA. It attempts to enhance the performance of DRE by giving a chance to the readers that experience collision to access the channel. Unlike GDRA, readers listen to the channel in $(k - 1)^{th}$ slot to check of its inactivity beforehand. If the channel was busy, each reader would calculate its distance D to the readers those occupied the channel based on the received signal strength:

- If $D > 2 \times d_{rt}$, the reader has a second chance to access the channel with $k = k+1$ and selects randomly a new channel and follows the same procedure of GDRA.
- Otherwise, the reader leaves the channel and waits a new AC paquet.

Equation (2) is used to measure received signal power:

$$P_{r,r} = P_r \frac{G_r G_r}{K_0 D^\alpha} \tag{2}$$

Where:

- $P_{r,r}$ represents the received signal power from the other reader that occupes the channel.
- P_r is the transmit signal power of the reader.
- D is the distance between two readers.
- G_r is the reader antenna gain.
- α is the path loss exponent.
- K_0 is a coefficient of channel path loss and power ratio in the band width.

2.3   Beacon Analysis Based Collision Prevention

Beacon Analysis based Collision Prevention (BACP) [11] is an enhancement to the GDRA protocol by allocating the maximum of available resources to readers in such manner those readers remains active without the occurrence of any collisions.

In contrast with GDRA, each slot is split into several sub-slots. Readers send their *Preference_Code* in the beacon message in their one of its sub-slot which is chosen randomly and listen to the medium to receive beacon messages from neighboring readers during the slot. This *Preference_Code* contains two information: *Prev_state* that refers to the status of the reader in the previous round and *Reader_ID*. *Prev_state* used to increase fairness in access to resources through avoiding repetitive access of the same readers that have the biggest *Reader_ID*, in a way that if the reader has successfully read the tags, the *Prev_state* is set to zero else is one. Unlike DRCA, if the channel is occupied and $D > 2 \times d_{rt}$, the reader is allowed to select randomly a channel only from the remaining channels with $k = k + 1$. If the selected channel is idle then reader sends its *Preference_Code*, there are two possible situations:

- If the reader did not receive *Preference_Codes* from its neighboring readers, it would enter the R-T Communication phase as he is alone on the channel and set its *Prev_state* to zero.
- Otherwise, the reader would compare its *Preference_Code* with the received *Preference_Codes* to allow the biggest *Reader_ID* reads tags and minimizes its chances to access the medium in the next turn by setting its *Prev_state* value to zero. Nevertheless, if that reader has not the big one, it will select again a new channel with $k = k + 2$ if its distance $D > 2 \times d_{rt}$.

The novelty of BACP protocol compared to GDRA and DRCA protocols is that BACP uses the maximum available resources and tries to make the largest possible number of active readers in each turn.

2.4 Distributed Efficient and Fair Anti-collision for RFID

Distributed Efficient & Fair Anti-collision for RFID (DEFAR) [6] is a distributed multichannel algorithm based on TDMA approach. It aims to improve throughput to be the most efficient possible by avoidance the inactivity of readers in collision and increases the fairness of RFID system by allocating three priority levels to readers: NEURAL as the priority of all readers at beginning, PUMPED UP for readers that failed to read tags and raise their priority for the following round and LAZY for readers that successfully read tags and lower their chance for the next round. In this protocol, the readers send its beacon message that contains ID reader and priority level, on its selected channel. According to receive beacon message, the readers decide to read tags or not and commute its priority level for the next round. If a reader do not receive any beacon during T_{beacon}, it can access the channel to read tags during R-T communication as he is the only one on this channel and commutes its priority to LAZY to decrease its chances of accessing the channel in the next turn. Otherwise, DEFAR favors access of reader that has the PUMPED UP priority level and/or the smallest ID reader.

In this protocol, the permissible distance to exchange beacon message between readers is limited to $d_{com} = 2 \times d_{AC} = 2 \times 3.3 \times d_{rt}$ with d_{AC} the adjacent channel interference range [12].

3 Evaluation Parameters

Many performance metrics can be used to evaluate RFID anticollision algorithms. However, there is no agreement on the most effective metrics for evaluating the performance of RFID reader-to-reader anti-collision protocols [13]. In this study, we focused on the most commonly used parameters which are described below:

Throughput: presents the total number of readers who successfully read the tags. Multiple research studies consider the throughput as good parameter for evaluating the reader-to-reader anti-collision protocols [13].

Collision: is used to define the total number of failed query sections [13]. It includes the channel access collisions, that occur when multiple readers in the interference range of each other choose the same slot and channel, and reading collisions that occur too when multiple readers access the tags in their interrogation range at the same time.

Efficiency: is the percentage of the total number of readers who successfully read the tags over the total number of readers who attempted to query the tags [13]. Network efficiency is computed by this given formula (4):

$$Eff = \frac{SQS}{AQS} \tag{3}$$

with:

$$AQS = SQS + FQS \tag{4}$$

Where:

- SQS presents successful query sections.
- FQS presents failed query sections.
- AQS presents attempted query sections.

SQS and FQS are counted every time a reader tries to get access to the channel.

Jain's Fairness Index (JFI): This metric shows how fairly the successful query sections is distributed among the readers in a dense RFID environment. It is computed as:

$$JFI = \frac{|\sum_{k=1}^{n} SQS_k|^2}{n \times \sum_{k=1}^{n} SQS_k^2} \tag{5}$$

Where n is the quantity of readers. Note that, if the all readers have the same SQS then JFI will be equal to one (fair behavior). In the worst case, it is equal to $\frac{1}{n}$ when a single reader has all the SQSs (unfair behavior). Therefore, the more SQS are equitably distributed, the more the value of the JFI will be closer to one.

4   Results and Discussions

In this section, we show simulations to analyze the performance of each protocol and establish the comparison between them. The performance of RFID reader anti-collision protocols was evaluated according to different criteria cited in the previous section.

4.1   Simulation Setting

For evaluate the performance of the anti-collision protocols, we consider a quantity of static readers, that reach to 500 readers (from 100–500 readers), deployed randomly and uniformly in a surface of $1000 \times 1000\,\mathrm{m}^2$. According to the reader maximum transmission power, the reader-to-tag read range is set to $d_{rt} = 10\,\mathrm{m}$ [8]. In addition, Reader-to-Tag collisions occur when readers are close to each others ($<20\,\mathrm{m}$). All protocols have been implemented on MATLABR2017b and were simulated 100 time after each changing the number of readers. To facility implementation of protocols on MATLAB, we consider that all readers have tags to read in each round. The table below presents the main evaluation parameters and their values (Table 1).

Table 1. Evaluation parameters

Parameter	Value
Simulation range	$1000 \times 1000\,\mathrm{m}$
Readers number	100–500
Number of slots	4
Number of frequency channels	4

Results are shown in throughput, collision, efficiency and fairness index, which they are presented in the following section.

4.2   Evaluation

Throughput: Figure 4 shows the comparison throughput between all protocols cited on Sect. 3.

When number of readers increases, throughput increases too because more readers deployed, more we have successful access to the channel (SQS). Also, we constant that there are a growing gap between centralized protocols and distributed protocols due to the way of contention resolution: In GDRA protocol, all readers in contention are disabled. Furthermore, the throughput BACP is better than GDRA and DRCA due to BACP resolve a problem of disabled readers. The main reason that DEFAR show a throughput higher than others is that in distributed algorithms, the readers for which interference range and communication range are reduced which favors the increase of the number of activated readers in a single round.

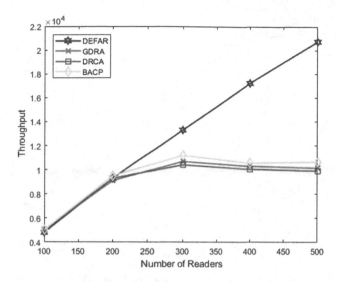

Fig. 4. Throughput of the evaluated protocols in $1000 \times 1000\,\mathrm{m}^2$ area with varying number of readers.

Collision: In Fig. 5, we illustrate the comparison of the collision respect to the number of readers, of all protocols viewed in Sect. 3.

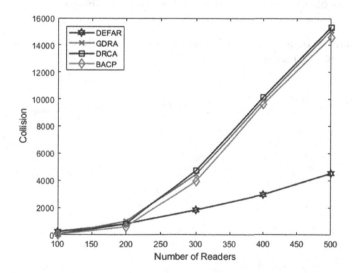

Fig. 5. Collision of the evaluated protocols in $1000 \times 1000\,\mathrm{m}^2$ area with varying number of readers.

When the number of readers increases, the collision occurs more frequently in all protocols but in distributed protocols it is very low comparing to centralized ones due to these last suffer from the high number of disabled readers at each round.

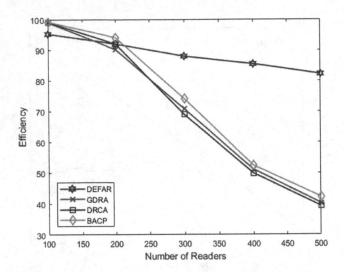

Fig. 6. Efficiency of the evaluated protocols in $1000 \times 1000 \, m^2$ area with varying number of readers.

Efficiency: In Fig. 6, we represent the results obtained for the efficiency evaluation of network respect to the number of readers for all compared protocols. According to this figure, distributed protocols present more efficiency than centralized protocols due to these latter register more collisions.

Jain's Fairness Index: Figure 7 shows the results of the fairness network with different number of readers.

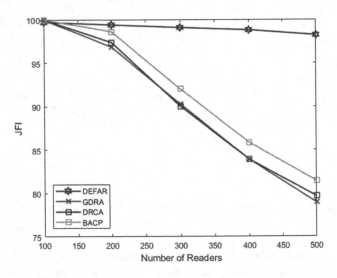

Fig. 7. Fairness of the evaluated protocols in $1000 \times 1000 \, m^2$ area with varying number of readers.

As shown in Fig. 7, all algorithms have a fair access to channel since they have an index of fairness higher than 75%, which mean that the medium is well shared among readers in the network. However, DEFAR and BACP algorithms are better since they define a priority levels for DEFAR and a priority code for BACP.

5   Conclusion

Radio interference is one of the most important challenges of RFID networks that affect its efficiency negatively. Many various protocols have been proposed to reduce it in dense RFID environments.

This paper presents a review of the most known RFID reader anti-collision protocols. It describes how reader collisions can occur in a dense deployment and the operating algorithms, which developed to mitigate it. Then, these algorithms are compared and discussed based on evaluation parameters to identify how they perform on different parameters. In this paper, we focused on evaluating the performance of these anti-collision protocols with variation in the number of readers, as many tags can be read at once by the reader without collision if they are in the activated interrogation range and do not exceed its limited number of readings. The Overall results obtained in this paper are:

– performances of distributed protocol are better than others due to that distributed protocol registers less collions even if the number of neighboring readers is higher.
– performances of protocols those have priority levels shows fairness in access to resources through avoiding repetitive access of the same readers. Finally, we have concluded that the distributed protocols present more efficiency than the centralized protocols.

References

1. Mbacke, A., Mitton, N., Rivano, H.: RFID reader anticollision protocols for dense and mobile deployments. Electronics **5**(4), 84 (2016). https://doi.org/10.3390/electronics5040084
2. Mohammed, A., Wang, Q., Li, X.: A study in integrity of an RFID-monitoring HMSC. Int. J. Food Prop. **20**(5), 1145–1158 (2016). https://doi.org/10.1080/10942912.2016.1203933
3. Ramudzuli, Z.R., Malekian, R., Ye, N.: Design of a RFID system for real-time tracking of laboratory animals. Wirel. Pers. Commun. **95**(4), 3883–3903 (2017). https://doi.org/10.1007/s11277-017-4030-9
4. Najera, P., Lopez, J., Roman, R.: Real-time location and inpatient care systems based on passive RFID. J. Netw. Comput. Appl. **34**(3), 980–989 (2011). https://doi.org/10.1016/j.jnca.2010.04.011
5. Rouan, E., Safi, S., Boumezzough, A.: An automated parking access control system based on RFID technology. In: 2020 15th Design & Technology of Integrated Systems in Nanoscale Era (DTIS), April 2020. https://doi.org/10.1109/dtis48698.2020.9080913

6. Mitton, N., Mbacke, A.A., Rivano, H.: Distributed efficient & fair anticollision for RFID protocol. In: 2016 IEEE 12th International Conference on Wireless and Mobile Computing, Networking and Communications (WiMob), October 2016. https://doi.org/10.1109/wimob.2016.7763249
7. Bueno-Delgado, M.V., Ferrero, R., Gandino, F., Pavon-Marino, P., Rebaudengo, M.: A geometric distribution reader anti-collision protocol for RFID dense reader environments. IEEE Trans. Autom. Sci. Eng. **10**(2), 296–306 (2013). https://doi.org/10.1109/tase.2012.2218101
8. Mbacké, A.: Collecte et remontée multi-sauts de données issues de lecteurs RFID pour la surveillance d'infrastructures urbaines, Thesis defended on 18 October (2018)
9. Ferrero, R.: Network Modeling and Interference Analysis in Pervasive Technology, Thesis defended (2012)
10. Golsorkhtabaramiri, M., Issazadehkojidi, N.: A distance based RFID reader collision avoidance protocol for dense reader environments. Wirel. Pers. Commun. **95**(2), 1781–1798 (2017). https://doi.org/10.1007/s11277-016-3918-0
11. Assarian, A., Khademzadeh, A., HosseinZadeh, M., Setayeshi, S.: A beacon analysis-based RFID reader anti-collision protocol for dense reader environments. Comput. Commun. **128**, 18–34 (2018). https://doi.org/10.1016/j.comcom.2018.06.006
12. Mbacke, A.A., Mitton, N., Rivano, H.: RFID anticollision in dense mobile environments. In: 2017 IEEE Wireless Communications and Networking Conference (WCNC), March 2017. https://doi.org/10.1109/wcnc.2017.7925493
13. Gandino, F., Ferrero, R., Montrucchio, B., Rebaudengo, M.: Evaluation Criteria for Reader-to-Reader Anti-collision Protocols, Technical Report n. 01–2011-UC

Game Theoretic Approaches to Mitigate Cloud Security Risks: An Initial Insight

Abdelkarim Ait Temghart[1]([✉]), M'hamed Outanoute[1], and Mbarek Marwan[2]

[1] TIAD Laboratory, FST, Sultan Moulay Slimane University, Beni Mellal, Morocco
[2] Smart Systems Laboratory, ENSIAS, Mohammed V, Rabat, Morocco

Abstract. Cloud computing is one of the most promising innovations impacting data storage and processing. And with it, clients rely on the IT solutions offered by an external provider instead of on-premise applications. Despite its enormous impacts, customers are still reluctant to outsource their business processes because of security concerns. As data is typically stored and governed by cloud vendors, users need to deal with security issues linked to the loss of control over their sensitive data. Cloud providers need to implement the appropriate security measures that might attract more clients while making the minimum investment. While used in various disciplines, game theory has recently expanded to investigate the effect of the defenders' and attackers' behaviors on strategic decision-making. This study aims to develop insights into how game theory can develop better security policies in cloud computing. First, we perform threat modeling to identify the potential threats facing cloud. Second, we identify the limitations of existing game based solutions and then suggest an improved model define an adequate strategy that would figure out the right balance between the required security level and the profit margins. Besides, we present future directions that can be explored to build highly reliable and optimal strategies for cloud services.

Keywords: Cloud computing · Security issues · Threats · Game theory

1 Introduction

It is commonly agreed that cloud computing has revolutionized the IT sector by pushing companies to opt a pay-per-use model for access on-demand services [1]. Concretely, a third-party vendor manages a huge range of functions from storage to processing. Despite its successes, cloud faces numerous obstacles that could affect its large-scale deployment and effective integration.

On the one hand, security concerns remain one of the most significant challenges facing public cloud [2, 3]. Note that traditional security measures focus considerably more on an access control and a secured communication channel between two sides (clients and cloud servers). Unfortunately, these encryption approaches are not suitable

The original version of this chapter was revised: The author Driss Ait Omar was included erroneously in the author list of the original publication and the name has now been removed. The correction to this chapter is available at https://doi.org/10.1007/978-3-030-76508-8_33

M. Fakir et al. (Eds.): CBI 2021, LNBIP 416, pp. 335–347, 2021.
https://doi.org/10.1007/978-3-030-76508-8_24

for the cloud environment since the end-users may lose control over their outsourced data. Such situation would reduce the ability of cloud providers to run the arbitrary computations and execute queries over encrypted data as well [4]. Besides, cloud storage may confront all kinds of attacks, i.e. insiders and outsiders [5]. Hence, robust data security strategies are deployed before uploading data to the cloud computing [6]. Based upon this consideration, cloud providers should strive to protect their clients' privacy through standards that take into account the following factors: confidentiality assurance, integrity protection, availability, secure access, compliance with laws and regulation, service audition [7]. Obviously, the implementation of security measures is necessary to boost the large-scale adoption of this new technology. In fact, it is clearly evident that there is an economic gain for cloud providers when building customer trust and loyalty [8].

On the other hand, the required cybersecurity measures to tackle cloud threats are often considered a large investment with a low profit margin. In practice, developing efficient countermeasures that creates a win-win situation comes with new challenges linked to the exponential increase of cyber attacks. The adopted policy is usually updated by comparing the payoff with the cost of the chosen defense mechanisms. In this case, risk analysis and the decision-making process is applied in cloud environment to answer the question regarding how the cloud providers will react to the attacker actions and then select appropriate security policy.

Based on the circumstances described above, game theory models are used to address potentially conflictual issues between cloud providers and attackers by exploring and analyzing the interactions among [9–11]. Notably, these models can successfully overcome many limitations of traditional solutions as they are characterized by the following attributes: proven mathematics, reliable defense, timely action, distributed solutions [12]. Logically, threats modeling process is done at an early stage to determine the potential loss or damage when adversaries exploit cloud vulnerabilities to gain unauthorized access to an asset. In fact, the most difficult challenges influencing the adoption of the cloud multi-tenancy and distributed nature of this environment. In the next step, game-theory models are extensively applied in cloud computing to analyze the interactions between attackers and defenders [13]. This stage is very crucial in choosing the defense mechanisms that ensure the right balance between security and profit [14]. In this chapter, we concentrate on two major constraints during the problem formulation. First, robust security measures are required to attract more consumers and generate higher returns on investment. Second, this strategy may necessitate a heavy investment in terms of the operating costs of security countermeasures. In light of this fact, this study suggests a game theory model based on minimum expenses made on security measures, as well as high-profit margin services. Additionally, we show how existing models in each field are being adapted to deal with security and privacy concerns and their limitation as well. In this respect, we propose an improved attacker-defender game model to elaborate an optimal strategy that figures out the right balance between data security measures and the profitability. Unlike existing models, we suggest a hybrid approach based on threats modeling and zero-sum games for decision-making in cloud threats.

The rest of this paper is organized as follows. Section 2 examines and analyzes the related work in this field. Section 3 provides background information on cloud threats and the principle of game theory, and we formulate and discuss the proposed game model. In Sect. 4, we highlight the major limitations of existing models as well as the fundamentals of the improved approach to tackle cybersecurity threats in cloud computing. Conclusion and perspectives are given in Sect. 5.

2 Related Work

This section briefly discusses the existing body of other researches related cyber-security game, and mentions how this method is used to analyze cloud market. In [12, 15], a comprehensive survey is provided especially the main difference between the existing game models applied to cyberspace security issues. To deal with complex systems, agent-based modeling and game theory are used to find the appropriate solution for attackers/defenders strategy [16]. In dynamic systems, stochastic game theoretic method and weighted directed graphs are used to detect intrusions in the case of computer networks [17]. Usually, game theory is used in risk analysis for strategic decision-making in the presence of adversarial attacks [18, 19]. In this respect, this technique is used for defending against a distributed denial of service attack in cloud computing [20]. In the same line, stackelberg games [21, 22] evolutionary game theory [23] are used to establish an optimal protection strategy by using game model of attack-defense scenarios. The authors in [24] rely on game theoretical analysis to determine the minimum verification requirement for cloud resources, and the optimal strategy of security policy. To this aim, they use Stackelberg game for deterministic verification model, and a probabilistic model as well. Alternatively, fuzzy inference method is used for selecting storage solutions in cloud [25]. In [26], the authors suggested a stackelberg game to find an optimal strategy that checks the outsourced data regarding its integrity and availability. In [27], game theory is used to define an optimal strategy for pricing scheme and resource allocation in cloud computing. They use mainly stackelberg equilibrium to obtain near optimal state for system's performance. Kamhoua in [28] use game theory to deal with cyber security interdependency in cloud computing. Basically, they analyze the impact of successful attacks on VM availability as well as the required expense associated with countermeasures. Besides, game theoretic model is used to ensure truth-telling ones among cloud providers. Tosh et al. use in [29] an evolutionary game theoretic model to share, among cloud providers, solutions that limit cyber attacks. More specifically, they use both dynamic cost scheme and distributed learning heuristic method to analyze how security investment affects the cybersecurity information sharing decisions.

As outlined above, significant progress has been realized in the analysis of cyber-security in cloud computing by using game theoretic models. However, the majority of existing approaches attempt to find the most generic solution that will work with all of the cloud threats. Besides, rigorous cloud threats evaluation methodology was not used. Unlike existing models, this study considers the importance of probabilistic risk models to analyze cloud threats. Additionally, we extend the traditional attacker-defender model by using a zero-sum game-theoretic to capture the dynamic nature of safety and security in cloud computing.

3   Background and Problem Statement

Overall, cyber-security is a critical process to successful identification of essential vulner-abilities and threats in cloud services so as to protect them. It is recognized that traditional methods usually involve a never-ending cycle of detection and response to new attacks taxonomy and countermeasures. Because of the cost involved in deploying security mea-sures, we suggest game theory to make the analysis and then put forward suggestions for cloud providers to invest in optimal security strategies. In this section, we first introduce the case study that illustrates security risks in cloud computing. Afterwards, we discuss how to model these threats and potential countermeasures.

3.1   Scenario

As stated in the ISO/IEC 27005 guidelines [30], a formal risk management process is fundamental to identify, estimate and evaluate the potential attacks that affect data confidentiality, integrity, and availability. In general, one of the underlying principles of security is ensuring that all entities involved in data management should strongly consider compliance with regulations and privacy standards. Hence, data protection is crucial for a successful implementation and use of remote cloud applications.

Basically, the entire cloud ecosystem consists of four different entities [31]: (1) the data owner, who is also a cloud user and has large amount of data to be stored in the cloud computing. (2) The cloud consumer, who is authorized by the data owner to access his data. (3) The cloud server, which is managed by cloud providers to offer remote services like storage, data sharing and processing. (4) The third party auditor (TPA), which is the trusted entity that assesses the cloud storage security on behalf of the data owner upon request. The system architecture for cloud services can be depicted as in Fig. 1.

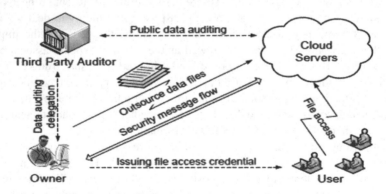

Fig. 1. The main parties involved in cloud ecosystem

In general, cloud service providers implement the necessary data protection mecha-nisms for online services to ensure compliance with cyber security regulations and the fulfilment of Service Level Agreement (SLA). Unfortunately, users and cloud providers may also cause vulnerabilities to distributed systems. In fact, user vigilance, information leakage by social engineering, and unawareness of security threats can lead to damage [32]. Thus, users should be aware of risk factors that can lead to potential cloud vulnerabilities; among these are poor access management, data breach and data leak, data loss, insecure API, misconfigured cloud storage, DoS attacks, malicious insiders, account hijacking, hared dangers, etc. [33]. In this context, Table 1 provides the main threats and its impacts in cloud computing.

Table 1. The major cloud threats [34]

Threat	Property violated	Threat definition
Spoofing identity	Authentication	Pretending to be something (clients)
Tampering with data	Integrity	Change or modify the original data
Repudiation	Non- repudiation	Reject the responsibility of an action
Information disclosure	Confidentiality	Disclosure secret information
Denial of service	Availability	Prevent access online services
Elevation of privilege	Authorization	Execute or access cloud resources

Briefly, security risk can be formally defined as the possibility of a threat to exploit particular cloud vulnerability. Based on the nature of countermeasures, each risk correctly represents the relationship between the probability of security threat event and magnitude of its consequence, using their product [35].

$$\text{Risk} = \text{Threat probability} \times \text{Potential loss}$$

Let's consider two events X_1 and X_2 associated with two threats with different attack densities $\lambda1$ and $\lambda2$ respectively.

The probability of two attacks, which are observed in t time units, can be estimated as follows [36].

$$X(t) = X_1(t) + X_2(t) \text{ and for } t \geq 0$$

$$P\{X_1(t) = n_1\} = e^{-\lambda_1} \frac{(\lambda_1 t)^{n_1}}{n_1!}$$

$$P\{X_2(t) = n_2\} = e^{-\lambda_2} \frac{(\lambda_2 t)^{n_1}}{n_2!}$$

As the X (t) is a Poisson process, the probability of these attacks is.

$$P\{X(t) = n\} = \sum_{n_2}^{n} P\{X_1(t) = n = n_2\}P\{X_2(t) = n_2\}$$

$$= \sum_{n_2=0}^{n} e^{-\lambda_1 t}\frac{(\lambda_1 t)^{n-n_2}}{(n-n_2)!} e^{-\lambda_1 t}\frac{(\lambda_2 t)^{n_2}}{(n_2)!}$$

$$= e^{(\lambda_1 t + \lambda_2)t}\frac{\{(\lambda_1 + \lambda_2)t\}^n}{(n)!} \tag{1}$$

Hence, Hyper-Erlang distribution is used to model the effect of cloud attacks [36]:

$$G(x) = \sum_{i=0}^{n} P_i E_i^r(x) \; 0 < P_i \leq 1 \tag{2}$$

Where E_i^r is the r-stage Erlang distribution with the probability P_i.

We suppose that security attacks on cloud computing assumed to be a Poisson distribution with the effect parameters λ. Hence, the effect of security attacks in cloud computing can be expressed as [36]:

$$P\left\{G_t = \frac{r}{t}\right\} = \frac{e^{-\lambda t} \lambda t^r}{r!} \text{ where } \lambda, \, r \geq 0 \tag{3}$$

3.2 Game Theory for Modeling Cloud Attacks

Basically, we can perform risk analysis more accurately if we place the situation in the form of a 'game' with two players, i.e. attackers and defenders. The central idea behind using game analysis is to find out how the players choose their strategies in different situations that maximize their own utility. At its simplest level, game theory would help cloud providers define their security policy and understand the return of an investment compared to its risk.

Formally, there is two players P1 and P2 (attacker and defender) in the cloud security game. In this case, the player i selects a security strategy from S_j that generates the payoff/utility u_j. In this context, the combination of the selected strategies of the attacker or defender is considered as a strategy profile. Cyber-security game in cloud can therefore be formulated as follows:

$$Game = (P, (S_j)_{i \in P}, (u_j)_{i \in P}). \tag{4}$$

Obviously it is impossible for cloud providers to act against all the defensive attacks at all times. Moreover, each player's payoff is affected by both the actions taken by him and the other player. In fact, the attacker is trying to maximize the amount of damage and effect of security attacks in cloud computing, which are expressed in Eq. 3. In the same line, cloud providers need to invest in securing their data by implementing robust measures to reduce the risk of cyber attacks.

As a consequence, a good investment decision rule has to maintain a fair balance between the cost of each security policy and its generated revenue. The maximum utility is achieved if all players are expected to converge to the state represented by the Nash equilibrium. At this steady state of cloud market, there is no player who can improve its profit by unilaterally modifying its strategies if the actions of the other are fixed. Mathematically, the Nash equilibrium strategy is a best reaction for both services provider and attacker. A strategic form of a state is formalized as follows:

$$U_j\left(\gamma_j^*, \gamma_{-i}^*\right) \geq U_j\left(s_j, \gamma_{-i}^*\right) \; \forall s_j \in S_j, \forall j \tag{5}$$

Where the strategy γ^* is the Nash equilibrium.

For the sake of simplification, the used security game models in this study share the following key assumptions: (1) protection costs are identical for all cloud entities, (2) attackers launch only one kind of attack at a time, (3) all players' decision are made simultaneously.

It may be the case when a cloud provider wants to generate the expected profit B by choosing the appropriate measures that maintains trust level φ, where $0 \leq \varphi \leq 1$. In this case, a successful cyber-attack can cause major damage to cloud, which is denoted L. Additionally, the most common cybersecurity attacks carried out against cloud have the probability distribution p, where $0 \leq \rho \leq 1$.

Based on the circumstances of this case, the generic utility function [37] has the following structure:

$$U = B - pL(1 - \varphi) \tag{6}$$

Note that the defender and the attacker strategy is limited by cost. Hence, we use a cost function f (x) [16] to define the relationship between security expenses and trust level φ.

$$[0\ 1] \longrightarrow \mathbb{R}$$
$$\varphi \longrightarrow f(x) = \frac{ax^2}{bx-1}$$

Where a and b are the scaling and asymptote parameters, respectively.

The attacker aims to maximize the amount of damage he causes, while the defender is trying to protect the targets and minimize the loss as well. Besides, this occurs because the attacker and the defender play simultaneously. An attacker-defender game represents the situation of a perfectly competitive in which one player can win something only by causing another player to lose it. The interactions of the attacker and the defender are typically modelled with a two-player zero-sum game aimed at more accurately predicting appropriate countermeasures against threats.

4 Improved Attacker-Defender Game Model

On the whole, attacker-defender game is a fairly new methodology for modeling security in highly complex distributed systems. The main feature of this modeling approach is the presence of a set of entities with different attributes and behaviors, and also the relationships between attackers and defenders.

In this case study, we will use a set of simple strategies for a non-cooperative game between a cloud provider and attacker. Logically, cloud providers can choose investing in protecting their data from cyber-attacks, recognizing that their choices will affect the expected profit. That is, each defender has two strategies: Invest in security (IS) or Not to Invest (NIS). In the same vein, the attacker has two strategies: launch an attack (A) on the cloud provider or Not launch an attack (NA). As security is the most principal factor in influencing the take-up of cloud services, end-users have two variants: trust (T) or (2) distrust (D) the cloud provider. After analyzing the interactions among players involved in cloud security, we use game theory models to produce an optimal decision-making for the adopted security policy.

Based on the circumstances described above, considering the case of a three-player game with a 3-tuple (T, IS, A) that represent a strategy profile, the utility function can be described as follows [38]:

The cloud payoff:

$$U(T, IS, A) = G - \lambda + a\lambda \tag{7}$$

Where

G: refers to the total costs of the delivered services
λ represents the total loss after attacks, which is calculated using Eq. 3
α is the probability of these attacks, which is calculated using Eq. 1

Naturally, each cloud provider would expect to get an outcome R obtained after investing in security measures and also incurs a cost e for the security expenses. Here the attacker can cause damage λ with a probability α by launching attacks against a particular cloud resource.

For Nash equilibria, it can help predict the behavior of the cloud provider wanting to maximize their payoff in the game [38].

$$R - e - \lambda(1 - \alpha) > R - \lambda, \text{ where } e < \lambda\alpha \tag{8}$$

It means that investing in security measures is the best option for the cloud provider in order to reduce risk of users' data leakage and improve the profit margin.

A survey of current game models applied in cloud demonstrates the importance of understanding the relation between the two parties involved in data security, i.e. providers and attackers. In general, the attacker's objective is in conflict with that of the cloud provider that aims at minimizing the damage of cyberspace. Thus, the issue between the provider and the attacker is commonly formulated as a non-cooperative model so as to capture the dynamic nature of the two parties' behavior. However, despite the importance of the existing models, most of them did not take into account other requirements linked to security's assumptions and constraints.

Based on this consideration, we are interested in dynamic non-cooperative model [39] because it is very useful in modelling and evaluating multiple actions taken by individual players. More precisely, we suggest zero-sum games for decision-making problems involving competition between cloud providers and attackers; where there is one winner and one lose. In fact, cloud providers have a strong interest in defending their facilities against intentional attacks, assuming the attacker will adopt a best response to defender's strategy. In other words, each cloud provider's gain or loss of utility is exactly balanced by the loss or gain of the attackers. Consequently, there is a single optimal strategy that is preferable to cloud provider to prevent data disclosure, while maintaining a reasonably high security level at low cost. This work introduced the idea that this conflict of interests could be formally expressed and analyzed by zero-sum games [40, 41]. For cloud threats, we rely on the study carried on the Sect. 3.1. For simplicity reasons, we only consider three elements to model the relation or competition in cloud market game: players, actions for implementing security strategy, and payoff matrix. Before giving the formal description of the game model, we first define the role of each player:

- **Players.** We consider two players: (1) attackers which benefit from harming cloud services (2) cloud provider utilize countermeasures to reduce the impact of attacks so as to protect clients' data.
- **Action sets.** It represents the sets of actions for each player. For cloud provider, we consider $D = \{d_1, d_2, ..., d_m\}$ as m refers to the number of the implemented security mechanism to protect outsourced data. Meanwhile, the attacker can conduct malicious actions that exploit n vulnerabilities in cloud computing; these actions are represented as $A = \{a_1, a_2, ..., a_n\}$.
- **Payoff matrix.** We use a reward function to essentially model the attacker's motivation and provider's payoff for an action Υ_{ij} (the action i for attacker and reaction j of cloud provider). This function is n × m matrix; each row represents the attacker action while each column is the possible cloud provider's reaction corresponding to this action. The payoff matrix can be expressed in the following way in mathematical terms:

$$\Upsilon = \begin{bmatrix} \Upsilon_{11} & \cdots & \Upsilon_{1m} \\ \vdots & & \vdots \\ \Upsilon_{n1} & \cdots & \Upsilon_{nm} \end{bmatrix} \tag{9}$$

The payoff function regarding cybersecurity in cloud computing can be expressed as follows [39]:

$$\Upsilon_{ij} = s_j - (s_i \times e_{ij}) + U \times p_i \times (1 - e_{ij}) + r_i \times (1 - e_{ij}), \forall i, j \tag{10}$$

Where

- s_i is the cost of implementing a defense strategy
- e_{ij} is the metric to measure efficacy of a defense mechanism
- U is the total production (the availability) of cloud services
- p_i is the rate of the loss in the cloud services caused by attack type i
- r_i is the recovery cost associated with action i due to a specific attack

To efficiently predict the behavior of attacker and defender, threat detection has evolved from static to dynamic behavioral analysis. Accordingly, the proposed approach to secure cloud computing is essentially composed of two stages: (1) threats modeling and (2) game theory optimization, as illustrated in Fig. 2.

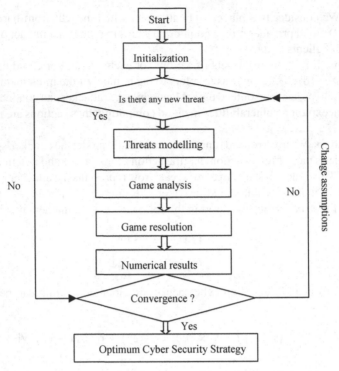

Fig. 2. The flowchart of the proposed approach

5 Conclusion

Despite still being in its infancy, the usage of game theory in cloud computing is a promising approach to promote this new paradigm. There have been several attempts to adopt game theory to ensure a healthy balance between productivity and security. Originally, game theory can successfully model the interaction between attackers and services providers. This work summarizes the advantages and limitations of existing game models to address cyber security risks in the cloud environment. Even though these models are extremely useful to find the most optimal solution, there are still improvements that need to be made to meet the expectations of both the providers and the end-users. In this respect, threats modeling methodology is an initial stage towards improving the elaboration of any security policy. Besides, it is important to select the most appropriate game theory model that well capture the essential characteristics of cloud computing and cyber security features. To this aim, we suggest zero-sum game model for selecting the most appropriate countermeasure options to a given threat in argue-scale cloud data centers. More precisely, the proposed strategy would undoubtedly help cloud providers take decisive and appropriate actions against adversary attacks. In the future work, we plan to analyze cybersecurity issues in cloud computing using zero-sum game with rational players and the proposed threats model [42]. Besides, we intend to study the influence of the suggested countermeasures on each player's payoff through numerical simulation and experimental validation. Another interesting and promising approach is the utilization of evolutionary games [43, 44] to analyze the dynamic relationship between attacker and defender, a process which clearly follows the evolutionary models more closely. Besides, stackelberg game is also a promising technique to simulate attacker/defender behavior for situation where the defender only knows the prior probabilities of different types of attackers.

References

1. Mell, P., Grance, T.: The NIST definition of cloud computing. Technical report, National Institute of Standards and Technology, vol. 15, pp. 1–3 (2009)
2. Cloud Security Alliance, Security guidance for critical areas of focus in cloud computing V4.0, Cloud Security Alliance, Seattle, WA, USA (2017)
3. Mazhar, A., Samee, U.K., Athanasios, V.: Security in cloud computing: opportunities and challenges. Inf. Sci. **305**, 357–383 (2015)
4. Fernandes, D., Soares, L., Gomes, J., Freire, M.M., Inácio, P.: Security issues in cloud environments: a survey. Int. J. Inf. Secur. **13**(2), 113–170 (2013). https://doi.org/10.1007/s10207-013-0208-7
5. Pearson, S., Benameur, A.: Privacy, security and trust issues arising from cloud computing. In: Proceedings of the IEEE Second International Conference on Cloud Computing Technology and Science (CLOUDCOM), Washington, DC, pp. 693–702. IEEE Computer Society (2010)
6. Xiao, L., Xu, D., Mandayam, N.B., Poor, H.V.: Cloud storage defense against advanced persistent threats: a prospect theoretic study. IEEE J. Sel. Areas Commun. **35**(3), 534–544 (2017)
7. Singh, A., Chatterjee, K.: Cloud security issues and challenges: a survey. J. Netw. Comput. Appl. **79**, 88–115 (2017)
8. Radwan, T., Azer, M.A., Abdelbaki, N.: Cloud computing security: challenges and future trends. Int. J. Comput. Appl. Technol. **55**(2), 158–172 (2017)
9. Kwiat, L., Kamhoua, C.A., Kwiat, K.A., Tang, J., Martin, A.P.: Security-aware virtual machine allocation in the cloud: a game theoretic approach. In: Proceedings of the 8th IEEE International conference on Cloud Computing, CLOUD 2015, New York City, NY, USA, pp. 556–563 (2015)
10. Pillai, P.S., Rao, S.: Resource allocation in cloud computing using the uncertainty principle of game theory. IEEE Syst. J. **10**(2), 637–648 (2016)
11. Li, Y.P., Tan, S.Y., Deng, Y., Wu, J.: Attacker-defender game from a network science perspective. Chaos: Interdisc. J. Nonlinear Sci. **28**(5), Article ID 051102 (2018)
12. Do, C.T., et al.: Game theory for cyber security and privacy. ACM Comput. Surv. **50**(2), Article 30 (2017)
13. Wu, H., Wang, W., Wen, C., Li, Z.: Game theoretical security detection strategy for networked systems. Inf. Sci. **453**, 346–363 (2018)
14. Cheng, L., Ma, D.H., Zhang, H.Q.: Optimal strategy selection for moving target defense based on markov game. IEEE Access **5**, 156–169 (2017)
15. Roy, S., Ellis, C., Shiva, S., Dasgupta, D., Shandilya, V., Wu, Q.: A survey of game theory as applied to network security. In: Proceedings of the International Conference on 43rd Hawaii International Conference, pp. 1–10 (2010)
16. Alan Nochenson, C.F., Heimann, L.: Simulation and game-theoretic analysis of an attacker-defender game. In: Grossklags, J., Walrand, J. (eds.) GameSec 2012. LNCS, vol. 7638, pp. 138–151. Springer, Heidelberg (2012). https://doi.org/10.1007/978-3-642-34266-0_8
17. Nguyen, K.C., Alpcan, T., Basar, T.: Stochastic games for security in networks with interdependent nodes. In: Proceedings of the International Conference on Game Theory for Networks, GameNets 2009, pp. 697–703. IEEE (2009)

18. Meng, S., Wiens, M., Schultmann, F.: A Game-theoretic approach to assess adversarial Risks. WIT Trans. Inf. Commun. Technol. **47**, 141–152 (2014)
19. Musman, S., Turner, A.: A game theoretic approach to cyber security risk management. J. Defense Model. Simul. Appl. Methodol. Technol. **15**(2), 127–146 (2018)
20. Wang, B., Zheng, Y., Lou, W., Hou, Y.T.: DDoS attack protection in the era of cloud computing and software defined networking. Comput. Netw. **81**, 308–319 (2015)
21. Jakóbik, A., Palmieri, F., Kołodziej, J.: Stackelberg games for modeling defense scenarios against cloud security threats. J. Netw. Comput. Appl. **110**, 99–107 (2018)
22. Jakobiki, A.: Stackelberg game modeling of cloud security defending strategy in the case of information leaks and corruption. Simul. Model. Pract. Theory **103**, Artile ID 102071 (2020)
23. Sun, P.J.: Research on the optimization management of cloud privacy strategy based on evolution game. Secur. Commun. Netw. **2020**, 18, Article ID 6515328 (2020)
24. Djebaili, B., Kiennert, C., Leneutre, J., Chen, L.: Data integrity and availability verification game in untrusted cloud storage. In: Poovendran, R., Saad, W. (eds.) GameSec 2014. LNCS, vol. 8840, pp. 287–306. Springer, Cham (2014). https://doi.org/10.1007/978-3-319-12601-2_16
25. Esposito, C., Ficco, M., Palmieri, F., Castiglione, A.: Smart cloud storage service selection based on fuzzy logic, theory of evidence and game theory. IEEE Trans. Comput. **65**(8), 2348–2362 (2016)
26. Ismail, Z., Kiennert, C., Leneutre, J., Chen, L.: Auditing a cloud providers compliance with data backup requirements: a game theoretical analysis. IEEE Trans. Inf. Forensics Secur. **11**(8), 1685–1699 (2016)
27. Jalaparti, V., Nguyen, G.D.: Cloud resource allocation games (2019)
28. Kamhoua, C.A., Kwiat, L., Kwiat, K.A., Park, J., Zhao, S.M., Rodriguez, M.: Game theoretic modeling of security and interdependency in a public cloud. In: Proceedings of IEEE 7th International Conference on Cloud Computing, pp. 514–521 (2014)
29. Tosh, D.K., Sengupta, S., Kamhoua, C.A., Kwiat, K.A.: Establishing evolutionary game models for cyber security information exchange (CYBEX). J. Comput. Syst. Sci. **98**, 27–52 (2018)
30. ISO/IEC 27005: Information technology security techniques information security risk management (2008)
31. Chang, V., Kuo, Y., Ramachandran, M.: Cloud computing adoption framework: a security framework for business clouds. Future Gener. Comput. Syst. **57**, 24–41 (2016)
32. Ravi Kumar, P., Herbert Raj, P., Jelciana, P.: Exploring security issues and solutions in cloud computing services: a survey. Cybern. Inf. Technol. **17**(4), 3–31 (2016)
33. Birje, M.N., Challagidad, P.S., Goudar, R.H., Tapale, M.T.: Cloud computing review: concepts, technology, challenges and security. Int. J. Cloud Comput. **6**(1), 32–57 (2017)
34. Julian, J.J., Surya, N.: A survey of emerging threats in cybersecuirty. J. Comput. Syst. Sci. **80**(5), 973–993 (2014)
35. Saripalli, P., Walters, B.: QUIRC: a quantitative impact and risk assessment framework for cloud security. In: Proceedings of International Conference on Cloud Computing, Miami, FL (2010)
36. Meetei, M.Z.: Mathematical model of security approaches on cloud computing. Int. J. Cloud Comput. **6**(3), 187–210 (2017)
37. Grossklags, J., Christin, N., Chuang, J.: Secure or insure? A game-theoretic analysis of information security games. In: Proceedings of International Conference World Wide Web Conference (WWW 2008), Beijing, China, pp. 209–218, April 2008
38. Njilla, L.Y., Pissinou, N., Makki, K.: Game theoretic modeling of security and trust relationship in cyberspace. Int. J. Commun. Syst. **29**, 1500–1512 (2016)
39. Lv, K., Chen, Y., Hu, C.: Dynamic defense strategy against advanced persistent threat under heterogeneous networks. Inf. Fusion **49**, 216–226 (2019)

40. Al Mannai, W.I., Lewis, T.G.: A general defender-attacker risk model for networks. J. Risk Finan. **9**(3), 244–261 (2008)
41. Halevy, N.: Resolving attacker-defender conflicts through intergroup negotiation. Behav. Brain Sci. **42**, E124 (2019)
42. Zarreha, A., Saygina, C., Wana, H., Leea, Y., Brachoa, A.: A game theory based cybersecurity assessment model for advanced manufacturing systems. Procedia Manuf. **26**, 1255–1264 (2018)
43. Cressman, R., Apaloo, J.: Evolutionary game theory. In: Başar, T., Zaccour, G. (eds.) Handbook of Dynamic Game Theory, pp. 461–510. Springer International Publishing, Cham (2018). https://doi.org/10.1007/978-3-319-44374-4_6
44. Khalifa, N.B., El-Azouzi, R., Hayel, Y., Mabrouki, I.: Evolutionary games in interacting communities. Dyn. Games Appl. **7**(2), 131–156 (2017)

Comparative Study on the McEliece Public-Key Cryptosystem Based on Goppa and QC-MDPC Codes

Es-said Azougaghe[1]([✉])[iD], Abderrazak Farchane[1], Idriss Tazigh[3],
and Ali Azougaghe[2][iD]

[1] Polydisciplinary Faculty, Sultan Moulay Slimane University, Beni Mellal, Morocco
a.es-said@usms.ma, essaidazougaghe@gmail.com, a.farchane@gmail.com
[2] ENSIAS, Mohammed V University in Rabat, Rabat, Morocco
aliazougaghe@gmail.com
[3] Regional Center for Education and Training Professions, Marrakech, Morocco
i.tazigh@gmail.com

Abstract. In recent years, much research has been conducted on quantum computers – machine that exploit the phenomena of quantum mechanics to solve difficult or insoluble mathematical problems for conventional computers. If large-scale quantum computers are built, they will be able to break many of the public key cryptosystems currently in use. This would seriously compromise the confidentiality and integrity of digital communications on the internet. Post-quantum cryptography aims to develop secure cryptographic systems against both conventional as well as quantum computers for interacting with existing protocols and communication networks. In this paper we present a public key cryptosystem of McEliece based on the correcting codes, using two types of correcting codes; QC-MDPC and Goppa correcting codes. This latter seems very interesting considering its two characteristics, namely the power of correction and the efficient decoding algorithm which resistant to quantum attacks due to difficulty of decoding a linear code. On the other hand, QC-MDPC cryptosystem code is rapid and more secure than Goppa cryptosystem.

Keywords: McEliece cryptosystem · QC-MDPC codes · Goppa code · Post-quantum cryptography · Bit flipping algorithm

1 Introduction

Over the last decades, public key cryptography has become an indispensable part of our global digital communication infrastructure [1]. These networks support a multitude of important applications for our economy, security and lifestyle, such as mobile phones, e-commerce, social networks and cloud computing. In such a connected world, the ability of individuals, businesses, and governments to communicate securely is of the highest importance.

© Springer Nature Switzerland AG 2021
M. Fakir et al. (Eds.): CBI 2021, LNBIP 416, pp. 348–360, 2021.
https://doi.org/10.1007/978-3-030-76508-8_25

Many of our critical communication protocols rely primarily on three key cryptographic features: public-key encryption, digital signatures, and key exchange. Currently, these features are mainly implemented using Diffie-Hellman key exchange [2], RSA cryptosystem (Rivest-Shamir-Adleman)[3] and elliptic curve cryptosystems. The security of these depends on the difficulty of certain theoretical problems relating to numbers, such as the factorization of integers or the problem of the discrete journal on various groups.

In 1994, Peter Shor of Bell Laboratories showed that quantum computers, a new technology exploiting the physical properties of matter and energy to perform calculations [4], can effectively solve each of these problems that base their security on the difficulty of factorization problem or the problem of discrete logarithm in polynomial time, thus rendering all public key cryptosystems based on such assumptions helpless. Thus, a sufficiently powerful quantum computer will jeopardize many modern forms of communication, from key exchange to encryption and digital authentication.

This threatens most, if not all, public key cryptosystems deployed in practice, such as RSA. Cryptography based on the difficulty of decoding a linear code, on the other hand, is resistant to quantum attacks and therefore considered a viable substitute for these systems in future applications [5]. Yet, regardless of their so-called "post-quantum" nature, code-based cryptosystems offer other benefits, even for current applications, because of their excellent algorithmic efficiency [6].

The first cryptosystems relying on coding theory was proposed by J.R. McEliece in 1978 [7]. It uses binary Goppa codes as a basis for the construction. The McEliece cryptosystem has some distinctive advantages over other public-key cryptosystems. One of the advantages is that the cryptosystem incorporates in each encryption a random element to improve security; it is a randomly generated error vector. RSA and other modern cryptographic systems do not incorporate such randomness into the encryption process. This random element contributes to the current status of the McEliece cryptosystem as a strong candidate for secure post-quantum computer encryption; RSA and other modern encryption methods can be easily manipulated by a quantum computer. The main disadvantage of a McEliece cryptosystem lies in the size of the public key. The public key of this cryptographic system (a matrix $nx(n-k)$) can be expensive to store, especially on a low capacity device such as a smartphone. This is one of the main reasons why McEliece cryptosystems are not more widely implemented.

To reduce the size of the public key, several works are carried out based on the family of LDPC codes. They have been repeatedly suggested for the McEliece scheme [8–10] The main problem of using LDPC codes in this context is that their low weight parity-check rows can be seen as low weight codewords in the dual of the public code.

In 2013, the use of quasi-cyclic MDPC (Moderate Density Parity Check) codes was suggested to instantiate the McEliece scheme in [11]. This version of McEliece enjoys relatively small key sizes (a few thousand bits) and its security provably reduces to the hardness of deciding whether a given quasi-cyclic code contains a word of small weight or not. Moreover, the decryption essentially consists in decoding and can be achieved with the same iterative algorithms as for LDPC codes. In particular, a low cost implementation, suitable for embedded systems, can be achieved using a hard decision bit flipping iterative algorithm, as demonstrated in [12]. Other works [12–14] are focused on how to choose the threshold in order to reduce the average number of iterations to successfully decode a given number of errors by using the bit flipping decoding algorithm.

Unfortunately, these codes have been attacked. To counter these attack, [15–17] used and approach in which she determined a variant of the bit flipping algorithm which favors the worst case for a particular set of parameters. If this variant is combined with intensive simulation, it can provide some guidelines to engineer the QC-MDPC-McEliece decryption that can resist timing attacks.

2   Preliminaries

2.1   McEliece Cryptosystem

The McEliece cryptosystem takes advantage of the error correction codes previously described as an encryption mechanism. The idea is that one can intentionally add errors to a codeword to hide/encrypt the message. The effectiveness of the cryptosystem depends on two difficult problems in the field of coding theory: the general problem of decoding and the problem of codewords of a given weight.

Problem 1 (closest codewords). Let C be a linear code [n, k] in F_q and $y \in F_q^n$. Find a code word $x \in C$ where d(y, x) is minimal. it is difficult to determine the x code word that is closest to y. This will serve as a basis for encryption in a McEliece cryptosystem.

Problem 2 (Finding weight w of code words). Let C be a linear code [n, k] in F_q et $w \in N$. Find a code word $x \in C$ where x has the weight w.

The first problem is important for decryption to be sufficiently difficult to constitute a cryptosystem, and the second is important because it assures us that a closely related problem is also difficult. But how difficult are these problems? In particular, we must be assured that they are strong enough to form the basis of a secure cryptosystem. Fortunately, in [18], the following principle is proved:

Theorem 1. *The general problem of decoding and the problem of finding weights are both NP-hard.*

With this assurance that the problems on which the McEliece cryptosystem is based are difficult enough to solve, we go on to the details of the system itself. To configure a McEliece cryptosystem, it is necessary to select a linear code $[n, k]$ capable of correcting t errors (via an efficient decoding method - the Patterson algorithm in the case of Goppa codes [19]. A disguised binary generator matrix for G serves as a public key (we will see how this "disguise" is constructed in the next section). The encryption process involves multiplication by the public key and adding an error vector to the result. Without the knowledge of G, the previous problems underline the difficulty of decryption. But with knowledge of

G and an efficient decoding algorithm, the receiver can efficiently decrypt the message.

Encoding and Encryption: One of the important differences to respect is the difference between the words "encode" and "encrypt". The word "encode" refers to the process of adding redundancy to a word to enable error correction. The term **"encryption"** refers to the process of hiding a message to be sent. In other words, coding is a function of a linear code, whereas encryption is a function of a cryptosystem. The same goes for the terms **"decode"** and **"decrypt"**. Certainly, this difference is rather subtle in McEliece cryptosystem. But with this distinction in mind, we will now address the additional details of the cryptosystem.

2.2 McEliece Cryptosystem Algorithm

Like all public key systems, this cryptosystem is made up of three algorithms:

Algorithm 1. Key generation

Input: Two integers n and t.
Output: The public key $p_k = $ (G', t) and the associated private key $s_k = $ (S, G, P, C)

1: Choose a linear code C t-corrector of dimension k and length n
2: Take the generator matrix $G \in M_{k,n}(F_2)$ in C;
3: Randomly choose an invertible matrix $S \in M_{k,k}(F_2)$;
4: Randomly choose a permutation matrix $P \in M_{n,n}(F_2)$;
5: Calculate the generator matrix $G' = S.G.P$;
6: **return** p_k, s_k

Algorithm 2. Encryption

Input: The public key $p_k = $ (G', t) and a message to encrypt $m \in F_2^k$.
Output: $c \in F_2^n$ the cipher text associated with m.

1: Encode the message $c' = m.G'$;
2: Randomly generate an error vector $e \in F_2^n$ weight $w_H(e) = t$;
3: Calculate $c = c' \oplus e$;
4: **return** c

Algorithm 3. Decryption

Input: The private key $s_k = $ (S, G, P, C) and cipher text $c \in F_2$.
Output: $m \in F_2^k$ the clear text associated with c.

1: Calculate $c'_p = c.P^{-1}.$;
2: Decode c'_p to find $m.S.G$;
3: Recover $m' = m.S$ from $m.S.G$;
4: Calculate $m = m'.S^{-1}$;
5: **return** m.

The following Fig. 1 summarizes the three McEliece system algorithms when transmitting a message m from Alice to Bob.

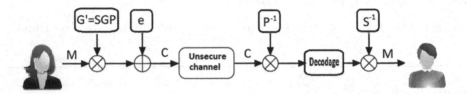

Fig. 1. McEliece schema.

Assuming that Alice wants to be able to send encrypted messages to Bob using a McEliece cryptosystem. Bob must first choose a linear code, generated by a matrix G, with a sufficient decoding algorithm, D, to correct the errors. He also chooses at random two matrices, S and P. Bob then publishes $G' = SGP$ and t and keep the rest of the secret information. Alice can now search for Bob's public key and send a message by transmitting the encrypted message to Bob:

$$y = mG' + e$$

where e is a random vector of weight t that Alice generates again each time she encrypts a message. Upon receiving this message, Bob can decrypt Alice's message by making $D(yP^{-1}) = D([mG' + e]P^{-1}) = D(mSGPP^{-1} + eP^{-1}) = D(mSG + e') = mSG$ where $e' = eP^{-1}$. Bob can then solve the resulting matrix for m by multiplying right by $(SG)^{-1}$ or another more efficient method of his choice.

Matrix S or the Scrambling Matrix: is an invertible matrix that plays a key role in the disguise of the structure of G. Note the effect of S on a particular message m; multiplying m on the left by S before multiplication by G produces a codeword different from mG. We can consider this as a method to hide the structure of G.

Switching Matrix: The permutation matrix provides another level of darkness to the public key. A permutation matrix is a matrix with exactly one non-zero bit in each row and column. In other words, a permutation matrix can be obtained by permuting the columns of an identity matrix.

public Key G': Thus, the public key published by the recipient is the binary matrix $G' = SGP$ and t, the number of errors that must be added to an encoded message. According to McEliece in, the astronomical number of choices for S and P are trying to recover G from infeasible G'.

Private Key: The matrices S and P as well as the decoding function D constitute the private key of a McEliece cryptosystem.

2.3 Existing Attacks

we have two types of attacks: Generic attacks and algebraic attacks each contains different methods which depend on the family of codes chosen for the schema, the Table 1 describes some codes used with the McEliece cryptosystem and the attacks they have suffered. The first family of codes proposed by McEliece is that of classic binary Goppa codes which currently remains good resistant to attacks.

Table 1. Comparison of the proposed code families with McEliece encryption

Code	Paramètres	Key	Security	Attacks
Goppa code Binaries	$[1024, 524, 101]_2$	67 Ko	2^{62}	
McEliece 1978	$[2048, 1608, 48]_2$	412 Ko	2^{96}	
RSG Code Niederreiter 1986	$[256, 128, 129]_2$	67 Ko	2^{95}	Attacks on θ^3
Reed Muller Code Binaries	$[1024, 176, 128]_2$	22.5 Ko	2^{72}	Attacks under exponential
Sidelnikov 1994	$[2048, 232, 256]_2$	59.4 Ko	2^{93}	
Geometric Codes Janwa Moreno 1996	$[171, 109, 61]_2$	16 Ko	2^{66}	Attacks polynomials

2.4 Definitions

In order to unify the notations and definitions, we will present a few definitions of the necessary concepts. All considerations will be conducted on finite field F_2.

Definition 1. *The **Hamming weight** (or simply weight) of vector $x \in F_2^n$ is the number of nonzero components denoted as $wt(x)$. A binary (n, r)-linear code C of length n and dimension r is an r-dimensional vector subspace of F_2^n. It is spanned by the rows of a matrix F_2^n, called a generator matrix of C. Also, it is the kernel of a matrix $H \in F_2^{n-r}$ called a parity-check matrix of C. The codeword $c \in C^n$ of a vector $m \in F_2^r$ is $c = mG$. The syndrome $s \in F_2^r$ of a vector $e \in F_2^n$ is $s = H.e^T$.*

Definition 2. *An (n, r)-linear code is a quasi-cyclic code (QC) if there is some integer n_0 such that every cyclic shift of a codeword by n_0 places is again a codeword. Additionally when $n = n_0.p$ for some integer p, it is possible to have generator and parity check matrices composed by $p \times p$ circulant blocks which are completely described by their first row (or column).*

Definition 3. *An (n, r, w)-LDPC or MDPC code is a linear code of length n, dimension r which admits a parity-check matrix of constant row weight w. LDPC and MDPC codes differ in the magnitude of the row weight w. We assume for MDPC codes row weight whose scale is $O(\sqrt{n \log n})$. On the other hand, the constant row weight is usually less than 10 for LDPC.*

2.5   MDPC and QC-MDPC Code

The construction of our codes is as follows:

$(n,\ r, w)$-MDPC code construction

1. Generate r vectors $(h_i \in F_2^n)_{0 < i < r}$.
2. The (n, r, w)-MDPC code is easily generated by selecting a random parity-check matrix $H \in F_2^{r \times n}$ of $i - th$ row weight h_i.

$(n,\ r, w)$-QC-MDPC code construction

1. Generate a vectors $h \in F_2^n$ of weight w at random.
2. The (n, r, w)QC-MDPC code is defined by a quasi-cyclic parity-chek matrix $H \in F_2^{r \times n}$ of row h.
3. The other $r - 1$ rows of H are obtained from the $r - 1$ quasi-cyclic shifts of h

The general definition of MDPC codes can be found in [11, 16]. For the purpose of this article, construction using $n_0 = 2$ will be discussed.

The (n, r, w)–QC-MDPC codes where $n = 2p$ and $r = p$. $H = [H_0 | H_1]$ So then the parity check matrix has the form where H_i is a $r \times r$ circulant block. To define the parity-check matrix H we pick up a random first row of weight w and the other $r - 1$ rows are obtained from $r - 1$ shift of the first row.

A generator matrix G in the row reduced echelon form can be easily derived from the H_i's blocks. Assuming that the block H_1 is non-singular (which particularly implies row h_i of matrix H_1 has $wt(h_i)$ odd) we construct a generator-matrix

$$G = \left[I | (H_1^{-1}.H_0)^T \right] \tag{1}$$

2.6   McEliece Cryptosystem Using QC-MDPC Codes

Let t denotes the number of errors which can be corrected by a bit flipping iterative decoder of an (n, r, w)-MDPC code with $n = 2r$. Typically, we expect that $t_w = O(n)$ and thus as $w = O(sqrt(n))$, we obtain $t = O(sqrt(n))$. The McEliece cryptosystem instantiated with (QC-)MDPC codes works as follow.

1. **Key Generation.** The key Generation procedure consists of two steps. First we generate a parity-check matrix $H \in F_2^{r \times n}$ of a t-error-correcting $(2r, r, w)$–QC-MDPC code by choosing the first row of the parity-check matrix. The second step is to generate the corresponding generator matrix $G \in F_2^{r \times n}$ in the row reduced echelon form.

   The public-key of this system is the tuple (G, t) and the private-key is matrix H.

2. **Encryption.** In order to encrypt message $m \in F_2^r$ we need to generate random vector $e \in F_2^n$ of $wt(e) < t$. The cipher-text is $c = mG + e \in F_2^n$ where $x \in F_2^r$ is a cipher-text.

3. **Decryption.** Decode the cipher-text c to get a codeword mG. The plaintext is obtained by truncating the first $n - r$ bits of that codeword.

3   Decoding Algorithms

3.1   Bit Flipping Algorithms

The decoding algorithms for MDPC codes are mainly divided into two families. The first class proposed by Berlekamp and Al in 1978 offers better error correction capacity but it is more complex in computation than the second family. In particular when processing large codes, the second family, called bit flipping algorithms introduced by Gallager in 1962 [20], dedicated to the LDPC code, seems to be more appropriate [21] are iterative and probabilistic algorithms, since they are made up of a series of operations which repeat themselves several times in order to return the right solution in a certain probability. They are based on the following principle:

– Calculate received cipher text syndrome $s = xH^T$.
– Count the number of unsatisfied parity check equations associated with each bit of cipher-text.
– Reverse every bit of cipher-text that causes more than b unsatisfied parity equations.
– Update the syndrome calculus.

This process is repeated until either the syndrome becomes zero or a predefined maximum number of iterations on which a decoding error is returned is reached. We name this Algorithm 4.

Algorithm 4. Modified Gallager's Bit Flipping Algorithm [16]

Input: $x \in F_2^n$, $H \in F_2^{r \times n}$, $r_{max} \in Z_+$
Output: $m \in C$ lub error
1: $s \longleftarrow x.H^T$

2: **for** $r \in \{0, ..., r_{max-1}\}$ **do**
3:    **for** $i \in \{0, ..., n-1\}$ **do**
4:       $\sigma_i \longleftarrow <s, h_i> \in Z$ ▷ *
5:       **if** $\sigma_i >= b$ **then**
6:          $x_i \longleftarrow x_i \oplus 1$ ▷ **
7:          $s \longleftarrow s \oplus h_i$
8:       **end if**
9:    **end for**
10:   **if** $s = 0^+$ **then**
11:      **return** x
12:   **end if**
13: **end for**
14: **return** *error*

---

* $<s, h_i>$ − means scalar product of two vectors
h_i − means i-th column of matrix H
** x_i − means $i_t h$ position in vector x

3.2 Security and Parameter Selection

Theoretical security of the QC-MDPC McElice cryptosystem has been presented [11, 16, 22–24]. In particular, an analysis of the safety and impact a quasi-cyclic structure on the security was presented. Recently, new attacks on system using those codes have been proposed. The most important is very powerful attack using a quasi-cyclic form of the parity check matrix [25]. The attack leverages the fact that there is some probability, termed the Decoding Failure Rate (DFR), that the decoding may fail to compute the errors.

Parameters for the examined systems are based on analyzes carried out in the work [26]. The suggested parameters are presented in Table 2. The tests have been carried out using these values. As shown in Table 3, the use of QC-MDPC codes allows to reduce the size of the keys for the cryptographic primitives which involve them compared to McEliece based on the Goppa codes. However, the quasi-cyclic character does not modify the behavior of these codes during decoding. Regarding its complexity, the encryption and key-generation are reduced to simple QCblock products, and the decryption takes less than 3 ms in implementation.

Table 2. Choice of the parameters of the codes $[2r, r, w]$-QC-MDPC for the McEliece cryptosystem with an initial error weight t [16]

Classical security	Quantum security	n	r	w	t	Public key size	Private key
80	58	9602	4801	90	84	4801	9602
128	86	19714	9857	142	134	9857	19714
256	154	65542	32771	274	264	32771	65542

Table 3. Key-size comparison [11].

Level security	QC-MDPC	Goppa
80	4801	460647
128	9857	1536537
256	32771	7 667 855

4   Simulation and Results

The implementation of McEliece's encryption algorithm with the Goppa codes and the QC-MDPC code using files of size 1Ko, 2Ko, 3Ko, 4Ko and 5Ko led us to these results (Fig. 2 and Fig. 3):

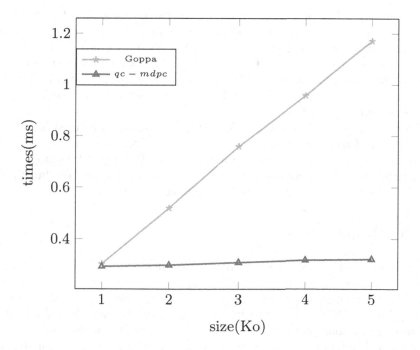

Fig. 2. Encryption time of files according to size

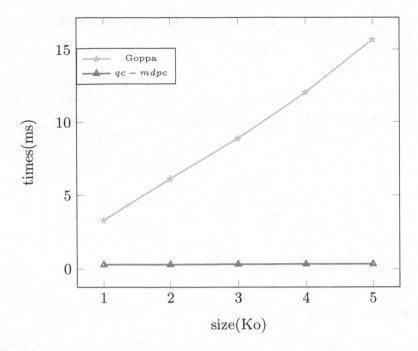

Fig. 3. Decryption time of the files according to the size

It is clear that McEliece's encryption by the two codes is almost with a linear time equal this fact is due to the same encryption procedure which is the message multiplication by the code generator matrix. For decryption, we observe that the Goppa code takes a longer time than the QC-MDPC code, but in spite of it we can consider the McEliece encryption with the QC-MDPC code is better than the Goppa code because the first is rapid and secure, while the second is deficient seen its decoding algorithm.

We believe that McEliece public key cryptosytem could be used in the next decades, even if no quantum computer is available. The advantage of cryptosystems based in QC-MDPC is that encoding and are faster which reduce the size of the public key. As a perspective of this work, we will improve the bit flipping decoding algorithm by seeking to reduce the decoding failure rate.

5   Conclusion

In this work we have explored the properties of a McEliece cryptosystem by comparing the two McEliece versions, the classic version based on goppa codes using the Paterson algorithm and the recent version based on QC-MDPC codes by adopting the bit flipping algorithm, then we used the predefined libraries for the fundamental tools, notably the applied mathematical formalism of the SAGE mathematical programming language. We conclude that the bit algorithm can be successfully applied to resist to potential quantum computer attacks. for the next work we will test large files on performance machines.

References

1. Chen, L., et al.: Report on post-quantum cryptography, vol. 12. US Department of Commerce, National Institute of Standards and Technology (2016)
2. Steiner, M., Tsudik, G., Waidner, M.: Diffie-Hellman key distribution extended to group communication, pp. 31–37 (1996)
3. Rivest, R.L., Shamir, A., Adleman, L.: A method for obtaining digital signatures and public-key cryptosystems, **26**(1), 96–99 (1978)
4. Costello, C., Jao, D., Longa, P., Naehrig, M., Renes, J., Urbanik, D.: Efficient compression of SIDH public keys. In: Coron, J.-S., Nielsen, J.B. (eds.) EUROCRYPT 2017. LNCS, vol. 10210, pp. 679–706. Springer, Cham (2017). https://doi.org/10. 1007/978-3-319-56620-7_24
5. Overbeck, R., Sendrier, N.: Code-based cryptography. In: Bernstein, D.J., Buchmann, J., Dahmen, E. (eds.) Post-Quantum Cryptography, pp. 95–145. Springer, Heidelberg (2009). https://doi.org/10.1007/978-3-540-88702-7_4
6. Bos, J., et al.: CRYSTALS-kyber: a CCA-secure module-lattice-based KEM. In: 2018 IEEE European Symposium on Security and Privacy (EuroS&P), pp. 353–367. IEEE (2018)
7. McEliece, R.J.: A public-key cryptosystem based on algebraic. Coding Thv **4244**, 114–116 (1978)
8. Monico, C., Rosenthal, J., Shokrollahi, A.: Using low density parity check codes in the McEliece cryptosystem. In: IEEE International Symposium on Information Theory (ISIT 2000) (2000)
9. Baldi, M., Chiaraluce, F.: Cryptanalysis of a new instance of McEliece cryptosystem based on QC-LDPC codes. In: 2007 IEEE International Symposium on Information Theory, pp. 2591–2595. IEEE (2007)
10. Baldi, M., Bodrato, M., Chiaraluce, F.: A new analysis of the McEliece cryptosystem based on QC-LDPC codes. In: Ostrovsky, R., De Prisco, R., Visconti, I. (eds.) SCN 2008. LNCS, vol. 5229, pp. 246–262. Springer, Heidelberg (2008). https:// doi.org/10.1007/978-3-540-85855-3_17
11. Misoczki, R., Tillich, J.P., Sendrier, N., Barreto, P.S.: MDPC-McEliece: new McEliece variants from moderate density parity-check codes. In: 2013 IEEE International Symposium on Information Theory, pp. 2069–2073. IEEE (2013)
12. Heyse, S., von Maurich, I., Güneysu, T.: Smaller keys for code-based cryptography: QC-MDPC McEliece implementations on embedded devices. In: Bertoni, G., Coron, J.-S. (eds.) CHES 2013. LNCS, vol. 8086, pp. 273–292. Springer, Heidelberg (2013). https://doi.org/10.1007/978-3-642-40349-1_16
13. Von Maurich, I., Güneysu, T.: Lightweight code-based cryptography: QC-MDPC McEliece encryption on reconfigurable devices. In: 2014 Design, Automation & Test in Europe Conference & Exhibition (DATE), pp. 1–6. IEEE (2014)
14. Von Maurich, I., Oder, T., Güneysu, T.: Implementing QC-MDPC McEliece encryption. ACM Trans. Embed. Comput. Syst. (TECS) **14**(3), 1–27 (2015)
15. Chaulet, J., Sendrier, N.: Worst case QC-MDPC decoder for McEliece cryptosystem. In: 2016 IEEE International Symposium on Information Theory (ISIT), pp. 1366–1370. IEEE (2016)

16. Janoska, A.: MDPC decoding algorithms and their impact on the McEliece cryptosystem, pp. 1085–1089 (2018)
17. Liva, G., Bartz, H.: Protograph-based quasi-cyclic MDPC codes for McEliece cryptosystems (2018)
18. Berlekamp, E., McEliece, R., Van Tilborg, H.: On the inherent intractability of certain coding problems (corresp.). IEEE Trans. Inf. Theory **24**(3), 384–386 (1978)
19. Patterson, N.: The algebraic decoding of Goppa codes. IEEE Trans. Inf. Theory **21**(2), 203–207 (1975)
20. Gallager, R.G.: Low-density parity-check codes, p. 8 (1962)
21. Maurich, I.V., Oder, T., Güneysu, T.: Implementing QC-MDPC McEliece encryption, **14**(3), 1–27 (2015)
22. Chaulet, J., Sendrier, N.: Worst case QC-MDPC decoder for McEliece cryptosystem, pp. 1366–1370 (2016)
23. Tillich, J.-P.: The decoding failure probability of MDPC codes. In: 2018 IEEE International Symposium on Information Theory (ISIT), pp. 941–945. IEEE (2018)
24. Sendrier, N., Vasseur, V.: About low DFR for QC-MDPC decoding. In: Ding, J., Tillich, J.-P. (eds.) PQCrypto 2020. LNCS, vol. 12100, pp. 20–34. Springer, Cham (2020). https://doi.org/10.1007/978-3-030-44223-1_2
25. Guo, Q., Johansson, T., Stankovski, P.: A key recovery attack on MDPC with CCA security using decoding errors. In: Cheon, J.H., Takagi, T. (eds.) ASIACRYPT 2016. LNCS, vol. 10031, pp. 789–815. Springer, Heidelberg (2016). https://doi.org/10.1007/978-3-662-53887-6_29
26. Yamada, A., Eaton, E., Kalach, K., Lafrance, P., Parent, A.: QCMDPC KEM: a key encapsulation mechanism based on the QCMDPC McEliece encryption scheme. NIST Submission (2017)

Optimization of Leach Protocol for Saving Energie in Wireless Sensor Networks

Abderrahim Salhi$^{(\boxtimes)}$, Mariam Belghzal, Brahim Minaoui, and Abdellatif Hair

Information Processing and Decision Laboratory, Faculty of Sciences and Technology, Sultan Moulay Slimane University, B.P. 523, Beni Mellal, Morocco

Abstract. In low powered Wireless Sensors Networks (WSNs), the optimization of energy consumption is major challenge facing researchers in this field.

A new clustering protocol for WSNs based on the existing LEACH protocol, is developed in this paper. The main idea in this work is to reduce energy consumption for sending the sensed data from sensors to the base station. In the existing LEACH protocol, in every cluster, sensed data are collected from all sensors, aggregated and directly sent to the base station by the cluster head. But our protocol differentiates the cluster heads which are far and near to the base station. Depending upon the distance the far node from the base station will send its aggregated data to an intermediate. Above cycle will be repeated until the entire data is delivered to the base station by its nearest intermediate node.

This technique, by bypassing long distance transmissions which are energetically greedy, has shown its great ability to preserve, for a long time, the energy and the number of nodes in WSNs with large areas, and low density of nodes. These results are confirmed by some different tests performed on a set of networks of different parameters.

Keywords: WSN · Hierarchical routing · LEACH · PEGASIS · Optimization of energy · MOD-LEACH

1 Introduction

Wireless Sensor Networks (WSNs) are Ad Hoc networks with a lot of numbers of nodes that are micro-sensors able to collect and transmit environmental data in an autonomous way to the base station [1–3]. These sensors are randomly dispersed in a geographical area. In fact, a sensor is powered by a battery, this power source is limited and it's usually irreplaceable. Therefore, energy is the most valuable resource in a WSN, since it directly influences their lifetime, and consequently the entire network.

Data routing is a very important factor in the energy saving management, that's why several research works [4, 5] have been done to bring to the strategies of routing some new approaches, techniques and improvements giving a better preservation of the energy of WSNs.

This article presents the main phases of the development of our approach which aims to optimize energy consumption in a WSN, beginning with the existing works, which

M. Fakir et al. (Eds.): CBI 2021, LNBIP 416, pp. 361–375, 2021.
https://doi.org/10.1007/978-3-030-76508-8_26

describes the different methods used to solve the problem of the energy of WSNs. Then comes a second section to present the energy model of sensor nodes. Finally, the last two sections are dedicated, to explain in detail, the principle of the proposed "MOD-LEACH" approach, and its energy calculation model, also, to present the results of the various tests and measurements that had been done to validate its reliability and performances.

2 Related Works

This document is not the first one that evaluated the various techniques of clustering. Singh [5] gave an overview on several algorithms and based on the LEACH protocol, aimed at reducing energy consumption in a WSN, such as: LEACH-A, LEACH-B, LEACH-C, LEACH-Cell, LEACH-E, LEACHEEE, LEACH-F, LEACH-K, LEACH-M, LEACH-Multihop, LEACH-S, LEACH-TL, LEACH-V, but the author concluded in this work that these algorithms still fail in prolonged WSNs. Heinzelman et al. In work [6] proposed the Low-Energy Adaptive Clustering Hierarchy "LEACH" algorithm which is also based on clustering. "LEACH" allows to form clusters of sensor nodes based on the areas where there is a strong signal received, then use Cluster Heads which are randomly selected and they communicate directly with the base station, to minimize power consumption and to reduce the amount of information sent to the base station.

In [7], the node having the highest energy should have the highest probability to become a Cluster Head. Indeed, each node must have an estimate total energy of all nodes in the network to calculate the probability that it becomes a Cluster Head. Therefore, each node cannot make the decision to become a CH if only its local information is known. In this case, the scalability of this protocol will be influenced.

Sh. Lee et al. have proposed a new "CODA" clustering algorithm [5] that divides the entire network into a few groups based on, the distance between a node and the base station, and the routing strategy. Each group has its own number of clusters and member nodes. As long as the distance between the nodes and the base station is large, more clusters will be built. However, the work of CODA is based on the global information of the position of the node, actually it's not scalable.

Fatima Es-sabery et al. [8], proposed a hybridization of two protocols "LEACH" and "PEGASIS". First of all, each node transmits the data to its next neighbor until it reaches the leader node that subsequently sends the data to the Cluster Head. Secondly, the same principle was applied but this time between the Clusters Heads. This approach is very effective in medium-sized networks, but it loses its performance in large-area networks.

3   Energy Model

The energy E_c consumed by a sensor is defined by the following equation:

$$E_c = E_{c-capture} + E_{c-traitement} + E_{c-communication} \qquad (1)$$

Such as:

- $E_{c-capture}$: Energy consumed by the capture unit;
- $E_{c-traitement}$: Energy consumed by the treatment unit;
- $E_{c-communication}$: Energy consumed by the communication unit;

The capture energy is very low, although the data processing energy is calculated as follows:

$$E_{DA} = 5nj/bit/signal \qquad (2)$$

$E_{c-communication}$ Means in two energies E_{tx} and E_{rx} as shown in [3].
Such as:

- E_{tx}: Transmission energy;
- E_{rx}: Receiving energy;

$$E_{c-communication} = E_{tx} + E_{rx} \qquad (3)$$

Or,

$$E_{tx}(s, d) = E_{tx-elec}(s) + E_{tx-amp}(s, d) \qquad (4)$$

$$\begin{cases} E_{tx}(s, d) = (E_{elec}.s) + \left(E_{fs}.s.d^2\right) \ ifd < d_0 \\ E_{tx}(s, d) = (E_{elec}.s) + \left(E_{amp}.s.d^4\right) \ elseif \end{cases} \qquad (5)$$

$$E_{rx}(s) = (E_{elec}.s) \qquad (6)$$

Such as:

- s: Size of the data in bits;
- d: Distance between the sending node and the receiving node;
- E_{elec}: Energy of electronic transmission;
- E_{fs}: Amplification energy for $d \leq d_0$;
- E_{amp}: Amplification energy for $d > d_0$;
- d_0: Limit distance for which the transmission factors change their value;

The distance d_0 is given as follows:

$$d_0 = \sqrt{\frac{E_{fs}}{E_{amp}}} \qquad (7)$$

If $d \geq d_0$, we use E_{amp}, otherwise we use E_{fs}.

The geographical distance "d", between the transmitter and the receiver, and the size of the data provided "s" are dominant factors in the energy consumption (see Fig. 1).

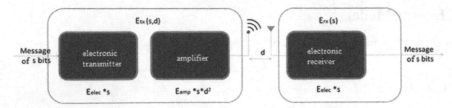

Fig. 1. Energy model of a wireless sensor node.

4 Proposed Approach

4.1 Operating Principle

Compared to direct transmissions in a WSN, the existed protocol "LEACH" has a good performances in saving energy and the number of nodes over the time, these performances are further improved by the hybridization approach of LEACH and PEGASIS, applied intra-cluster and inter-cluster [7]. However, these two techniques still present energy-intensive transmissions which are made over distances d greater than d_0, $d > d_0$.

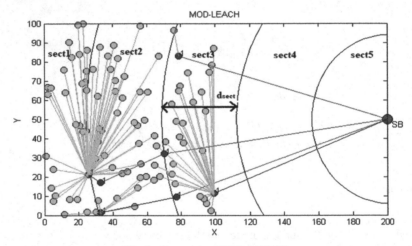

Fig. 2. Sectors in «MOD-LEACH».

During a transmission over a long distance $d > d_0$, the energy consumed is multiplied by a coefficient d^4 according to the energy model given above, which exhausts the energy of the node considerably, and accelerates the death of it.

The approach proposed in this work is to reduce the energy consumed by avoiding these transmissions over long distances. This is achieved by inserting the intermediate nodes acting as proxies between the cluster head and the base station.

In order to properly choose these intermediate nodes, the area of the WSN is divided into circular sectors concentric around the base station, and distant by a distance d_{sect} (see Fig. 2). For Cluster Heads furthest from the base station, our protocol chooses an intermediate node in each sector separating the Cluster Head from base station, and the transmission instead of being direct, it will be proxied by these intermediate nodes.

To prevent that the planning of WSN becomes a multi-hop planning, an intermediate node is selected once, and Cluster Heads cannot be intermediate nodes to not accelerate the depletion of their batteries.

4.2   Improvement of Routes

Our algorithm "MOD-LEACH" (LEACH MODifier) is used to improve the routes calculated by the "LEACH", and it has four main steps:

The first step comes to initialize the parameters, and to build the network.
The second step is to apply "LEACH" on the WSN, which is composed of the following sub-steps:

- Verification of dead nodes;
- Election of nodes that can be Cluster Heads; according to the conditions of "LEACH" protocol.
- Cluster formation, which means, associating each node with the nearest Cluster Head;

The third step aims at modifying the routes formed by "LEACH", to eliminate the transmissions with distances exceeding d_0, it relates to 2 sub-steps:

- Find the set L of nodes that will make transmissions at distance $d > d_0$;
- Find, for each node of the set L, recursively, an intermediate node (previously not used as intermediate and not Cluster Head) in the previous sector, until we arrive at the base station, if we do not find a relay node in a sector, we look in the sector that precedes it.

The fourth step completes the loop of the algorithm by calculating the energy of the network, and updating the energy of each node.

The flow chart of "MOD-LEACH" is shown in Fig. 3.

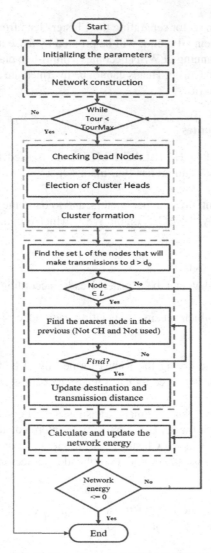

Fig. 3. Flow chart of «MOD-LEACH».

4.3 Energy Calculations

Technically "LEACH" and "MOD-LEACH" allow to calculate for each node of the WSN, the transmission distance to the destination node, the number of packets to receive and the number of packets to send.

In the case of application of "LEACH", a normal node n sends a single packet of k bits, towards its destination, the energy E_{c-n} consumed by this node is given by the following equation:

$$\begin{cases} E_{c-n} = k.(E_{tx} + E_{fs}.d^2) \ if \ d \leq d_0 \\ E_{c-n} = k.(E_{tx} + E_{amp}.d^4) \ if \ d > d_0 \end{cases} \tag{8}$$

A Cluster Head, receives nodes of its Cluster, s packets of k bits, gathers and compresses $(s + 1)$ packets in a single packet, and transmits it to the base station, the expression of its consumed energy E_{c-ch} is given by the following equation:

$$\begin{cases} E_{c-ch} = k.(E_{rx}.s + E_{DA}.(s + 1) + E_{tx} + E_{fs}.d^2) \ if \ d \le d_0 \\ E_{c-ch} = k.(E_{rx}.s + E_{DA}.(s + 1) + E_{amp}.d^4) \ if \ d > d_0 \end{cases} \qquad (9)$$

In the case of application of "MOD-LEACH", an intermediate node n_i, receives a single packet of k bits, from the source, groups and compresses two packets (the one received + its own packet) and transmits a single packet to its destination. The energy consumed E_{c-ni} is given by the following equation:

$$E_{c-ni} = k.(E_{rx} + 2.E_{DA} + E_{tx} + E_{fs}.d^2) \qquad (10)$$

The choice of d_{sect} affects the energy consumed by the network, a comparison of the consumption according to the distance was made for the sending of a single data packet by a Cluster Head to Base Station, with a direct transmission, and with a proxied transmission, as it shown in graph of Fig. 4,

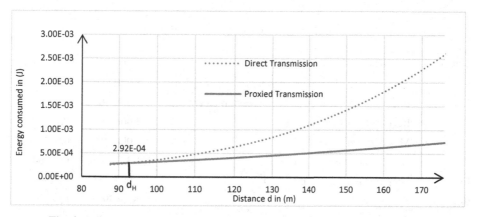

Fig. 4. Energy consumed by a direct transmission and an indirect transmission

It can be seen that in the direct case, the transmission energy increases exponentially with the distance d, while the proxied transmission shows a small linear increase according to the distance d.

We also note that for $d < d_H$ the direct transmission energy is slightly lower than the proxied transmission energy, this is explained by the fact that the terms of the energy of reception of the data packet and the energy of aggregation of two packets, added to the expression of the proxied transmission energy consumption, which unnecessarily increases consumption. And for $d \ge d_H$, we find that direct transmission becomes greedy energy.

Hence the choice of:

$$d_{sect} = \frac{d_H}{2} \qquad (11)$$

this choice of d_{sect} ensures that the maximum distance between two nodes in two neighboring sectors does not exceed d_H, and subsequently the transmissions between them are made at short distance, which avoids the use of amplification circuits and saves the energy of nodes.

5 Simulation Results

The development and implementation of our algorithm "MOD-LEACH" is made using MATLAB. A coding of "LEACH" algorithm and "DT" proved necessary in order to use it to compare the performances of our approach.

5.1 Paramètre De Simulation

To test the performances of our algorithm "MOD-LEACH", and to compare it with the other algorithms, we took for parameters of the WSNs with a small, medium, large and extra-large areas, and with a variable node densities as shown the following tables (Tables 1, 2, 3, 4 and 5):

Table 1. Energy parameters taken for simulation.

Symbol	Parameter	Value	Unit
Etx	Transmission electronics energy	5,00E−08	j/bit
Erx	Energy of aggregation and compression electronics	5,00E−08	j/bit
EDA	Receive electronics energy	5,00E−09	j/bit
k	Message size	2,00E+03	bits
Eamp	Amplification energy if d ≥ d0	1,30E−15	j/bit/m4
Efs	Amplification energy if d < d0	1,00E−11	j/bit/m^2
d_{sect}	Sector distance	46,3529	m

Table 2. Parameters taken for the simulation in the small network.

Symbol	Parameter	Value	Unit
(X, Y) = S	Area	(50 × 50) = 2500	m^2
(sx, sy)	Base station coordinates	(150 × 25)	m
Density	Density of nodes	0,04	Nudes/m^2
N	Number of nodes	100	Node
d_{sect}	Sector distance	46,3529	m

Table 3. Parameters taken for the simulation in the medium network.

Symbol	Parameter	Value	Unit
(X, Y) = S	Area	$(100 \times 100) = 10000$	m^2
(sx, sy)	Base station coordinates	(200×50)	m
Density	Density of nodes	0,04	Nudes/m^2
N	Number of nodes	400	Node
d_{sect}	Sector distance	46,3529	m

Table 4. Parameters taken for the simulation in large network.

Symbol	Parameter	Value	Unit
(X, Y) = S	Area	$(350 \times 100) = 35000$	m^2
(sx, sy)	Base station coordinates	(450×50)	m
Density	Density of nodes	0,0114	Nudes/m^2
N	Number of nodes	400	Node
d_{sect}	Sector distance	46,3529	m

Table 5. Parameters taken for the simulation in the extra-large network.

Symbol	Parameter	Value	Unit
(X, Y) = S	Area	$(400 \times 200) = 80000$	m^2
(sx, sy)	Base station coordinates	(440×100)	m
Density	Density of nodes	0,0075	Nudes/m^2
N	Number of nodes	600	Node
d_{sect}	Sector distance	46,3529	m

5.2 Results

Small Network Figure 5, Fig. 6 and Fig. 7 show the results obtained by applying algorithms DIRECT, LEACH, Hybridization of LEACH and PEGASIS, MOD-LEACH to the small network.

Medium Network. Figure 8, Fig. 9 and Fig. 10 show the results obtained by applying the four planning algorithms to the medium network.

Large Network. Figure 11, Fig. 12 and Fig. 13 present the results obtained by applying the four planning algorithms to the large network.

Extra-Large Network. Figure 14, Fig. 15 and Fig. 16 present the results obtained by applying the four planning algorithms to the extra-large network.

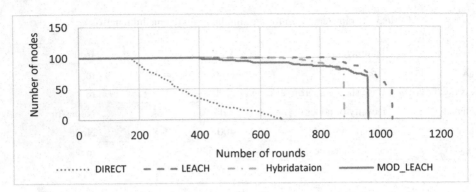

Fig. 5. Number of live nodes according to the number of turns, case of the small network.

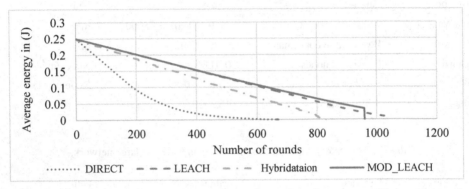

Fig. 6. Average energy of the network according to the number of turns, case of the small network.

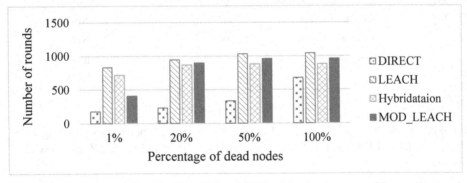

Fig. 7. Number of rounds for 1%, 20%, 50% and 100% of dead nodes, case of the small network.

5.3 Analyze

In the case of small network, with a small area (50 m × 50 m), and a relatively high density, about 0.04 Nodes/m², "LEACH" presents the best performances in terms of energy saving and network nodes, compared to Hybridization and "MOD-LEACH".

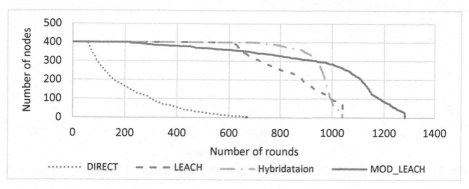

Fig. 8. Number of live nodes according to the number of turns, case of the medium network.

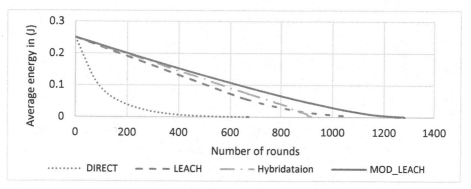

Fig. 9. Average energy of the network according to the number of turns, case of the medium network.

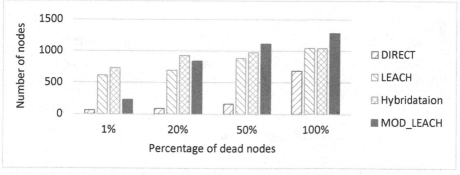

Fig. 10. Number of rounds for 1%, 20%, 50% and 100% of dead nodes, case of the medium network.

This is explained by the fact that these two last algorithms are unnecessarily overloaded with receptions and aggregations of data packets that increase the energy consumption of in the network. Then "LEACH" shows its efficiency in preserving energy in networks of

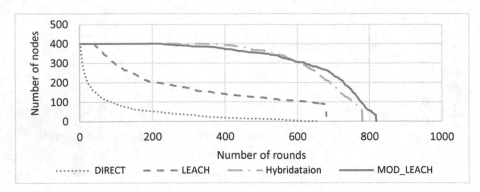

Fig. 11. Number of live nodes according to the number of turns, case of the large network.

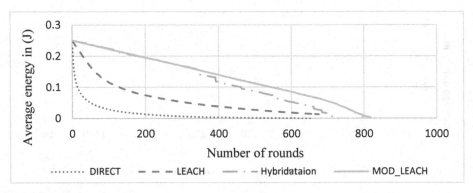

Fig. 12. Number of live nodes according to the number of turns, case of the large network.

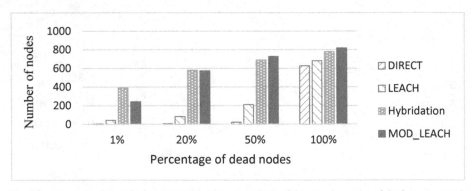

Fig. 13. Number of rounds for 1%, 20%, 50% and 100% of dead nodes, case of the large network.

sensors with small area and high density of nodes, while the choice of direct transmission "DIRECT" in such a network, appears as an energetic genocide for the nodes.

By increasing the area of the WSN (100 m × 100 m), while maintaining the density of the nodes constant (0.04 Nodes/m²), the case of the medium network, we note a

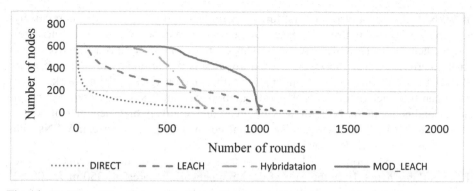

Fig. 14. Number of live nodes according to the number of turns, case of the extra-large network.

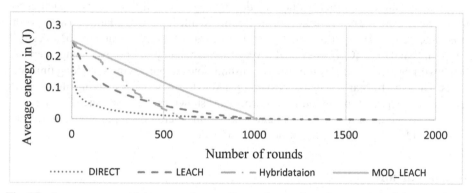

Fig. 15. Average energy of the network according to the number of turns, case of the extra-large network.

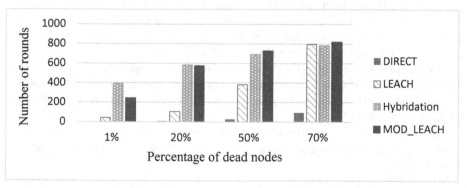

Fig. 16. Number of rounds for 1%, 20%, 50% and 70% of the dead nodes, case of the extra-large network.

change of roles between the candidate algorithms. "LEACH" reveals its inability to preserve the energy of the network, because the limits of the managed distances are so far. While the Hybridization algorithm presents the best performances in these kind of WSN, and especially in the first turns where the density of the nodes is relatively high. As we advance in time (increase of the turns), the number of dead nodes increases and the density of the nodes decreases. This decrease in density has a negative effect on the performance of the Hybridization algorithms which ends with the give in to "MOD_LEACH" in terms of saving nodes lifetime and energies of the network.

It can be concluded that the Hybridization algorithm, has a good preservation of energy and nodes in WSNs with medium areas, and relatively high densities of nodes. While "MOD-LEACH" has high reliability and competitive performance, in WSNs with large areas, and low nodes densities.

This last conclusion is validated by the tests carried out on the large network and the extra-large network. In the case of the large network, an area of (350 m × 100 m) is used, with a node density higher than 0.0114 Nodes/m^2. In the first moments when density of nodes remain higher, the Hybridization algorithm is more efficient and more competitive, until the moments when the density begins to fall, il loses its capacities to preserve energy of the network and nodes lifetime. While in the extra-large network, with a too large area (400 m × 200 m) and a very low density of nodes, about 0.0075Nodes/m^2 are used, "MOD-LEACH" presented a very high efficiency of energy management of the network by saving largely the energy consumption and preserving for a long time the nodes. These performances are the best compared to those of the competing algorithms.

The new approach proposed in this work, has shown this performances by solving problems of energy optimisation encountered by some hierarchical planning algorithms in WSN with a large areas and low densities of nodes.

6 Conclusion

In a WSN, transmissions over long distances activate the amplification circuits in the micro-sensors, which quickly depletes their power. This constraint motivated us to propose a new algorithm "MOD-LEACH", to optimize the standard LEACH algorithm for saving energy in WSN by avoiding these energy-intensive transmissions. Routes calculated by "LEACH" are replaced by short-distance transmissions using well-chosen intermediate nodes between Cluster Heads and Base Station.

In order to validate the approach in terms of extended network life and the efficient management of energy consumption, we have implemented it using MATLAB and compared the results obtained with protocols "DT", "LEACH", and the Hybridization "LEACH + PEGASIS".

The results of the various tests performed, prove that our protocol offers better energy saving compared to concurrent protocols, in WSN with large areas and low nodes densities.

References

1. Merck, M.: The icecube detector: a large sensor network at the south pole. IEEE Pervasive Comput. **4**, 43–47 (2010)
2. Akyildiz, I.F., Su, W., Sankarasubramaniam, Y., Cayirci, E.: Wireless sensor network: a survey. Comput. Netw. **38**(4), 393–422 (2002)

3. Elappila, M., Chinara, S., Parhi, D.R.: Survivable path routing in WSN for IoT applications. Pervasive Mob. Comput. **43**, 49–63 (2018)
4. Hitesh, M., Amiya, K.R.: Fault tolerance in WSN through PE-LEACH protocol. IET Wirel. Sensor Syst. **9**(6), 358–365 (2019)
5. Singh, K.: WSN LEACH based protocols: a structural analysis. In: International Conference and Workshop on Computing and Communication (IEMCON), pp. 1–7 (2015)
6. Heinzelman, W.: An application -specific protocol architecture for wireless microsensor networks. IEE Tran. Wirel. Commun. **1**(4), 600–670 (2002)
7. Lindsey, S., Raghavendra, C.: PEGASIS: power-efficient gathering in sensor information systems. In: Proceedings of IEEE Aerospace Conference, pp. 1125–1130 (2002)
8. Es-sabery, F., Ouchitachen, H., Hair, A.: Energy optimization of routing protocols in wireless sensor. Int. J. Inform. Commun. Technol. (IJ-ICT) **6**(2), 76–85 (2017)

Big Data, Datamining, Web Services and Web Semantics (Poster Papers)

Brain Cancer Ontology Construction

Fatiha El Hakym and Btissame Mahi[✉]

Information Processing and Decision Support Laboratory, Department of Computer Science,
Faculty of Science and Technology, PO Box. 523, Beni Mellal, Morocco

Abstract. An ontology is an integral part of a semantic web. Ontology can be designed and create the necessary metadata elements to develop a semantic web applications. The evolution of semantic web has encouraged creation of ontologies in many domains. This work aims to create an ontology of brain cancer to do this we need to define the steps of creation of an ontology; first, we need a lot of information about brain cancer. This article describes the different steps involved in creating a brain cancer ontology. We have used a Protege 4. 2 to for the construction of a brain cancer ontology.

Keywords: Semantic web · Ontology · Protege · Brain cancer · Owl

1 Introduction

The semantic web aimed to enable a much sophisticated management system by organizing knowledge in conceptual spaces using automated tools [1]. The idea of the semantic web is to present web content in a form that is easier to use for machine processes, and to intelligent take advantage of this rendering. To represent information in a form that can be interpreted syntactically and semantically with the computer, we use ontologies in the semantic web. Ontologies have a hierarchical structure of terms that describe a domain that can be used as the basis for a knowledge base [2, 3]. In this article, we have shown a method of ontological construction for brain cancer using protégé 4.2. The ontology of brain cancer can be used as a guide to understanding how brain cancer works. We have shown that brain cancer works with different classes. The different steps involved in creating a brain cancer ontology are described in the fourth part.

The newspaper was divided into five sections. The introductory section provides an initial discussion of some aspects of the paper, such as its purpose, scope, methodology, and system design: a description of Brain Cancer Ontology. It also deals with the search techniques within the knowledge base; there have also been discussions about integration with an ontological browser, which is technically known in information science as a faceted web browser. The final section presents the results achieved during the various stages of ontological construction, along with provisions for future work.

2 Literature Review

Many studies based on artificial intelligence and knowledges share has been developed to help physician understanding and improving the diagnosis results in several medical

© Springer Nature Switzerland AG 2021
M. Fakir et al. (Eds.): CBI 2021, LNBIP 416, pp. 379–387, 2021.
https://doi.org/10.1007/978-3-030-76508-8_27

fields, such as brain cancer [4], breast cancer [5, 6], diabetic retinopathies [7, 8], etc. In this section, we are focusing of brain cancer previous work. Several studies have been performed on the detection of brain tumors, and different researchers have raised different ideas based on their opinions and methods used to detect them, so that a diversified conclusion about this disease and its causes and remedies can be achieved. Different techniques are currently used that are based on neural network, Convolutional Neural Network CNN is then used to group and segment the images and finally to detect the cancer [9]. [10] have come up with an improved implementation of Brain tumor detection, segmentation, and classification are three processes that can be used to produce computer-aided methods for diagnosing tumors from magnetic resonance imaging. Diagnosis by means of MRI and MRS is the most important way to detect brain tumors. MRI in Morocco is an expensive affair. Brain scans using magnetic resonance imaging, computed tomography, as well as other imaging modalities, are fast and safer methods for tumor detection [11].

Our developed ontology will primarily benefit physicians and researchers, but its main purpose is to assist novice neurosurgeons. By introducing the symptoms into the system, surgeons may be able to recover the information corresponding to a particular disease. In addition, the system can also help these physicians identify the disease and consider available treatments. Our system is smartly equipped to show possible medical options available for the disease in question.

3 Purpose

The objective of this work is the construction of an information retrieval system based on ontology. To create this ontology, information has been obtained from a wide variety of information sources in the brain cancer literature. This collected information has been analysed to provide a standard, reliable and relevant information base for our proposed system.

4 Methodology

We will see in this section the different steps of the construction [12] of an ontology that we would like to create [13]. However, before creating our domain ontology, we need to have an idea of what ontology can deal with:

- Which doctors have specialized knowledge about brain cancer, and in which hospitals are they available?
- What are the different symptoms of brain cancer?
- What types of brain cancer?
- What are the different hospitals in the country specialized in the treatment of brain cancer?
- Who are the brain cancer specialists in Morocco?
- Which brain cancer are most common in men and women between the ages of 40 and 70?

4.1 Problems of Current Information Retrieval Systems

The various search engines rely mainly on keyword research mechanisms. The results retrieved in this way will show all documents wherever that particular term appears many unnecessary documents are retrieved [14]. For example, someone looking for information on a hospital specializing in the treatment of brain cancer in flap, he will have a lot of useful and useless information. For this, we have tried to develop an information search system based on an ontology.

4.2 Steps of Building Ontology

Step 1 Identification of Terminology: We have relied for our work on technical terms that have been collected from various sources published by different brain cancer associations. We have selected "about brain cancer: a primer for patients and caregivers" by the American Brain cancer Association®, as main sources. The American Association of Neurological Surgeons (AANS) has also taken terms from a classification.

Step 2 Analysis: The formal concepts collected in the previous step were analysed to identify their commonalities and differences. They were also analysed to identify the concepts that will serve as an instance, category or property. Than these terms have been used as basic elements to construct a brain cancer ontology.

Step 3 Synthesis: The synthesis involves the order of the facets according to their similarity in the properties and the labelling of the categories. After the division into classes, they were organized in hierarchical order. Step 4 standardization;
   Brain cancer types

- Primary brain cancer
- Secondary brain cancer

   Brain cancer cause

- Environmental cause
- Genetic cause

Step 4 Standardization: Aims to conceptualize the results of the previous step. The terms to be kept are defined according to their context and from a definition in natural language. The concepts are then identified as well as the semantic relationships between them. They are alike in the form of a semantic network.
   For our work, we have standardized terms from SNOMED CT®.

Step 5 Ordering: The order of the terms should be based on the purpose, scope, and object of the ontology. We have organized a brain cancer class according to the AANS classification system. One limitation is that in our work we only have standardized terms that were used to construct a brain cancer type classes in addition to SNOMEDCT®.

4.3 Brain Cancer Ontology Using the Concept DERA

DERA is a faceted [15, 16]. Knowledge organization framework (Domain, Entity, Relation and Attribute). Domains consist of three elements in DERA [17], namely entity (E), attribute (A) and relation (R). Then we will describe the ontology of brain cancer from the DERA point of view. In this ontology, brain cancer is a domain that contains a class, the relationship between classes or objects, and the attribute or property.. Entity is an elementary component consisting of facets built of classes and their instances, having either perceptual correlates or only conceptual existence within a domain in context. An example of entity class:

- Glioma
- Glioblastoma
- Anaplastic astrocytoma
- Oligodendroglioma
- Ependymoma
- Ganglioglioma

Relation is an elementary component consisting of facets built of classes representing the relation between entities. An example of Relation: Doctors brain cancer specialize, in which hospital they are available. Attribute is an elementary component consisting of facets built of classes denoting the qualitative/ quantitative or descriptive properties of entities.

Table 1. Glossary of terms

Term name	Description
Brain cancer	A tumor lesion that develops in the skull
Glioma	These are the most well-known cancers, so called because the damaged cell is the glial cell, which is found around neurons
Glioblastoma	This is grade IV gliomas, it is the most aggressive and the most common
Chordomas	Originating in embryonic cells of the spinal cord or the base of the cranial nerve
Hemangioblastomas	That originate in the blood vessels
Meningiomas	Beginning in the membrane covering the brain
Hospital	An institution where the sick or injured are given medical or surgical care
Benign Tumor	Are abnormal accumulations of cells that multiply slowly and usually remain isolated from the surrounding normal brain tissue
Patient	Sick person

We describe some terms on which on which we are based (Table 1) to arrive at the construction of the ontology.

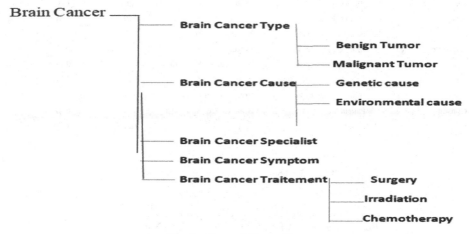

Fig. 1. Concept hierarchies.

A partition of a concept C is the set of subclasses of C, which do not share instances and cover C. Figure 1 represents the hierarchy of the concepts of our ontology.

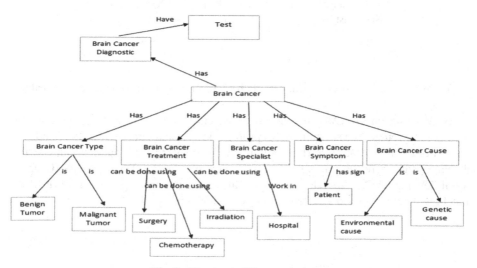

Fig. 2. Diagram of binary relations.

We describe the binary relationships between the source concepts and the target concepts with their cardinalities (Table 2), if we consider Brain cancer as a source concept and Brain cancer diagnostic as a target concept (Fig. 2) we will have a source cardinality of 0.n and target cardinality from 1.1 as shown in Table 2.

Table 2. Table of relationship

Name of relationship	Source concept	Target concept	Source card	Target card
Has	Brain Cancer	Brain Cancer Diagnostic	0..n	1..1
Has	Brain Cancer	Brain Cancer Specialist	0..n	1..n
Is	Brain Cancer Type	Malignant Tumor	1..1	1..1
can be done using	Brain Cancer Treatment	Surgery	0..1	0..1
Have	Brain Cancer Type	Brain Cancer Symptom	1..n	0..n
specialist	Brain Cancer Specialist	Brain Cancer Type	1..n	1..n
Causedby	Brain Cancer Type	Brain Cancer Cause	0..n	0..n
Consult	Patient	Brain Cancer Specialist	0..n	0..n
Do	Patient	Test	1..n	1..n

Table 3. The table of concept instances

Concept	Instance	Attribute
Benign Tumor	Chordomas	Name
Benign Tumor	Meningiomas	Name
Benign Tumor	Hemangioblastomas	Name
Malignant Tumor	Glioma	Name
Brain Cancer Specialist	Dr.LAZRAK Hicham	Name
Hospital	MedV	Name

Table 3 represents some instances of the concepts and the type of each attribute (Fig. 3).

By clicking on OWLVIZ for visualization, class and subclasses appear, and dot error may be found. The solution to this problem is to install Graphiviz 2.28 (Window). Then, in the Protégé 4.2 File (tab) preferences, OWLVIZ, a dot application, add the absolute path where this "dot.exe" file is located, e.g. in case Windows OS (C: \programFile\Graphviz2.28 \bin\dot.exe). The validity of the ontology was confirmed by matching the inferred and asserted ontology as shown in Fig. 4.

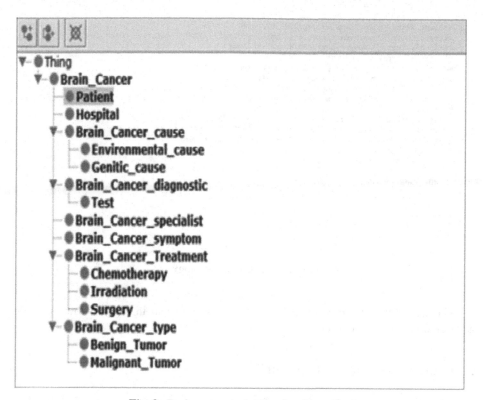

Fig. 3. Brain cancer ontology class hierarchy.

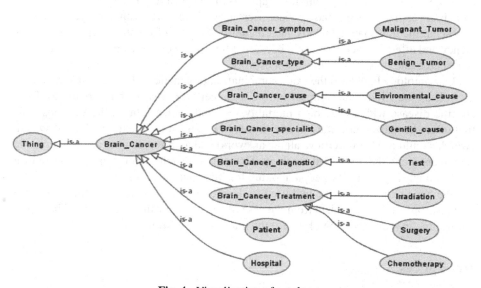

Fig. 4. Visualization of ontology.

```
<Ontology xmlns="http://www.w3.org/2002/07/owl#"
  xml:base="http://www.semanticweb.org/fatih/ontologies/2021/1/untitled-ontology-7"
  xmlns:rdfs="http://www.w3.org/2000/01/rdf-schema#"
  xmlns:xsd="http://www.w3.org/2001/XMLSchema#"
  xmlns:rdf="http://www.w3.org/1999/02/22-rdf-syntax-ns#"
  xmlns:xml="http://www.w3.org/XML/1998/namespace"
  ontologyIRI="http://www.semanticweb.org/fatih/ontologies/2021/1/untitled-ontology-7">
<Prefix name="" IRI="http://www.semanticweb.org/fatih/ontologies/2021/1/untitled-ontology-7#"/>
  <Prefix name="owl" IRI="http://www.w3.org/2002/07/owl#"/>
  <Prefix name="rdf" IRI="http://www.w3.org/1999/02/22-rdf-syntax-ns#"/>
  <Prefix name="xsd" IRI="http://www.w3.org/2001/XMLSchema#"/>
<Prefix name="rdfs" IRI="http://www.w3.org/2000/01/rdf-schema#"/>
<Declaration>
<Class IRI="#Benign_Tumor"/>
    </Declaration>
        <Declaration>
  <Class IRI="#Brain_Cancer"/>
  </Declaration>
  <Declaration>
  <Class IRI="#Brain_Cancer_Treatment"/>
  </Declaration>
```

Fig. 5. Brain Cancer ontology.

The RDF OWL code generated for classes is shown in in Fig. 5 show the code generated for object properties.

5   Conclusion and Future Work

Presenting active information about brain cancer is very important and very helpful. The computer-based brain cancer ontology supports the work of researchers in gathering data from brain cancer research and provides users around the world with intelligent access to new scientific information quickly and efficiently. Shared information improves the efficiency and effectiveness of research because it helps to avoid unnecessary redundancies in research and thus avoid duplication of work.

Our ontology facilitates the exact combination of genetic and environmental factors as well as their individual impact on brain cancer. We wish to build ontology for a specific disease with a very high mortality rate and gather data on the best hospitals that offer specialized hospitals. Our vision is to create an information retrieval system (which should be able to retrieve all the answers) and at the same time act as a semantic search engine in the developed ontology. This ontology helps to design all new medical professionals as well as lay people looking for information.

We have reviewed the method used to develop ontology; while in future work we will put concrete and real data from the case of Morocco for the creation and sharing of knowledge in the form of ontology for the diagnostic of Brain cancer.

References

1. Antonial, G., Harmelen, F.: Semantic Web Premier. MIT Press, Cambridge (2008)
2. Kapoor, B., Sharma, S.: A comparative study ontology building tools for semantic web applications. Int. J. Web Semant. Technol. **1**(3), 1–13 (2010)

3. Staab, S., Studer, R. (eds.): Handbook on Ontologies. Springer, Heidelberg (2010). https://doi.org/10.1007/978-3-540-92673-3

4. El-Dahshan, E.S.A., Mohsen, H.M., Revett, K., Salem, A.B.M.: Computer-aided diagnosis of human brain tumor through MRI: a survey and a new algorithm. Expert Syst. Appl. **41**(11), 5526–5545 (2014)

5. Elmoufidi, A., El Fahssi, K., Jai-andaloussi, S., et al.: Anomaly classification in digital mammography based on multiple-instance learning. IET Image Process. **12**(3), 320–328 (2017)

6. Elmoufidi, A.: Pre-processing algorithms on digital x-ray mammograms. In: 2019 IEEE International Smart Cities Conference (ISC2). IEEE (2019)

7. Gulshan, V., et al.: Development and validation of a deep learning algorithm for detection of diabetic retinopathy in retinal fundus photographs. Jama **316**(22), 2402–2410 (2016)

8. Skouta, A., Elmoufidi, A., Jai-Andaloussi, S., Ochetto, O.: Automated binary classification of diabetic retinopathy by convolutional neural networks. In: Saeed, F., Al-Hadhrami, T., Mohammed, F., Mohammed, E. (eds.) Advances on Smart and Soft Computing. AISC, vol. 1188, pp. 177–187. Springer, Singapore (2021). https://doi.org/10.1007/978-981-15-6048-4_16

9. Jijja, A., Rai, D.: Efficient MRI segmentation and detection of brain tumor using convolutional neural network. Int. J. Adv. Comput. Sci. Appl. **10**(4), 536–541 (2019)

10. Abd-Ellah, M.K., Awad, A.I., Khalaf, A.A.M., et al.: A review on brain tumor diagnosis from MRI images: practical implications, key achievements, and lessons learned. Magn. Reson. Imaging **61**, 300–318 (2019)

11. Tandel, G.S., Biswas, M., Kakde, O.G., et al.: A review on a deep learning perspective in brain cancer classification. Cancers **11**(1), 111 (2019)

12. Horridge, M., Jupp, S., Moulton, G., et al.: A practical guide to building owl ontologies using protégé 4 and co-ode tools edition1. 2. The University of Manchester, vol. 107 (2009)

13. Giunchiglia, F., Dutta, B., Maltese, V., et al.: A facet-based methodology for the construction of a large-scale geospatial ontology. J. Data Semant. **1**(1), 57–73 (2012). https://doi.org/10.1007/s13740-012-0005-x

14. Campos, P.M.C., Reginato, C.C., Almeida, J.P.A., et al.: Building an ontology network to support environmental quality research: First steps. In: CEUR Workshop Proceedings (2018)

15. Giunchiglia, F., Dutta, B.: DERA: a faceted knowledge organization framework (2011)

16. Giunchiglia, F., Dutta, B., Maltese, V.: From knowledge organization to knowledge representation. KO Knowl. Organ. **41**(1), 44–56 (2014)

17. Naskar, D., Das, S.: HNS Ontology Using Faceted Approach. KO KNOWLEDGE ORGANIZATION **46**(3), 187–198 (2019)

Semantic Web for Sharing Medical Resources

Mohamed Amine Meddaoui[✉]

Laboratory of Information Processing and Decision Support, Department of Computer Science, Faculty of Sciences and Technics, Sultan Moulay Slimane University, Beni Mellal, Morocco

Abstract. The classical web is very helpful for using in different domain, especially in medicine. But still trying to improve results for this, we propose to use semantic web for sharing medical resources. In this article, we will discover the principle of the semantic web, which consists understanding the true meaning of the concepts in order to simplify the presentation of the data, which then serves to give the precise result during the research and the advantage of using the semantic web for sharing of medical resources. We use logical «Protege» to present the concepts and relation between them. And the limits that require innovation.

Keywords: Semantic web · Ontology · Protege · Medical resources · Owl

1 Introduction

The practice of modern and quality medicine cannot be dissociated from rational processing of medical resources. Indeed, the increasing complexity of current medicine naturally lead to the establishment of information systems capable of helping the practitioner in his daily care tasks of the patient. For this, medical resources draws on research from various fields such as knowledge engineering and the semantic web. Each of these research areas provide medical informatics with methods, techniques and tools to improve formalization of data and knowledge in information systems.

The information generated by a hospital can be extremely useful information in a healthcare context when it is shared, for example, by various hospitals. Identified, the most significant of which is certainly the difficulty we have in represent medical knowledge in a standardized and unique way, even within the same medical specialty. The structures for storing (and by extension sharing) information do not help either no more single and concrete answer to the problem of medical resources. It raises the question of the use of information technologies and communication as part of the improvement of medical practices, the discovery of new knowledge or research medical. We will discuss the concepts of sharing this information More generally we will try to clearly expose the importance of semantics in the paradigm of sharing medical information at this day.

We start by state of the art to understand the importance of semantic web and we use logiciel «Protege 5.5.0» for the presentation and we discuss and limits that require innovation and we finish by conclusion and future work.

© Springer Nature Switzerland AG 2021
M. Fakir et al. (Eds.): CBI 2021, LNBIP 416, pp. 388–398, 2021.
https://doi.org/10.1007/978-3-030-76508-8_28

2 State of the Art

We know well the very important role of IT in the medical field, the use and processing of data are easier and practicable. But we notice with the accumulation of numerous data, the storage and the use and processing of data become more difficult and complex, for that the specialists seek to optimize these operations without touching modifying the content but changing the way of processing data.

Given its great importance, usefulness and survivability, specialists seek to apply semantics in the field of medicine taking into account the complexity of gathering data and categorizing them according to criteria in order to facilitate the exact extraction of a precise data and avoid searching in waves, we are talking about optimizing data storage and data processing and their extractions.

We know that medical resources are ramifying in the semantic approach, we adopt linked data, which makes it possible to visualize at once the two foundations of the Web: the data and the link. And which in passing emphasizes the current lack of links to give meaning to the data. We distinguish two major objectives of the Semantic Web:

- Adding meaning to increase the relevance of content.
- The creation of a web-scale database.

The ultimate goal is to allow machines/applications to use these relationships in order to offer the best possible services to individuals, services also allowing creating new relationships, in order to facilitate and optimize the use and processing of data and sharing medical resources.

3 Related Work

By [1] even before ontology validation methods began to develop, many researchers were interested in validation criteria that make it possible to objectify the expected quality of an ontology. In this section, we present four major lists of criteria. Sabou and Fernandez [2012] resume in a book dedicated to the validation of ontologies, the major validation methods that have been developed and the criteria taken into account. They also note that the criteria, which make it possible to validate an ontology very often, depend on its use.

4 Terminologies and Classifications

Terminology, considered a science, is concerned with the identification of concepts of a domain and the terms that designate it to facilitate the exchange of knowledge in one language and from one language to another. Terminology involves the standardization of terms in a domain so that they can be organize in relation to each other. The main interest of a terminology is to reduce the ambiguity that exists between the terms of a domain and therefore to be able to better share information. Defines a classification as the act of distributing concepts by classes and categories of a domain. It is a systematic distribution by classes or categories having common characteristics in a specific context. The structure and the depth of classification depends on the designer's objective.

5 Ontologies

The term ontology in the field of information science has been used and disseminated, which will define an ontology as "a shared specification of a conceptualization". A refined definition of what an ontology is: "An ontology involves or includes certain view of the world in relation to a given domain. This view often conceived as a set of concepts entities, attributes, processes, their definitions and their interrelationships. This called a conceptualization. Ontology can take different forms, but it will necessary include a cabular vocabulary of terms and a specification of their meaning. Is a specification partially reflecting a conceptualization? This last definition clarifies the previous ones and introduces an important term, to our meaning (Fig. 1).

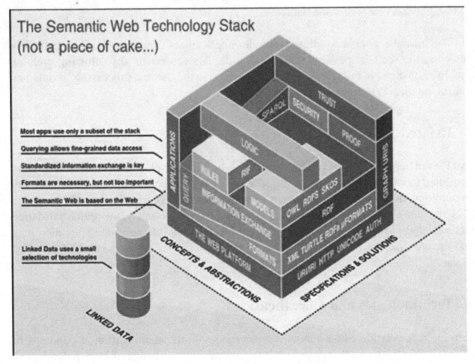

Fig. 1. Semantic web layers

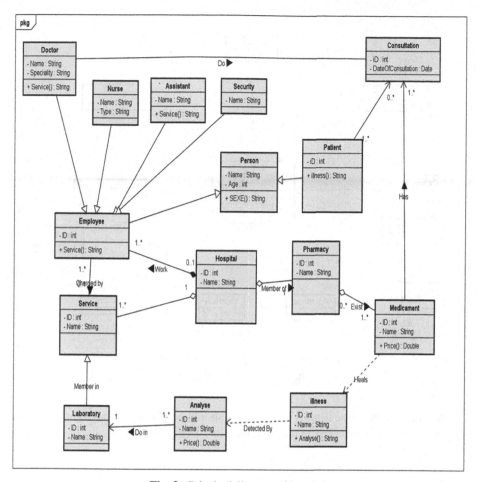

Fig. 2. Principal diagram of hospital

6 Schema

Hospital receive patients in the reception by assistant which will orient this patient in the appropriate service, nurse should prepare this patient to doctor for the consultation, some patients need analyze or scanner or radiology to know the degree of the disease to give in order to give the right medicament for this medicament (Figs. 2 and 3),

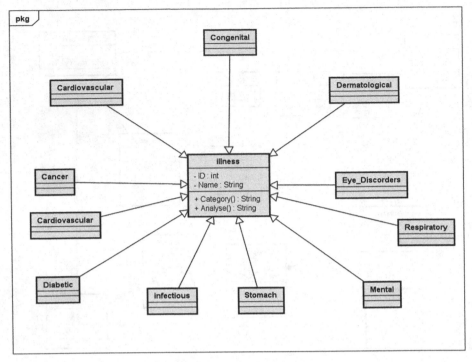

Fig. 3. Illness class diagram

This diagram describe illness, every illness should be classify in the right class, and describe the appropriate analyze to detect this illness (Fig. 4).

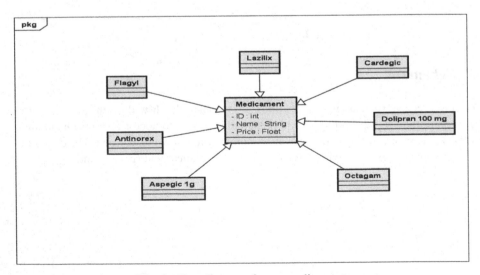

Fig. 4. Class diagram of some medicaments

This diagram describe medicament which presented by the appropriate attributes (Fig. 5).

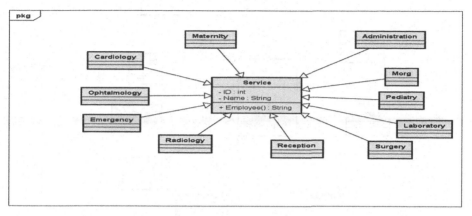

Fig. 5. Diagram of services

Lot of services constitutes the hospital; everyone specified to appropriate specialty. The hospital 'administration is managed by the administrator (Fig. 6).

A service joined employee (Doctor, Nurse, Assistant ...) of the same specialty.

The hospital has lot of services, every doctor is delegate only in one service abide by the appropriate specialty of doctor.

As exception, a doctor cardiologist can be Surgeon.

A service has also nurse's major leader and some trainee nurse. In addition, assistants.

We use **Protege-5.5.0** Logical to present the ontology of medical resources and relation between elements (Fig. 7).

Person should be just Man or Woman; we have disjunction, it is impossible to be Man and Woman in the same Time.

A Doctor should be a special just in one domain, doctor's specialty are independents. Some of Doctors have authority to practice Surgeon like Cardiologist. To have a perfect organization, it is necessary to classify Patient using Age criterion in order to make easy diagnostic and control (Table 1).

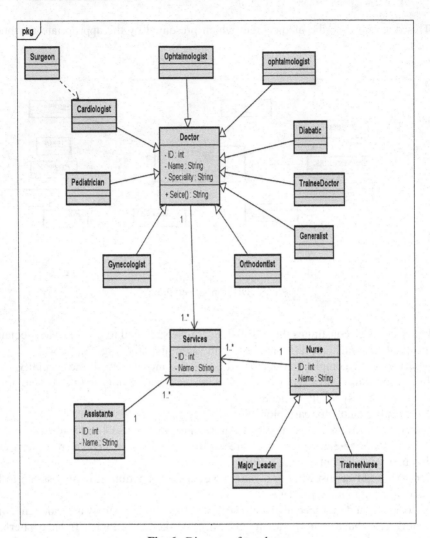

Fig. 6. Diagram of employee

6.1 Extract Information

We use SPARQL To find the specific data wanted, we give an example to extract data:
Find patients who have age more than fourteen years:

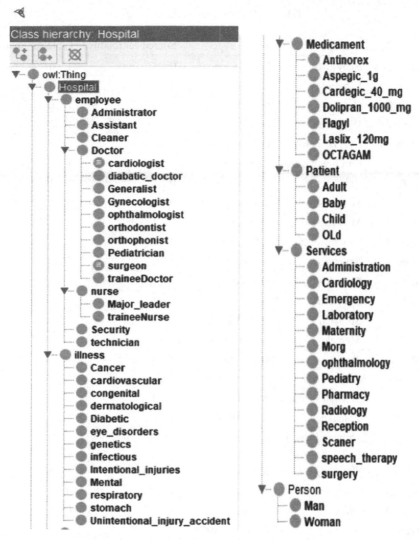

Fig. 7. Ontology of the hospital 'hierarchy

```
PREFIX pateint:    <http://xmlns.com/patient/>

SELECT ?name ?age

WHERE  {


        ?x patient:name ?name


        ?x patient:age age > 40


    }
```

Table 1. Description of relation ship

Relationship	Domain	Range	Source cardinality	Cible cardinality
Cared_By	Baby	Pediatrician	0...1	1...n
Doit_By	Laboratory	Analyse	1...n	1...n
Detected_By	Cancer	Scaner	1...n	1...n
Detected_By	Diabatec	Analyse	1...n	1...n
Cared_by	Eye_Discorders	Ophthalmologist	1...1	1...n
Cared_by	Cardiovasculare	Cardiologist	1...1	1...n
Cared_by	Baby	Pediatrician	0...1	1...1
Done_by	Surgeon	Cardiologist	0...1	1...1
Exists	Medicament	Pharmacy	0...1	1...1
Managed_By	Administration	Administrator	1...1	1...1
Received_By	Reception	Assistant	1...2	0...n
Cared_by	Genetics	Gynecologist	1...1	1...n
Cared_by	Infectious	Generalist	1...1	1...n
Cared_by	Stomach	Orthodontist	1...1	1...n
Cared_by	Mental	Psychologist	1...1	1...n
Cared_by	Teeth	Orthophonist	1...1	1...n
Cared_by	Respiratory	Orthophonist	1...1	1...n
Cared_by	Tension sickness	cardiologist	1...1	1...n
Cared_by	Pregnancy	Gynecologist	1...1	1...n
Cared_by	Unintentional_injury_accident	Generalist	1...n	0...n
Cared_by	Cracked bones	Radiology	1...1	0...n
Managed_by	Umergency	Assistant	1...2	0...n
Managed_by	Morg	Assistant	1...n	0...n
Managed_by	Maternity	Gynecologist	1...n	0...n
Managed_by	Pharmacy	Assistant	1...2	0...n
Managed_by	Technicals Problems	Technician	1...2	0...n
Managed_by	Speech_therapy	Orthophonist	1...1	0...n
Surgery	Operations	Surgeon	1...3	0...n
Traineeship	TraineeDoctor	Major_leader	1...n	0...n
Traineeship	TraineeNurse	Major_leader	1...1	0...n
Consultation	Patient	Doctor	0...1	1...n

7 Discussion

By [8] the organization was rich in data but poor in information; actually this will be true when we don't have enough links between data, for that it is necessary to have sufficient description of data in order to specify the right link, but the question what is the criterion to do a right link ?

This question can be answered by [6] who propose since we assume that the info Sources exist within the data system independently of the conceptual layer, the entire system are going to be supported specific mechanisms for mapping the information at the sources to the weather of the ontology.

8 Limit

Machine is capable of performing trillions of operations on a chain. Then we understand that the machine must evolve in order to be able to face and treat the complexity of semantic networks. We are not talking about a power evolution, but rather how it should process information.

We know that there are several types of knowledge necessary for treatment some information. Knowledge of the location of data, of the structure storage, on the vocabularies used, on the query language. Then knowledge about the field, whether or not we will organize it according to the founding concepts (top-ontology). Once this knowledge is formalized, it must be linked. Use them for specific purposes. These uses are generally defined; we took the part to formalize the knowledge of data as best as possible, their location, and the domain. We then created applications capable of to use this knowledge for purposes defined by man. We could sometimes use formal knowledge to help interoperability, for example to deduce queries during the data integration process, or to adapting the results obtained to each source towards a common formalism.

We were also able to use the knowledge to make groupings of data. But much remains to be done.

Semantics is a necessary step in moving towards medical resources sharing but not enough. First of all, because there are as many semantics as there are observers or uses. Current representation languages (ontologies) although extremely powerful, do not seem to be sufficient to be able to model everything, to all uses. It will therefore be necessary to invent systems capable of using several models of the same world. This does not mean to stop the effort of convergence, because it often improves the quality of the modeled field. But he There will certainly never be a single ontology. Again, the machine must help man, not force him.

9 Conclusion

Today, medical resources are very rich data, but the cumulus of this data generate, for this, the semantic web is very helpful to describe medical resources. By proposing using ontology to understanding, concepts in order describe all organization of the hospital: Services, Doctors, Patients, illness Medicaments…the good understanding for the hospital's organigram will make easy do lot of task very quickly more than normal web.

This will make extract information too easy and very quickly ,because when we have a good comprehension of subject then we can find a perfect answer or result for the search , using searching by concept which give a precise result , **contrariwise** searching by keyword who give lot of results that can make it disturbed because of wave results. Also using semantic web will economize lot of storage space because of two important points:

Data are **clarified**: that means we have no ambiguity, because data are describe. Data are **simplify** in fact this because of good understanding of concepts we will minimize **redundancy** data. Finally, to complete and ensure the performance of using semantic web in medical resources, it is necessary to describe links between data. We will try to use a **big data** in order to assure the efficiency of the semantic web, especially when have many processes in the same time and test performance. In addition, we will make sure the security of data, in order to protect privacy.

References

1. Charlet, J., Declerck, G., Dhombres, F., Gayet, P., Miroux, P., Vandenbuscche, P.Y.: Building a medical ontology to support information retrieval: terminological and metamodelization issues (2013)
2. Lemoine, M.: Trois conceptions sémantiques des théories en médecine. Lato Sensu, revue de la Société de philosophie des sciences 1(1), 1–11 (2014)
3. Semantic Web - Wikipedia figure the semantic web - Not a piece of cake... - benjamin nowack's blog (https://bnode.org/)
4. Messai, R.: Ontologies et services aux patients: application à la reformulation des requêtes, Doctoral dissertation, Université Joseph-Fourier-Grenoble I (2009)
5. Du Chateau, S., Mercier-Laurent, E., Bricault, L., Boulanger, D.: Modélisation des connaissances et technologies du Web sémantique: deux applications au patrimoine culturel. Humanités numériques, (2) (2020)
6. Poggi, A., Lembo, D., Calvanese, D., Giacomo, G., Lenzerini, M., Rosati, R.: Linking data to ontologies. In: Spaccapietra, S. (ed.) Journal on data semantics X. LNCS, vol. 4900, pp. 133–173. Springer, Heidelberg (2008). https://doi.org/10.1007/978-3-540-77688-8_5
7. Bertaud-Gounot, V., Guefack, V.D., Brillet, E., Duvauferrier, R.: Les technologies du web sémantique pour un renouveau des systèmes experts en médecine. Principes, problèmes et propositions à partir de l'exemple du myélome dans le NCI-T. Systèmes d'information pour l'amélioration de la qualité en santé: Comptes rendus des quatorzièmes Journées francophones, 47 (2012)
8. Burgun, A., Bodenreider, O., Jacquelinet, C.: Issues in the classification of disease instances with ontologies. Stud. Health Technol. Inform. 116, 695 (2005)
9. Karami, M.: Semantic web: a context for medical knowledge discovering and sharing. Front. Health Inform. 7, 2 (2018)

Construction of Glaucoma Disease Ontology

Yousra Nahhale and Hind Ammoun[✉]

Faculty of Science and Technology, Sultan Moulay Slimane University, Beni Mellal, Morocco

Abstract. Nowadays, accessing and sharing available information about the medical field in a smart way is a major problem. Ontologies in the medical field provide a machine-understandable vocabulary as well as it associates basic concepts of this field and the relations between these concepts, which makes it possible to make the knowledge reusable. In this article we will build an ontology of glaucoma disease.

Keywords: Ontology · Glaucoma · Concept

1 Introduction

The ontology of the domain has been studied and used in multiple fields of research (such as medicine, biology, military applications, philosophy, linguistics…). The ontology allows the storage of domain knowledge in the common formalism. Unlike databases, they offer the possibility of storing domain semantics and of providing reasoning mechanisms. Over the past decades, research in the field of biomedical informatics has continued to grow. At the same time, various biomedical tools have been designed to perform data collection and management.

Nowadays, database systems are faced with a great deal of information management (for Examples of clinical records, patient diagnoses, laboratory examinations, medical images…). Therefore, it is very important:

- Correctly collect and store medical information provided during consultation, treatment and medical research;
- In addition, gather useful knowledge from this data in order to provide patients with the best treatment.

These efforts aim to create models that effectively standardize all knowledge in medical activities, so that it can be used later, and to make the detection and treatment of known pathologies more efficient. In the formal mechanisms used in recent years, we have discovered ontology.

They are interested in the creation of entities: attributes and relationships of concrete or abstract terms. Ontology is based on the establishment of a general vocabulary and it uses the representation and concepts defined by the terminological resources. To build ontology resources to meet specific needs. These resources enable data interoperability and help increase knowledge.

© Springer Nature Switzerland AG 2021
M. Fakir et al. (Eds.): CBI 2021, LNBIP 416, pp. 399–413, 2021.
https://doi.org/10.1007/978-3-030-76508-8_29

In every country or continent, multidisciplinary groups have sprung up, seeking to standardize medical knowledge and language in order to better use these resources.

Glaucoma covers a group of eye diseases responsible for excavation and atrophy. The gradual disappearance of the optic disc and the gradual disappearance of the visual field. Without treatment or late diagnosis, glaucoma can lead to blindness. Defining the early stages of the disease, detecting and treating it early is still a challenge. In the country Canada, glaucoma is the second leading cause of blindness in people over 50. In all cases, 90% are primary open-angle glaucoma (POAG) and it increases with age. It is more common in blacks than in blacks. In white. High intraocular pressure is an important risk factor for the disease, but its positive predictive value for the disease is low. The prevalence of primary narrow-angle glaucoma also increases with age, and the prevalence of Inuit and Asians is higher. The disease will also be more common in women. A variety of diagnostic methods can be used to detect the presence or absence of glaucoma, but unless routine screening is performed, no method can effectively identify the disease at an early stage. The common treatment for glaucoma is to reduce intraocular pressure. Usually, but not always, the development of the disease can be prevented [16].

In our work, we try to build an ontology for the pathology of glaucoma, which formalizes the concept of field. It includes the concepts, properties that characterize the patient and all relevant examinations and diagnostic tests.

2 Methodology

In this section, we give a brief overview of the ontology development methodologies. Gruninger and Fox [12] used the capacity problem to extract concepts, axioms and definitions of terms from ontology, and established the conditions that characterize the integrity of ontology. The methodology of the literature [11] provides step-by-step advice to perfect the components of the ontology. Its framework provides good interoperability between agent-based applications. According to [13], there is no one and only correct way to model domain ontology, because there are always alternative ways to model it. The method proposed in [14] focuses on the user domain. This work proposed four concepts of ontology, but did not study the interrelation between these concepts. Nanda et al. [15] developed their product family domain ontology methodologies. They use formal conceptual analysis to determine the similarities between design artefacts. We use methodologies [11] to develop our ontology, which defines the tasks to be performed to develop a coherent and complete conceptual model. These tasks gradually increase the complexity of the conceptual model. The progressive improvement of the components of the ontology makes the ontology dynamic and open to change and growth. In addition, the methodology helps to build ontologies from scratch and it can be applied to the reuse of existing ontologies.

3   The Notion of Ontology

Ontology is a term borrowed from philosophy and a branch of philosophy that involves the nature and organization of reality. In recent years, several definitions of ontology have been given, but one of the definitions that characterizes the essence of ontology is based on the related definition in: "Ontology is a formal specification and explicit of shared conceptualization. Conceptualization refers to the abstract model of certain phenomena in the world, which determines the appropriate concept of the phenomenon. Explicit means that the type of concept used and the constraints of its use must be clearly defined. Formalization refers to the fact that the machine must be able to understand ontology, that is, it can interpret the semantics of the information provided. Shared indicates that ontology supports consensus knowledge, and it is not restricted to certain individuals but accepted by a group [2].

3.1   Ontology Components

Ontology can be seen as a set of structured concepts and the relationships between these concepts, aiming to represent objects in the world in a form understandable by humans and machines. The components of the ontology are:

- Concept [3]: Or class, which defines a set of abstract or concrete objects, we want to model a model for a given domain. Knowledge carrier on the objects to which we refer through concepts. A concept can represent a material object, a notion and an idea. A concept is characterized by a set of properties:

  – If the extension is not accepted, the concept is universal. The truth, for example, there is no extension.
  – A concept has an identity property, if the attribute can achieve the following goals: Distinguish two instances of this concept.
  – If a concept cannot be an instance of other concepts, it is rigid. For example, a living being is a rigid concept, but "humans" is not a rigid concept, because humans are "a living being".
  – If a concept can be an instance of the others, then it is anti-rigid concept.

- Relations [3]: They reflect the (correlated) links that exist between the concepts that exist in the reality analysis part. These relationships include associations: The ≪ subclass of ≪, the ≪ part of ≫, with is associated with the instance ≪ of ≫, ≫. These relationships allow us to see the structure and interrelation of concepts, one related to the other. One or more terms, and a signature specifying the following the number of instances of the concept of relational link, its type and order Concept, how to read the relation, characterize them.

- Instances [3]: or individuals constitute the extensional definition of ontology (to represent specific elements).
- Function [3] Function is a special case of relation, where an element of a relation is defined from other elements.
- Axioms [3]: Ontology is further composed of axioms, which form the semantic constraints of inference are stored in a conceptualization. They take the form of logical theory. Axiom The form makes it possible to check the coherence of the ontology.

3.2 Life Cycle of an Ontology

Since the ontology is intended to be used as a software component among the following components: The development of a system that meets different operational objectives. Ontology must be seen as an evolving technical object and have the need to specify the life cycle. In this case, with on the one hand, the ontology is management activities. Including planning, control, and quality assurance. On the other hand, development-oriented activities brought together Pre-development activities, development and post-development the inspired software engineering life cycle (see Fig. 1) includes an initial phase Needs assessment, it becomes an idea, the realization of the idea Reflected in the design that has been published for its use. Then there is the evaluation stage, which generally passes to a Model evolution and maintenance stage. Reassess the ontology and requirements must be met after each major use. The ontology can be extended and partially reconstructed if necessary. The verification of the knowledge model is at the heart of the process and performs repeatedly.

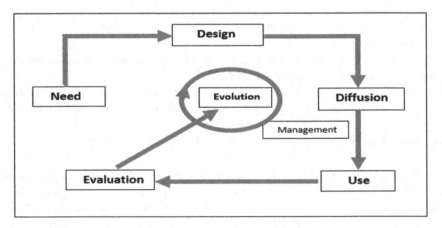

Fig. 1. Life cycle of an ontology [2]

4   Glaucoma Disease

4.1   Definition

Glaucoma is an eye disease that can cause progressive and irreversible damage to the optic nerve.

The most common cause is excessive pressure in the eye. Deterioration of the optic nerve leads to amputation of the visual field, in particular amputation of the peripheral (lateral) visual field. There are two types of conditions: open angle glaucoma (OAG) and iridocorneal angle closure glaucoma. Glaucoma is the second leading cause of blindness in the world after age-related macular degeneration (AMD). In France, nearly 800,000 people suffer from glaucoma, and it is estimated that around 500,000 people suffer from this disease and are unknown [1].

4.2   The Eye and Humor Aqueous

The eye or eyeball is a spherical organ that is specifically used to detect light signals, which are transmitted to the brain in the form of electrical signals to form images. The eyes are soft organs comparable to balloons. This is because it produces and contains a liquid called aqueous humor, which is spherical like the air in a balloon. This aqueous humor is produced in the ciliary body and drained through a filter called trabeculae. This fluid puts pressure on the eyes (eyes), and this pressure changes with disease. In glaucoma, the drainage or outflow of aqueous humor may be disturbed, resulting in high pressure in the eye [1] (Fig. 2).

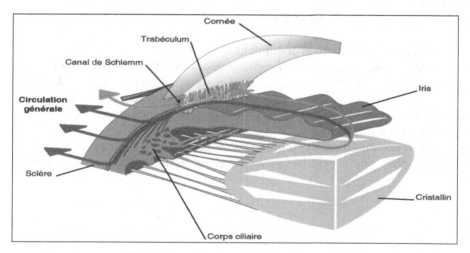

Fig. 2. Normal course of aqueous humor [1]

4.3   The Two Types of Glaucoma

The difference between the two types of glaucoma depends on the cause of the increased pressure in the eye.

- **Open Angle Glaucoma (GAO)**
  GAO is the most common form of the disease, accounting for almost 90% of glaucoma cases. This is due to the progressive deterioration of the drainage filter (trabeculae) of the aqueous humor. As the rate of fluid elimination slows, aqueous humor builds up in the eyes, which gradually increases intraocular pressure.
  High intraocular pressure can damage the origin of the optic nerve (head): the eye papilla. Damage to the nipple is progressive, but irreversible, resulting in loss of peripheral vision and irreversibility [1].
- **Iridocorneal close angle glaucoma**
  There is an angle between the iris (the color of the eye) and the cornea, called the iridocorneal angle. This angle is the passage from the aqueous humor to the trabecular. Closed-angle glaucoma is the closure of the iridocorneal angle, which causes sudden obstruction of the flow of aqueous humor. This interruption causes a sharp rise in intraocular pressure, which quickly and severely damages the optic nerve. Due to its severity, the disease is considered a medical emergency because its important prognosis for the eye is short-term [1].

5   Conception and Construction of Glaucoma Disease Ontology

After having a more complete understanding of the ontology, we will devote this part to the conception of the ontology of glaucoma pathology.

Step 1: Determine the domain of ontology
Our ontology is specially developed for glaucoma diseases. The objective of the ontology is to bring together the different concepts used in glaucoma diseases and their definitions and the properties that characterize patients. It also includes different factors, symptoms, possible diagnoses and associated tests. Different medical agents (doctors, nurses, etc.) or patients will use the ontology.
Step 2: List the important terms of the ontology
Glaucoma, diagnostic, patient, test, symptom, clinic, treatment (Tables 1, 2, 3, 4 and 5).

Table 1. Glossary of terms

Term name	Description
Glaucoma	A degenerative disease of the optic nerve
GAO	Chronic open angle glaucoma
GAFA	Angle closure glaucoma
Test	Medical exam
Watery humor	Aqueous humor
Symptom	Is a sign, a disorder observable by a patient
Treatment	Is a set of measures applied by a health professional to a person vis-à-vis a disease
Patient	Is a natural person receiving medical attention or to whom care is provided
Diagnostic	Is the process by which the doctor determines the condition from which the patient suffers, and which makes it possible to propose a treatment
Pressure	Abnormally high blood pressure in the blood vessels
Laser	Glaucoma operation by laser treatment
FO	Eye fundus
HTA	Arterial hypertension
HTO	Ocular hypertonia
NO	Optic nerve

Table 2. Definition of super classes

Classes	Descriptions
Glaucoma	It consists of a hierarchy of classes describing the pathology glaucoma. Can be considered the backbone of any ontology
Patient	Contains patient characteristics
Test	Understands the different devices and techniques to confirm a diagnosis of glaucoma
Treatment	Consists of the different treatments that will be offered to deal with this pathology
Diagnostic	Contains possible diagnoses

Step 3: Classes definitions

Step 4: Description of class properties

Step 5: Description of relationships

Table 3. Extract from class properties

Properties	Class	Type
CIN	Patient	String
Age	Patient	Integer
Date_of_Birth	Patient	String
High_risk	Factore_risk	Boolean
Id	All classes	Integer
....	...	...

Table 4. Extract from Class relations

Relationship	Domain	Range	source cardinality	Cible cardinality
Has_description	All classes	Description	0...1	1...1
Because of	Diagnostic	Symptom	0...n	0...n
Contains	Test	List _test	1...n	1...n
Has_ sign	Patient	Symptom	0...n	0...n
Confirmed_by	Diagnostic	Result	0...n	1...n
Need	Symptom	Test	0...n	1...n
Take_a_test	patient	List_test	1...n	1...n
Consult	Patient	Ophthalmologist	1...n	1...n
Is_of_type	Glaucoma	Type	0...n	1...1
Is a	Type	GAO	0...1	1...1
Is a	Type	GAFA	0...1	1....1
Contains	Glaucoma	Factor_risk	0...1	1...n
Of_type	Factor_risk	Disease	0...1	1...n
Of_type	Factor_risk	Age	0...1	1...n
Of_type	Factor_risk	Inherited	0...1	1...n
Of_type	Factor_risk	Ethnic_origin	0...1	1...n
Of_type	Factor_risk	Medication	0...1	1...n
A_risk_factor	Diagnostic	Factor_risk	0...n	1....n
A_risk_factor	Patient	Factor_risk	0...n	1...n
Treated by	Diagnostic	Treatment	1...n	1...1
Work in	Ophthalmologist	Hospital	1...1	1...n
...	...	...	...	...

Step 6: Extract from instances of classes

- **Explanation of some classes**

  The glaucoma class includes a set of subclasses, like the Factor risk subclass, which contains different types of risk factors, such as old age, medication, diseases (like diabetes)... As shown Figure, Type subclass include different types of glaucoma disease, namely GAO and GAFA (see Fig. 3).

Table 5. Extract from instances of classes

Class	Instance
Test	The fundus examination
Treatment	Laser treatment
Factor_risk	Disease (diabetes)
Factor_risk	Age (≥40)
Factor_risk	Ethnic origin (People with dark skin and people of African descent are more likely to develop glaucoma)
...	...

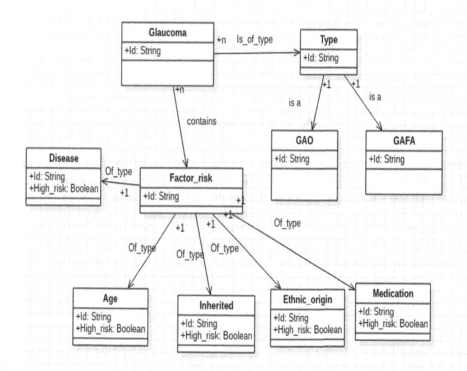

Fig. 3. Glaucoma class diagram

The test class contains a set of subclasses. For example, the "Test list" subclass contains a detailed test list (intracavity pressure measurement, The_iridocorneal angle check…) and the "Results" subclass contains all possible requested test values (See Fig. 4).

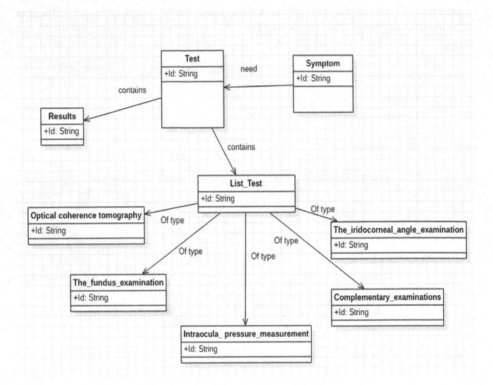

Fig. 4. Test class diagram

The "patient" class has multiple relationships. Every patient has a diagnosis and there are symptoms that help doctors diagnose the patient. Each patient can take a set of tests (Take test relation) and get their own results, and treatment based on the results (See Fig. 5).

The diagnostic class and the treatment class have a "treated by" relationship, that is, each diagnosis has a very specific treatment method. Each class of symptoms accompanied by a diagnosis also has a "because of" relationship (See Fig. 6).

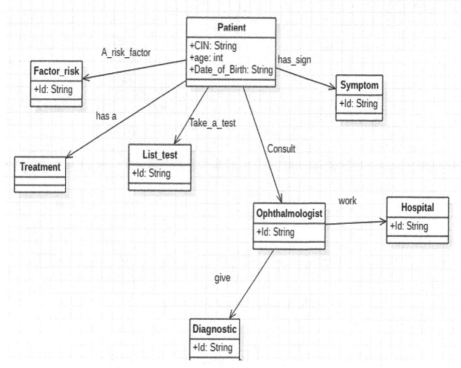

Fig. 5. Patient class diagram

6 Implementation

- Protégé

Protected is an authoring system used to create an ontology. It was created by Stanford University and it is very popular in the fields of the Semantic Web and Computer Research. Protégé is developed in Java. It is free and its source code is released under a free license (Mozilla Public License). Protégé can read and save ontologies in most ontology formats: RDF, RDFS, OWL, etc. It has several competitors, such as Hozo, OntoEdit and Swoop. He is known for his ability to handle a large ontology [4] (Figs. 7, 8, 9 and 10).

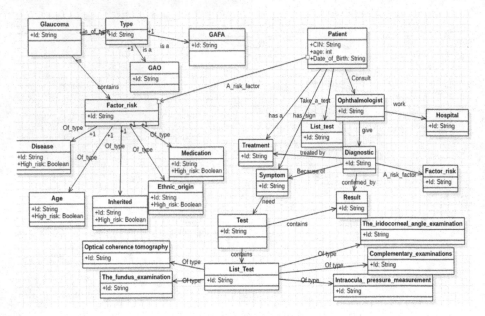

Fig. 6. Diagnostic class diagram

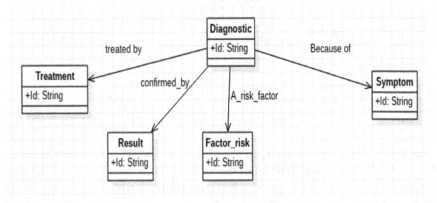

Fig. 7. **Global** diagram

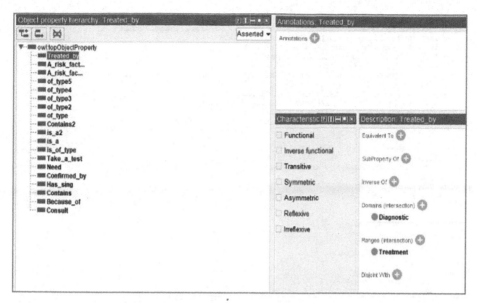

Fig. 8. Hierarchy of ontology

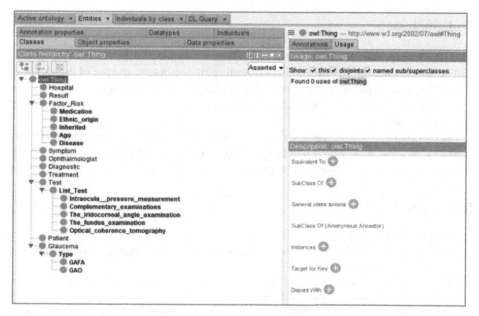

Fig. 9. Hierarchy of property

```
<Ontology xmlns="http://www.w3.org/2002/07/owl#"
    xml:base="http://www.semanticweb.org/hp/ontologies/2021/1/untitled-ontology-4"
    xmlns:rdf="http://www.w3.org/1999/02/22-rdf-syntax-ns#"
    xmlns:xml="http://www.w3.org/XML/1998/namespace"
    xmlns:xsd="http://www.w3.org/2001/XMLSchema#"
    xmlns:rdfs="http://www.w3.org/2000/01/rdf-schema#"
    ontologyIRI="http://www.semanticweb.org/hp/ontologies/2021/1/untitled-ontology-4">
    <Prefix name="owl" IRI="http://www.w3.org/2002/07/owl#"/>
    <Prefix name="rdf" IRI="http://www.w3.org/1999/02/22-rdf-syntax-ns#"/>
    <Prefix name="xml" IRI="http://www.w3.org/XML/1998/namespace"/>
    <Prefix name="xsd" IRI="http://www.w3.org/2001/XMLSchema#"/>
    <Prefix name="rdfs" IRI="http://www.w3.org/2000/01/rdf-schema#"/>
    <Declaration>
        <Class IRI="#Age"/>
    </Declaration>
    <Declaration>
        <Class IRI="#Complementary_examinations"/>
    </Declaration>
    <Declaration>
        <Class IRI="#Diagnostic"/>
    </Declaration>
    <Declaration>
        <Class IRI="#Disease"/>
    </Declaration>
    <Declaration>
        <Class IRI="#Ethnic_origin"/>
    </Declaration>
    <Declaration>
        <Class IRI="#Factor_Risk"/>
    </Declaration>
```

Fig. 10. Glaucoma ontology owl file

7 Conclusion

With the emergence of the Semantic Web and its commitment to the automated and unified use of data for the use and understanding of computers, the use of ontology to present knowledge has become a trend. Ontologies provide knowledge and the connections between them. Semantics are expressed in formal representations in an incomplete but simpler way.

In this paper, we have made a design as well as an ontology construction for the topology of glaucoma, which combines many concepts and the links between these concepts (classes and subclasses), properties and instances.

References

1. Vue, L.A.: GLAUCOME
2. Fatima, B.: Conception et implémentation d'une ontologie médicale Cas: insuffisance cardiaque. This de doctorate
3. Rahamane, S., Daouadji, M.: Les ontologies pour modéliser les processus de soins en établissement de santé. Thèse de doctorat
4. Bodenreider, O., Burgun, A.: Biomedical ontologies. In: Medical Informatics, pp. 211–236. Springer, Boston (2005)
5. Keita, A.K.: Conception coopérative d'ontologies pré-consensuelles : application au domaine de l'urbanisme. Institut National des Sciences Appliquées. Lyon, Ecole Doctorale Informatique et Information pour la Société (2007)
6. Noy, N., Mcguinness, D.L.: Développement d'une ontologie 101: Guide pour la création de votre première ontologie. Université de Stanford, Stanford, Traduit de l'anglais par Anila Angjeli (2000). https://www.bnf.fr/pages/infopro/normes/pdf/no-DevOnto.pdf

7. Espinasse, B.: Introduction aux ontologies. Notes de support de cours, Université Aix-Marseille, 48 p. (2010)
8. Bachimont, B.: Engagement sémantique et engagement ontologique: conception et réalisation d'ontologies en ingénierie des connaissances. Ingénieriedes connaissances: évolutions récentes et nouveaux défis, pp. 305–323 (2000)
9. Uschold, M., King, M.: Towards a methodology for building ontologies. In : Proceedings of the IJCAI-95 Workshop on Basic Ontological Issues in Knowledge Sharing, Montreal, Canada (1995)
10. Zghal, S.: Contributions à l'alignement d'ontologies OWL par agrégation de similarités. 2010. Thèse de doctorat. Artois (2010)
11. Fernández, M., Gómez-Pérez, A., Juristo, N.: Methontology: from ontological art towards ontological engineering. In: AAAI 97 Spring Symposium on Artificial Intelligence in Knowledge Management, Stanford University, California, USA, pp. 33–40 (1997)
12. Gruninger, M., Fox, M.S.: Methodology for the design and evaluation of ontologies. In: Proceedings of International Joint Conference of Artificial Intelligence Workshop on Basic Ontological Issues in Knowledge Sharing, Montreal, Canada, pp. 1–10 (1995)
13. Noy, N.F., McGuinness, D.L.: Ontology development 101: a guide to creating your first ontology. Knowl. Syst. Lab. **01**, 05 (2001)
14. Ahmed, S., Kim, S., Wallace, K.M.: A methodology for creating ontologies for engineering design. J. Comput. Inf. Sci. Eng. **7**(2), 132–140 (2007)
15. Nanda, J., Simpson, T.W., Kumara, S.R.T., Shooter, S.B.: A methodology for product family ontology development using formal concept analysis and web ontology language. J. Comput. Inf. Sci. Eng. **6**(2), 103–113 (2006)
16. Elolia, R., Stokes, J.: Série de monographies sur les maladies liées au vieillissement: XI. Glaucome. Maladies chroniques au Canada, vol. 19, no. 4, p. 172 (1999)

Signal, Image and Vision Computing
(Poster Papers)

Creation of a Callbot Module for Automatic Processing of a Customer Service Calls

Imad Aattouri[1]([⊠]) [iD], Mohamed Rida[1], and Hicham Mouncif[2]

[1] Hassan II University, Casablanca, Morocco
[2] Sultan My Slimane University, Beni-Mellal, Morocco
h.mouncif@usms.ma

Abstract. Bots are now more and more present in our daily life, we can find them in the form of chatbots, voicebots or FAQs. Since the number one source of customer demand is phone calls, callbot becomes a requirement in order to ensure a 24/7 presence and to minimize customer wait times in addition to demining charges. Callbots present a new kind of machine-human interface.

In this article, we are making a prototype call bot to manage incoming calls to a French-speaking call center specializing in customer relations. The first callbot is deployed for a period of 4 months, the first results are satisfactory and motivate for move on to new stages of setting up the callbot system.

Keywords: NLP · Human machine systeme · Callbot · French-speaking

1 Introduction

The first objective in the call centers that manage the customer relations department is mainly customer satisfaction. Customer satisfaction is generally lowered because of two points the bad response from the interlocutor or the long waiting time.

The callbot is software allowing communication between a machine and a human in a telephone call. It aims to respond to the user's customer and meet their needs. The callbot is a less well-known solution compared to other variant of automatic system (chatbots and voice bots).

The purpose of the callbot is to provide 24/7 processing, reduce processing time, lower processing costs, have only one response for the same request and everything is mainly to ensure customer satisfaction.

This article presents the realization of a prototype of a callbot capable of handling first level customer requests and only transferring calls where the callbot has difficulty in finding the answer.

The structure of this article is as follows: we will start with a study of the existing in the "Background" part, then we will detail the modeling of the callbot that we are going to set up in "Modelization" section. In the "Experiment" section, we will present the data set and the tools used in the implementation of this study. Then we will study the results in the "result and discussion" part.

© Springer Nature Switzerland AG 2021
M. Fakir et al. (Eds.): CBI 2021, LNBIP 416, pp. 417–424, 2021.
https://doi.org/10.1007/978-3-030-76508-8_30

2 Background

Many bots are used in daily life by many people like Siri in "Apple phone's", Alexa from Amazon, Cortana for Microsoft, and Bixby in Samsung phone's, theses bots are able to open apps, play music, set calendar events, and more actions (Dale 2016).

ELIZA was created in 1966 at MIT to imitate human conversation to emulate a physiotherapist in clinical treatment (Shawar and Atwell 2007). By using keyword matching, the Chatbot aimed to encourage users to disclose information about themselves and their family members. The ALICE chatbot was created in 1995 and was inspired by ELIZA (Shawar and Atwell 2015; Zahour et al. 2020).

3 Modelization

The presentation of the callbot framework is presented in the Fig. 1.

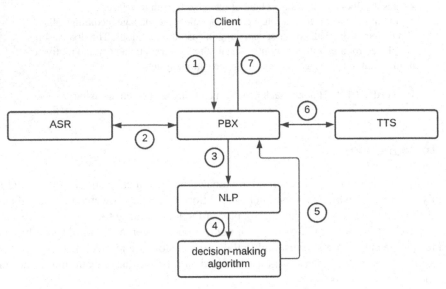

Fig. 1. Callbot framework presentation

- **Step 1:** Customer express his request.
- **Step 2:** The PBX picks up the client's request and sends it to the ASR system for converting to text. ASR sends text to PBX.
- **Step 3:** the PBX sends the customer request in text format to the NLP module.
- **Step 4:** The NLP module cleans the client's expression, retrieves the synonyms of the keywords detected and transfers it all to the decision-making algorithm module for decision making
- **Step 5:** the decision-making algorithm module finds the best result and sends the response to be provided to the client to the PBX in text format

- **Step 6:** the PBX transfers the text to the TTS, and retrieves it in voice form.
- **Step 7:** PBX converts voice into telephone signal and answers the customer.

- **Private Branch Exchange (PBX)**
  A private branch exchange (PBX), is a company telephone system whose role is to switch (redirect) incoming calls to users on local lines and allow everyone to share a certain number of external telephone lines (Jackson and Clark 2007).
- **Automatic speech recognition (ASR)**
  Automatic Speech Recognition (ASR) is the identification and processing of the human voice using computer hardware or/and software techniques for. The role of the ASR is to detect the words that a person has spoken.
  Automatic speech recognition is also known as automatic voice recognition (AVR), voice-to-text, speech-to-text or simply speech recognition.

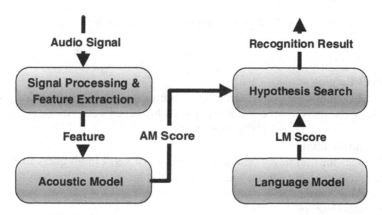

Fig. 2. Architecture of ASR systems (Yu and Deng 2014)

As shown in Fig. 2:

- The signal processing and feature extraction component takes as input the audio signal,
- enhances the speech by removing noises and channel distortions,
- converts the signal from time-domain to frequency-domain,
- extracts salient feature vectors that are suitable for the following acoustic models.

- **Text to speech (TTS)**
  Text to speech is the artificial production of human speech. A Text-to-Speech System (TTS) converts normal language text into speech; it works as follows:

  - **Normalization (or tokenization):** converts raw text containing symbols like numbers and abbreviations into the equivalent of written words.
  - **Text-to-phoneme conversion:** assigns phonetic transcriptions to each word, and divides and marks the text into prosodic units.

- Converts the symbolic linguistic representation into sound. In certain systems.
- **NLP:** Natural Language Processing (NLP) is a branch of artificial intelligence that enables machines to understand and process human languages. NLP enables interactions between humans and computers through natural language.

  This processing allows the analysis and understanding of natural language between computers and humans without going through computer languages but rather natural human languages instead of computer languages (Zhou et al. 2020; Chawla et al. 2018).

  The presentation of the NLP algorithm used is presented in Fig. 3

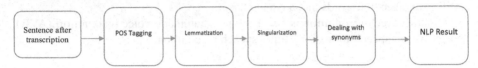

Fig. 3. Representation of the NLP algorithm

- **POS Tagging**

  In NLP, each word of a sentence is considered a token. The process of assigning one of the parts of speech (like nouns, verbs, adverbs etc.) to the given word is called Parts of Speech (POS) tagging.
- **Lemmatization**

  At times the user might enter different forms of verbs in his query like requires, required etc. But to frame the correct query, we need just the root form of the word i.e. 'require'.
- **Singularization**

  The user might enter the plural form or singular form. To resolve this issue, we need to singularize the item name before passing it on. In case the item name is already in singular form, it is passed on as it is otherwise its singular form is passed.
- **Dealing with Synonyms**

  Lists of all possible synonyms of all keywords were created and these lists were appended with all possible tense and singular/plural forms of the words already present in them (Chawla et al. 2018).

4   Experiment

A first step in setting up the callbot is to understand the customer's request and transfer it to the right service without giving an answer.

The purpose of this first part is to fill the knowledge base in order to know the customer requests and the way in which it is expressed and to see at what level the callbot manages to understand what the customer is asking.

- **Dataset**
  The data in this article are gathered in a call center for handling customer relations calls in a French company.
  The number of calls retrieved in this article is more than 194,000 calls. Gathered in a period of 4 months.
  The language for handling these calls is French.
- **Used tools**
  In this experiment, we used:

PBX: Asterisk
Asterisk is an open source PBX (Jackson and Clark 2007).
TTS: Google Cloud
Convert text into natural-sounding speech using an API powered by Google's AI technologies.
The price: Free for the first 0 to 1 million characters per month, after 1 million characters $16.00/1 million characters.
ASR: Google Cloud
Apply Google's most advanced deep learning neural network algorithms for automatic speech recognition (ASR). Supports more than 125 languages.
The price: Free for the first 0 to 60 min per month, after 60 min $0.006/15 s.
Flask
Flask is a lightweight WSGI web application framework. It is designed to make getting started quick and easy, with the ability to scale up to complex applications. It began as a simple wrapper around Werkzeug and Jinja and has become one of the most popular Python web application frameworks.
SpaCy
SpaCy is an automatic language processing Python software library developed by Matt Honnibal and Ines Montani. SpaCy is free software released under the MIT license.
The SpaCy library allows the following analysis operations to be carried out on texts in more than 60 languages (Explosion AI 2020).
NLTK
NLTK is a leading platform for building Python programs to work with human language data. It provides easy-to-use interfaces to over 50 corpora and lexical resources such as WordNet, along with a suite of text processing libraries for classification, tokenization, stemming, tagging, parsing, and semantic reasoning, wrappers for industrial-strength NLP libraries, and an active discussion forum (NLTK Project 2020).

5   Result and Discussion

This table presents the comparison of the operation of the voice IVR before the installation of the callbot and after.

The axes of comparison are the communication time, the percentage of calls transferred correctly and percentage of calls where the callbot was able to recover the customer's request.

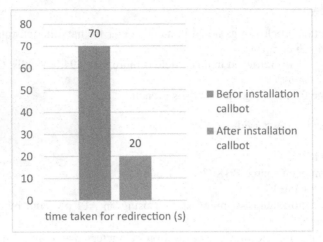

Fig. 4. Transfer time comparison

The first improvement noticed with the use of the callbot is the time to get in touch with the right service (Fig. 4), which went from 70 s on average to only 20 s when using the callbot. This lack of waiting time can be linked mainly to the time that the customer waited to listen to the voice message of the customer relations service to choose the right number to choose, which is done automatically using the callbot.

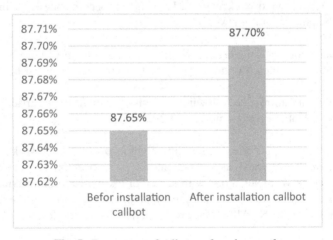

Fig. 5. Percentage of calls transferred correctly

The second factor compared is the percentage of calls transferred correctly to the right service (Fig. 5), the comparison shows that this figure is almost stable, with a slight improvement on the call bot side of 0.05%.

The third axis is the percentage of calls or the callbot to detect a response, this stat (Fig. 6) shows that the callbot manages to understand a large part of customer requests.

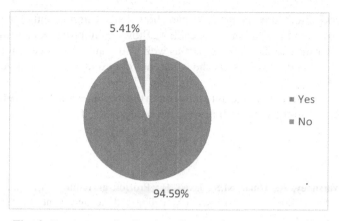

Fig. 6. Percentage of calls where the callbot detected a response

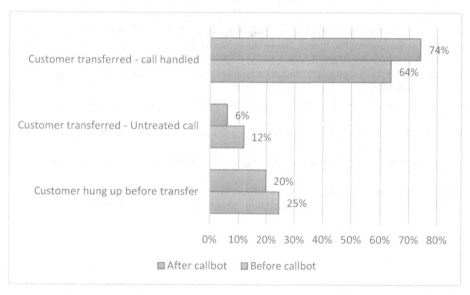

Fig. 7. Call processing statistics before and after setting up the callbot

In Fig. 7, we can see that the number of customers who hang up before the transfer to the right service has decreased by 5%, and the number of customers who hang up after the transfer has decreased by 6%.

The number of clients transferred and processed is increased by 10%.

6   Conclusion

In this work we have set up a framework model of a callbot. A first test step was to set up the callbot on the first level of call at customer relationship service level to ensure the transfer of customers to the right service.

The results presented in the previous part show that the implementation of the voice IVR (first step of the installation of the callbot) allows an improvement of the results in terms of transfer time to the right service, as well as the quantity of calls forwarded.

On the other hand, these results show that the robot understands a large number of requests.

Finally, the results encourage us to move on to the implementation of the final version of the callbot, where the robot will process the requests itself.

References

Chawla, A., Varshney, A., Umar, M.S., Javed, H.: ProBot: an online aid to procurement. In: 2018 International Conference on System Modeling and Advancement in Research Trends, Moradabad, India, pp. 268–273 (2018)

Shawar, B.A., Atwell, E.: Chatbots: are they really useful? In: LDV-Forum, pp. 29–49 (2007)

Shawar, B.A., Atwell, E.: ALICE chatbot: trials and outputs. Computación y Sistemas 19, 625–632 (2015)

Dale, R.: The return of the chatbots. Nat. Lang. Eng. 22(5), 811–817 (2016). Industry Watch

Yu, D., Deng, L.: Automatic Speech Recognition: A Deep Learning Approach. Springer, Seattle (2014). https://doi.org/10.1007/978-1-4471-5779-3

Explosion AI: Industrial-Strength Natural Language Processing in Python, 25 October 2020. spaCy: spacy.io

Jackson, B., Clark, C.: What is asterisk and why do you need it? In: Asterisk Hacking, pp. 1–20 (2007)

Zhou, M., Duan, N., Liu, S., Shum, H.Y.: Progress in neural NLP: modeling, learning, and reasoning. Engineering 6(3), 275–290 (2020)

NLTK Project: NLTK 3.5 documentation, 13 April 2020. Natural Language Toolkit. https://www.nltk.org/

Zahour, O., Eddaoui, A., Ouchra, H., Hourrane, O.: A system for educational and vocational guidance in Morocco: Chatbot E-Orientation. Procedia Comput. Sci. 17, 554–559 (2020)

Applied CNN for Automatic Diabetic Retinopathy Assessment Using Fundus Images

Amine El Hossi[1(✉)], Ayoub Skouta[2], Abdelali Elmoufidi[3], and Mourad Nachaoui[1]

[1] Mathematics Applications Laboratory, Faculty of Sciences and Technics, Sultan Moulay Slimane University, Beni Mellal, Morocco
elhossismiamine@gmail.com, nachaoui@gmail.com
[2] Computer Science, Modeling Systems and Decision Support Laboratory, Faculty of Science Aïn Chock, Hassan II University, Casablanca, Morocco
ay.skouta@gmail.com
[3] Information Processing and Decision Support Laboratory, Faculty of Sciences and Technics, Sultan Moulay Slimane University, Beni Mellal, Morocco
a.elmoufidi@usms.ma

Abstract. In the area of ophthalmology, diabetic retinopathy affects an increasing number of people. Early detection avoids severe diabetic proliferative retinopathy complications. In this paper, we propose a method for binary classification of retinal images using convolutional neural networks architecture. This method is formed to recognize and classify a retinal image as normal or abnormal retina. The paper setup is, first of all a preprocessing step is applied, next by data augmentation, and then a CNN formed, and applied. To train, validate and test the proposed model, we have used a public dataset "Resized version of the Diabetic Retinopathy Kaggle competition dataset" from Kaggle web site. Proposed model has trained using 4000 images of the normal retina and 4000 images of abnormal diabetic retina, and 500 images of the normal retina and 500 images of abnormal diabetic retina for testing. The accuracy Achieves 89% in 100 images of single prediction words.

Keywords: Convolutional neural networks · Binary classification of images · Deep learning · Diabetic retinopathy

1 Introduction

One of the most important sensors of the body is the eye, it is the responsible part of scanning, detection and analyzing that helps humans to interact with their environment [1]. The eye expresses the feelings, aliments of the body and it makes a huge difference in the beauty of the face. Nowadays, the eye undergoes a lot of risks and diseases [2] that requires conducting research and studies on the eye in order to understand it, know and facilitate methods of caring for it. In addition, among the most common dangers to the human eye is the DIABETIC RETINOPATHY (DR) because of the global spread of diabetes that reached epidemic proportions according to the International Diabetes Federation. Otherwise, ophthalmologists are studying this disease to avoid many cases

© Springer Nature Switzerland AG 2021
M. Fakir et al. (Eds.): CBI 2021, LNBIP 416, pp. 425–433, 2021.
https://doi.org/10.1007/978-3-030-76508-8_31

of blindness in order to recommend the appropriate treatment [3, 4]. In our study, we are going to focus on classifying a retinal image as a normal or abnormal retina. Therefore, although doctors are capable of it, but in some cases the lesions are too small and difficult to be detected. As we know, artificial intelligence is very useful in these cases [5], and convolutional neural network has proved itself effective in the automation procedure [6, 7]. It is able to study and recognize the important segment and features of an image and use it in the decisions [8, 9]. The structure of a convolutional neural network, layers and sequences, is capable to transform an inputted image to a value between 0 and 1 in the output layer that helps in what variety we place the image [9]. This transformation is the result of learning and validation from examples in the datasets. Our article will be structured as follows: the second section tackles the binary classification of diabetic retinopathy, and the next section about the definition of deep learning and convolutional neural networks. The fourth section explains and discuss our used method. In the final section, which is the conclusion, we summarize our results. This study is for classifying retinal images using convolutional neural network in one of this two categories, [0: abnormal retina candidate to be sick, 1: normal healthy retina].

2   Related Previous Work

In the medical domain, many computer-aided diagnoses are used to help in the detection of many diseases [10]. Even more, many diseases are early-detected using artificial intelligence [10]. In addition, one of the newest examples is the creation of two models, one to classify a COVID-19 test of a suspect patient as positive or negative, and the other to classify the hospitalization units of patients with COVID-19 [11]. Other diseases are breast cancer [12, 13]. In this part, we are focusing of the diabetic retinopathy previous works. Such as the evaluation study diagnosis of a decision support software for the detection of the diabetic retinopathy in Cameroon, where every patient (represented by two images) was classified by the ophthalmologist and the software as sick or not sick and the software was correct in 97,0% in the 538 images used in the study [14]. Moreover, in [15], an evaluation of retinal adults, with diabetes, images a machine learning algorithm had high sensitivity and proved specific for detecting referable retinas. An interesting example also is a deep learning model sensitivity was better than the general ophthalmologist's [16]. Also [17] published a classification method using deep learning. Another example in [18], were a computer-aided screening system (DREAM) that analyzes fundus images with varying illumination and fields of view and generates a severity grade for diabetic retinopathy (DR) using machine learning. Classifiers such as the Gaussian Mixture model (GMM), k-nearest neighbor (kNN), support vector machine (SVM), and AdaBoost are analyzed for classifying retinopathy lesions from non-lesions. GMM and kNN classifiers are found to be the best classifiers for bright and red lesion classification, respectively. Also [19] proposed a new learning model called granular support vector machines for data classification problems. Granular support vector machines systematically and formally combine the principles from statistical learning theory and granular computing theory.

The reasons why we used deep learning are the performance and the precision of CNN classification results in working with images. In addition, deep learning may allow to us continuous improvement after the start of using the model to work on live images of real patients.

3   BINARY Classification of Diabetic Retinopathy

The diabetic retinopathy is the leading cause of blindness among working-aged adults [9]. In addition, to limit the spread of this disease, we must know what the factors of it is and how to classify the symptoms for appropriate treatment. Many classifications have been used in DR [14–17], but in our paper, we are going to use a classification that allows us to predict either an image is for a normal or abnormal retina like in Fig. 1, the difference between the two cases of the retina is clear, without specifying the type of difference.

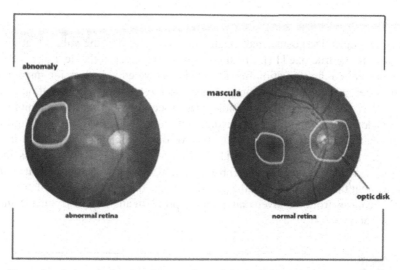

Fig. 1. Normal and abnormal retina (images from diabetic retinopathy (resized) dataset) (Color figure online)

In the Fig. 1 (images from "Resized version of the Diabetic Retinopathy Kaggle competition dataset"), normal retina contains just mascula and optic disc in clearly way, and a smooth presentation of the veins (Fig. 1). Otherwise, abnormal retina includes abnormal shapes and colors (Fig. 1), that we will consider it as abnormal in all disease types (Microaneurysms, Neovascularization, Vitreous/preretinal hemorrhage etc....).

4   Deep Learning and Convolutional Neural Network

The computer science engineers are always using human body to be inspired in lot of cases, and when it comes to intelligence, nothing in this world is more intelligent as human brain [20]. That why, to create an intelligent machine, engineers use the neural network (Fig. 2), which is the way that the brain works (feel, calculate, predict, decide, etc....).

A neural network is a group of neurons connected which each using linear equation AX + B. These coefficients (A and B) are randomly selected in the first try, then they are

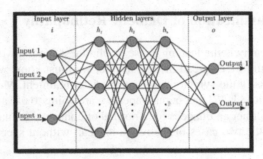

Fig. 2. Artificial neural network

calculated and optimized using mathematic algorithms to find the values that give the closest value to the true one [11]. The transition from a layer to the next one is decided by a function called the activation functions [11]. As we cited before, the inspiration is also done when it comes to analyzing images and the result was convolutional neural network. CNN are very inspired by biological process of seeing, it is a simple ANN (Fig. 2) in addiction to a number of convolutional layers, when we say convolution, we say filters that keeps only what we want survive on our image, and we give it to the ANN to optimize the A and B values [11]. The use of neural network is classified as deep learning, which is the process of providing a model the data and make it train and validate without the intervention of the programmer, and keep working on data (learn to extract the features from the given classification problem and to classify them) until we have a good accuracy (Fig. 3).

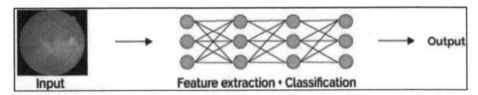

Fig. 3. Deep learning model

Otherwise, machine learning is the process of working in data statistic's, datamining, pattern recognition and predictive analyzes [12]. In general, machine learning works, learn and makes decisions based on what has been learned but the deep learning is about to create a model (ANN), and makes it learn and make intelligent decisions [12]. Deep learning remains better than machine learning, because experiences proved that the performance of deep learning is better the machine learning when it comes to big amount of data [12].

5 Proposed Method

In this work, we develop a model to classify a retina as normal retina or abnormal. First, we created two folders, training set and test set, and in each of these folders, we created the two sub-folders: normal and abnormal. Because we are going to work on the Keras library ImageDataGenerator that going to label whatever image in the folder with the name of the folder. For example, every image in the normal folder will be automatically classified as normal with 0 in the label, and either for abnormal folder. In this way, our convolutional neural network classes every image in the dataset with a numerical label (0: abnormal, 1: normal), so it will be easy in the final layer to predict the normality of a retina, that because in the output layer we will have one neural with 0 or 1 as a value. The architecture of our CNN (Fig. 4) is based on three convolution layers with 3×3 filter using ReLU as activation function followed by a maximum pooling layers of 2×2 bits. The number of filters is 16, 32, and 64 respectively in each layer. Next, we flat everything out and we apply two dense layers with sigmoid activation function since we have binary mode classification (Table 1).

Table 1. Parameters of the used CNN model

Layers	Layer type	Filters	Filters size	Activation	Param
1	Conv2D	16	3*3	ReLU	448
	Maxpool		2*2		0
2	Conv2D	32	3*3	ReLU	4640
	Maxpool		2*2		0
3	Conv2D	64	3*3	ReLU	18,496
	Maxpool		2*2		0
4	Flatten				73,856
5	Dense	512			3,392,265
Classifier	Dense			Sigmoid	1

Our model, as presented in the table, contain 3,489,971 parameters all Trainable.

The execution of our model has been adapted to 50 epochs. The result obtained is "98.1%" precision with two classes (normal or proliferative).

6 Dataset

The data we used in our experiment is extracted from a "Diabetic Retinopathy (resized)" dataset provided via Kaggle (Resized version of the Diabetic Retinopathy Kaggle competition dataset) on diabetic retinopathy in the training and testing. The dataset contains two folders, resized_train and resized_train_cropped, in each folder 35 100 images. The images are divided into five categories (normal: level_0, light: level_1, moderate: level_2, severe: level_3 and proliferating: level_4). In our study, we focused on only 4000 images from resized_train folder and we used resized_train.csv to move the images into normal and abnormal folders like follows. Every image with 0 in the label is moved to normal folder, else are moved to abnormal folder until we complete 2000 images in each folder.

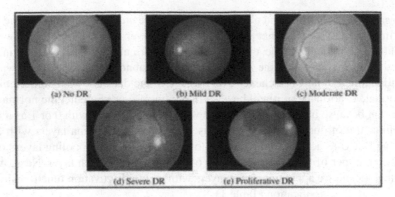

Fig. 4. Diabetic retinopathy (DR) levels with increasing severity [22]

7   Image Augmentation

To create a good classify method we need a much data for training and testing. For this reason, we are using a technique that makes us have more data from a limited size of images. This augmentation technique enlarges the quantity of the dataset used, by simply creation-modified images from the original versions. The process of increasing data can be used to train a model based on features in the image, regardless of their position in the input [21]. In our paper, we simply supplement images by a horizontal and vertical image of the old version, in (Fig. 4) a b and c are generated image by rotation from the original one.

8   Results

In this paper, we have created a model for binary classification of retina image as normal or proliferative. This model was created by implementing the deep learning approach, specifically we have built a model using on ImageDataGenerator library from Keras and giving it both normal and abnormal retina images from "Diabetic Retinopathy (resized)" dataset. The Fig. 4 below is a screenshot shows the last five obtained experimental epochs. The accuracy of the model achieved 98.88% as maximum value. Otherwise, experience has proofed the efficiently and the robustness of the used model.

The Fig. 5 and 6 represents model's accuracy and loss error details obtained in the experiment (Fig. 7).

For validation, we recorded 100 images for the testing step mixed between normal and proliferative retina. We are using the confusion matrix, which is a table that describes the performance of a classification model, to calculate the accuracy which the formula:

$$\text{Accuracy} = \frac{True\ Positive + True\ Negative}{True\ Positive + False\ Positive + True\ Negative + False\ Negative}$$

```
Epoch 45/50
53/53 [==============================] - 136s 3s/step - loss: 0.0374 - accuracy: 0.9900
Epoch 46/50
53/53 [==============================] - 136s 3s/step - loss: 0.1207 - accuracy: 0.9652
Epoch 47/50
53/53 [==============================] - 139s 3s/step - loss: 0.0569 - accuracy: 0.9864
Epoch 48/50
53/53 [==============================] - 135s 3s/step - loss: 0.0562 - accuracy: 0.9841
Epoch 49/50
53/53 [==============================] - 138s 3s/step - loss: 0.0755 - accuracy: 0.9740
Epoch 50/50
53/53 [==============================] - 136s 3s/step - loss: 0.0417 - accuracy: 0.9888
```

Fig. 5. Last 5 epochs

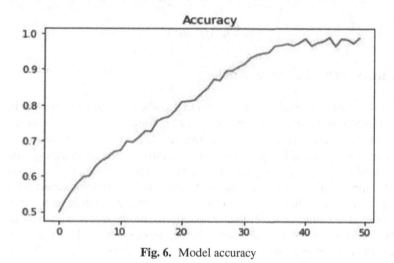

Fig. 6. Model accuracy

The accuracy is the number of images correctly classified a divided by the total number of images. The result obtained by our model is 98.1%; it correctly classified 98.1 images from 100 images. Therefore, the accuracy of our CNN is 98.1%.

The proposed model, despite its simplicity, can give desirable results, but it does not substitute for the intervention of the doctor. This is due to several reasons, the most important of which is that the classification does not Do not predict disease proactively, but only gives an expectation of whether the eye is sick or not.

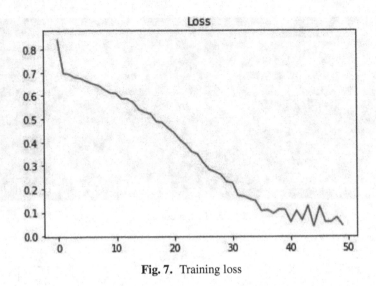

Fig. 7. Training loss

9 Conclusion and Future Work

In this proposal, we have created a model to help in diabetic retinopathy disease diagnosis using digital fundus images. The objective is to automate the diagnosis based on the CNN architecture and using digital fundus images. The model proved itself capable to differentiate normal retina image from the proliferative retina by detecting anomalous in the retina photography. This differentiation has 98.1% accuracy. This model could be used in the implementation of a rapid aid system of anomaly in retina screening. It could be very practical and helpful for the doctors if the screening retina image is inputted at the model at the time of the consultation. In the future work, we are focusing on improving the model to be able to classify the degree of severity of the retinopathy, our model could be developed and improved to be able to classify which type of anomaly we have in the retina, microaneurysms or Neovascularization, exudate and hemorrhage.

Reference

1. Martinez-Conde, S., Macknik, S.L., Hubel, D.H.: The role of fixational eye movements in visual perception. Nat. Rev. Neurosci. **5**(3), 229–240 (2004)
2. Behar-Cohen, F., Martinsons, C., Viénot, F., et al.: Light-emitting diodes (LED) for domestic lighting: any risks for the eye? Prog. Retinal Eye Res. **30**(4), 239–257 (2011)
3. Ennant, M.T.S., Greve, M.D.J., Rudnisky, C.J., et al.: Identification of diabetic retinopathy by stereoscopic digital imaging via teleophthalmology: a comparison to slide film. Can. J. Ophthalmol. **36**(4), 187–196 (2001)
4. Lisboa, P.J.G.: A review of evidence of health benefit from artificial neural networks in medical intervention. Neural Netw. **15**(1), 11–39 (2002)
5. Rynjolfsson, E., Mcafee, A.: The business of artificial intelligence. Harvard Bus. Rev. 1–20 (2017)

6. Charya, U.R., Oh, S.L., Hagiwara, Y., et al.: Deep convolutional neural network for the automated detection and diagnosis of seizure using EEG signals. Comput. Biol. Med. **100**, 270–278 (2018)
7. Weimer, D., Scholz-Reiter, B., Shpitalni, M.: Design of deep convolutional neural network architectures for automated feature extraction in industrial inspection. CIRP Ann. **65**(1), 417–420 (2016)
8. Uhrig, R.E.: Introduction to artificial neural networks. In: Proceedings of IECON 1995 - 21st Annual Conference on IEEE Industrial Electronics, vol. 1, pp. 33–37. IEEE (1995)
9. Krogh, A.: What are artificial neural networks? Nat. Biotechnol. **26**(2), 195–197 (2008)
10. Masood, A., Ali Al-Jumaily, A.: Computer aided diagnostic support system for skin cancer: a review of techniques and algorithms. Int. J. Biomed. Imaging **2013** (2013)
11. Freire, D.L., de Oliveira, R.F.A.P., Carmelo Filho, J.A.B., et al.: Machine learning applied in SARS-CoV-2 COVID 19 screening using clinical analysis parameters. IEEE Latin Am. Trans. **100**(1) (2020)
12. Elmoufidi, A., El Fahssi, K., Jai-Andaloussi, S., et al.: Anomaly classification in digital mammography based on multiple-instance learning. IET Image Proc. **12**(3), 320–328 (2017)
13. Elmoufidi, A.: Pre-processing algorithms on digital X-ray mammograms. In: 2019 IEEE International Smart Cities Conference (ISC2), pp. 87–92. IEEE (2019)
14. Bediang, G., Panpom, V.A., Koki, G., et al.: Utilisation d'un Logiciel d'Aide à la Décision pour le Dépistage de la Rétinopathie Diabétique au Cameroun. Health Sci. Dis. **21**(3) (2020)
15. Alam, M., Le, D., Lim, J.I., et al.: Supervised machine learning based multi-task artificial intelligence classification of retinopathies. J. Clin. Med. **8**(6), 872 (2019)
16. Ortiz-Feregrino, R., Tovar-Arriag, S., Ramos-Arreguin, J., et al.: Classification of proliferative diabetic retinopathy using deep learning. In: 2019 IEEE Colombian Conference on Applications in Computational Intelligence (ColCACI), pp. 1–6. IEEE (2019)
17. Gargeya, R., Leng, T.: Automated identification of diabetic retinopathy using deep learning. Ophthalmology **124**(7), 962–969 (2017)
18. Roychowdhury, S., Koozekanani, D.D., Parhi, K.K.: DREAM: diabetic retinopathy analysis using machine learning. IEEE J. Biomed. Health Inform. **18**(5), 1717–1728 (2013)
19. Lee, R., Wong, T.Y., Sabanayagam, C.: Epidemiology of diabetic retinopathy, diabetic macular edema and related vision loss. Eye Vis. **2**(1), 1–25 (2015)
20. Oth, G., Dicke, U.: Evolution of the brain and intelligence. Trends Cogn. Sci. **9**(5), 250–257 (2005)
21. Wang, S.-C.: Artificial neural network. In: Wang, S.-C. (ed.) Interdisciplinary Computing in Java Programming. The Springer International Series in Engineering and Computer Science, vol. 743, pp. 81–100. Springer, Boston (2003). https://doi.org/10.1007/978-1-4615-0377-4_5
22. Mola, A., Vishwanathan, S.V.N: Introduction to Machine Learning, vol. 32, no. 34. Cambridge University, Cambridge (2008)
23. Xiuqin, P., Zhang, Q., Zhang, H., et al.: A fundus retinal vessels segmentation scheme based on the improved deep learning U-Net model. IEEE Access. **7**, 122634–122643 (2019)
24. Skouta, A., Elmoufidi, A., Jai-Andaloussi, S., et al.: Automated binary classification of diabetic retinopathy by convolutional neural networks. In: Saeed, F., Al-Hadhrami, T., Mohammed, F., Mohammed, E. (eds.) Advances on Smart and Soft Computing, Advances in Intelligent Systems and Computing, vol. 1188, pp. 177–187. Springer, Singapore (2020). https://doi.org/10.1007/978-981-15-6048-4_16

RFID Based Security and Automatic Parking Access Control System

El hassania Rouan[1](✉) and Ahmed Boumezzough[2](✉)

[1] Faculty of Sciences and Technics, University Sultan Moulay Slimane,
Beni-Mellal, Morocco
[2] Faculty Poly-disciplinary, University Sultan Moulay Slimane,
Beni-Mellal, Morocco

Abstract. Nowadays, the most of existing parking management systems still rely on human interventions to perform thousands of parking transactions that can be more difficult in a huge parking, which receives many visitors. Problems that arise due this kind of parking management systems are access by unauthorized users, less public security and fraud to pay parking fees. In this paper, we address these problems by applying an automated technology at the terminal parking, which is Radio Frequency IDentification (RFID). With RFID technology, users cannot access to parking lots without RFID tag as identification and the check-in/ check-out can be done very fast by avoiding the congestion problem and decreasing the waiting time at every parking terminal. This system ensure a higher security as only the registered users are allowed to access into parking lot. In this paper, an automated parking access control system is designed and its implementation was made by simulating a miniature of the gate portal system using RFID technology. This system allows identification of vehicles, records at what time they enter or leave the parking lot and calculates the parking fee based on time duration between its arrival and its departure. The user is notified for the parking fee and payment receipt through an email.

Keywords: Parking fee · Radio Frequency IDentification · Smart parking system · Security · Unique IDentifier.

1 Introduction

Most of the existing parking systems need expensive equipment, infrastructure and deployment or they are semi-automated requiring human interventions for monitoring access of vehicles, which causes many parking problems. The solution for the parking problems is being raised. The smart parking system can be a good solution to minimize these problems and improving the utilization of parking facilities [1].

Recently, many research studies have concentrated on Radio Frequency IDentification (RFID) and have focused on different purposes to solve the parking

© Springer Nature Switzerland AG 2021
M. Fakir et al. (Eds.): CBI 2021, LNBIP 416, pp. 434–443, 2021.
https://doi.org/10.1007/978-3-030-76508-8_32

problems and improve the utilization of parking facilities. The employment of RFID technology in parking systems makes them full automatic with much lower costs and higher security by collecting vehicle information in real time using its unique identifier stored in its RFID tag. This technology has many benefits over previous identification technologies, such as smart cards and bar codes. RFID reader reads a data object without a physical contact in case of smart cards and does not need a direct line of sight as bar codes, which makes it able to read more than one item at the same time. Additionally, RFID can store information on the tag RFID according to its storage capacity.

In paper [2], authors proposed a prototype of an automatic barrier gate system controlled by mobile application. The system is developed by using RFID technology to check if the vehicles have permission to enter to the restricted area while the barrier gate is controlled using an application mobile. In this paper, the developed system is not full automatic because users need pressing the "open" button on their mobile phones to enter into the parking lot.

In another paper [3], authors designed an intelligent parking management system utilizing Radio Frequency IDentification. This system detects an empty parking slot by the RFID reader mounted on the drone and guides vehicles to it. In addition, this system is able to calculate the parking fee and charge it to the owners automatically. Although this paper did not mention anything about managing the parking access.

The authors of [4] proposed an automatic access control system using Arduino and RFID. This system is designed to distinguish between authorized and unauthorized users. The RFID reader reads RFID tag attached to user and checks it for matching with stored unique identification numbers on the Arduino Microcontroller for granting access. However, this paper did not include the attendance monitoring system for automatically registering vehicles those are entering and leaving the parking area.

This paper is an extended version of our work published in [5]. In this, an automatic and secure RFID-based access control system for parking lot is designed using NodeMCU V3 hardware and Arduino IDE software. The developed system combines RFID technology by using RFID module RC522 that acts as an RFID card reader and NodeMCU V3 board to accomplish the required task. The RFID reader installed at entrance and exit terminals of parking lot scans RFID card allotted to the registered user. The RFID reader captures the user Unique IDentifier (UID) and compares it with the stored UIDs for a match. If the captured user UID match with any of the stored UIDs, access is granted; otherwise access is denied. At the same time, the entrance time and exit time will record in database and vehicle parking fee will be generated based on it. The parking fee and payment receipt will be charge it to its owners automatically via an email.

The rest of this paper is organized as follows: Sect. 2 provides an overview on RFID technology including its main components and applications. Section 3 is dedicated to describe hardware and software components used in this project.

Section 4 describe the working prototype of the proposed system. Finally, some conclusions are drawn in Sect. 5.

2 An Overview on RFID Technology

Radio Frequency IDentification is an emerging technology with a vast array of applications in many industries and helps to identify animate or inanimate objects through radio waves. RFID is a wireless communication technology that allows a non-contact communication and identification of tagged objects without the need of a direct line of sight, which improves its processes efficiency, and ease of use [6].

RFID technology was invented during World War II but it was still unknown for commercial applications until the 1980s. One of its first applications is a British Radar system that used to identify allied or enemy planes through radiating a signal able to read an identification number which is used to distinguish allied aircrafts from enemy ones [7].

Nowadays, RFID play an important role in many fields due to its various benefits and features such as supply chain management, object tracking, inventory, smart parking systems, etc.

An RFID system consists mainly of three components, which are Reader(s), Tag(s) and RFID middleware. The readers identify tags attached to objects, animals, or persons within a given radio frequency range through radio waves without the need of human intervention or data entry for tracing purposes. The principle is to emit a radio frequency signal towards tags, which in return use the energy issued from the transmitted reader signal to reflect their response [8]. The RFID middleware presents the smart part of an RFID system. It is responsible for monitoring readers, managing the exchanged data and aggregating it to the dedicated applications [9].

In general, tags can be divided into three categories according to their power source: passives, semi-passives, and actives. An active tag has an integrated power supply that runs the integrated circuitry and the transmitter. It sends a radio frequency signal without being called by a RFID reader. A passive tag has not an embedded power source which makes it rely only on the received signal from RFID reader to generate its own quick response. A semi-passive tag has an internal power source that keeps only its own microchip and other integrated circuits such as sensors activated at all times but it is based on the same principle as passive tag for transmitting data to the RFID reader [9].

3 Overview on RFID Design

The main objective of this research paper is to design and implement a fast and secure controller of access for parking system. This system is equipped with a website programmed in order to display and register user accounts, receive payment notifications via email. This system includes many features:

– Authorize access only for the vehicles those recognized as registered.
– Record vehicles login and logout in a database.
– Calculate the parking fees.
– Notify users about their parking prices and payment receipt via an email.

To successfully design and implement an RFID based security and automatic parking access control system both hardware and software are being used.

3.1 Hardware

Various components were used to implement an RFID based parking access control system. The required hardware components for this project are described below in detailed manner:

Node MCU V3 (ESP8266): Is an open source firmware and development kit that plays a vital role in designing an IoT product. In this project, Node MCU V3 acts as an interface between the software part and the hardware part.

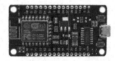

RFID Reader (RC522): Is a highly integrated RFID card reader that supports contact less 13.56 MHz communication.

RFID Tag: Is a passive RFID tag operating at 13.56 MHz frequency, which is allotted to registered users and contains a Unique IDentifier.

Servo Motor: Is a device that has a rotating axis. It is used in this project as a parking barrier gate controller.

LCD Display: Is an electronic visual display that displays 16 characters by two lines. It is used to display parking access information.

Buzzer and LEDs: Buzzer is an audio signaling device that used in this project to indicate an unsuccessful access of visitors. LEDs are used to show a successful (green light) or an unsuccessful (red light) access of visitors.

Block Diagram: The design of proposed RFID system is shown in Fig. 1. The proposed system is simulated using MF RC522 RFID module. The MF RC522 reader is connected to the NodeMCU V3 board in order to establish the communication between the RFID module and the program. The others components are also connected to NodeMCU V3 board to make them acting depending on the situation using a specific order of connection.

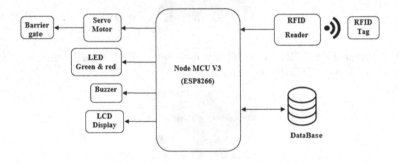

Fig. 1. The overall system block diagram

3.2 Software

We have used three main software programs in order to monitor the entrance and exit terminals of parking area and keep a record of user history, which are Arduino IDE, phpMyAdmin and MySQL. The proposed system is comprised of three phases: registration phase, recognition phase and payment phase.

- Registration phase: When the RFID reader detects the RFID card, the reader retrieves the UID and saves it in the database. From which it is retrieved and used to accord access into parking area. The user information are stored manually in a database using an interface web, which is administered by an admin.
- The recognition phase comes when the user wants to enter or leave the parking lot. This phase aims to check the user authenticity at two points: parking entrance and parking exit.

– In the payment phase, parking system generates the parking fees based on duration between arrival date and departure date. The user will be notified via an email about the details of its parking fee and payment receipt.

4   System Operation

This section describes the working procedure of the developed system and its implementation as a proof of concept. The procedures are pointed out using the flow diagram shown in Fig. 5. In this system, when a vehicle arrives at the parking entry or parking exit, the RFID reader detects if the presence vehicle has an RFID tag or not. If this vehicle has not an attached tag on it, the barrier gate will still closed. On the other hand, the RFID reader reads the UID stored in RFID tag and transfers the captured UID to the software part for checking its registration.

After getting user UID, as it came in the range of few millimeters from RFID reader, the software part will be able to compare the captured UID with the stored UIDs in database and generates one of the two messages: "access authorized" corresponding to registered user and "access refused" corresponding to a non-registered user. In case of registered user, access will be granted into parking area, the barrier gate will automatically open without pressing buttons or using remote controls and an authorized access message will be displayed on the LCD as shown in Fig. 3. On the other hand, if the UID of RFID tag does not match, the barrier gate will remain closed and the alarm will turn on to indicate unsuccessful access (Fig. 4). At the same time, information provided by the system are maintained in a database which keeps record of user history including check-in time and checkout time as shown in Fig. 5. Based on time duration between the arrival and the departure of each vehicle, the parking system can also calculate the parking fees and charge it to its owner automatically through an email (Fig. 6) but users still need to pay manually at each departure (Fig. 2).

Figure 3 shows that user has a valid UID, hence the authorized access is displayed on the LCD screen and the barrier gate open automatically.

Figure 4 shows that user has a non registered UID which prohibit him to have an access and an unauthorized access message is displayed on the LCD screen and the barrier gate stills closed.

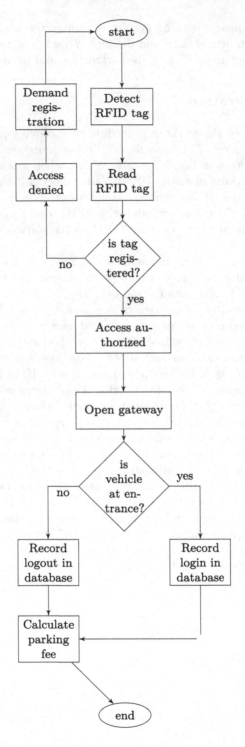

Fig. 2. Flow diagram of the introduced system.

Fig. 3. Authorized access.

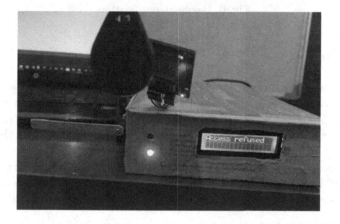

Fig. 4. Refused access.

RFID PARKING SYSTEM

Home

Registration

Users Data

Logs

Users Log Table

Select the date log:

jj/mm/aaaa

Select Date

N°	CardID	User Name	Date login	Time In	Date logout	Time Out	User Status	Payment (DHs)	Payment Status
1	459FEC20	El hassania ROUAN	2020-10-01	15:17:08	2020-10-01	16:55:00	Exit	0 Pay	Paied
2	047A399A	Mouhcine ROUAN	2020-10-01	16:34:12	2020-10-01	21:52:41	Exit	106 Pay	Non paied

Fig. 5. Login and logout interface.

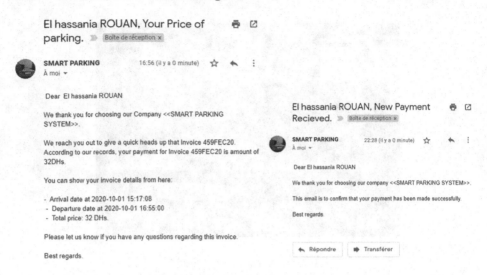

Fig. 6. Payment notification.

5 Conclusion

In this work, we have designed and implemented an automatic and secure RFID-based access control system for parking area that can be installed at its entrance and exit terminals. This system aims to monitor the parking entry/exit, keeps a record of user history, generates the parking fees and charges it to their owners automatically through an email.

The proposed system relies mainly on the employment of RFID technology and NodeMCU V3 board to distinguish between registered and non-registered users. This system helps to improve utilization of a parking facilities by:

- avoiding the wait time at each entry and exit terminals;
- increasing its security by allowing access only for the registered commuters.
- providing a reliable means of granting or denying access in a restricted area.
- providing a real time information about users: time of arrival, time of departure and parking fees.

Finally, the implementation of RFID technology in parking systems make them full automatic with higher security. It provides a low cost prototype for smart parking system, efficient monitoring system and better utilization of the overall parking system.

References

1. Tsiropoulou, E., Baras, J., Papavassiliou, S., Sinha, S.: RFID-based smart parking management system, Cyber-Phys. Syst. **3**(1–4), 22–41 (2017)

2. Kasym, K., Sarsenen, A., Segizbayev, Z., Junuskaliyeva, D., Ali, M.H.: Parking gate control based on mobile application. In: International Conference on Informatics, Electronics & Vision, pp. 399–403 (2018)
3. Wu, H.P.: Intelligent parking management system utilizing RFID. In: ACM MobiSys 2019 Rising Stars Forum (RisingStarsForum 2019), Seoul, Republic of Korea, 5 pages. ACM, New York (2019)
4. Orji, E.Z., Oleka, C.V., Nduanya, U.I.: Automatic access control system using arduino and RFID. J. Sci. Eng. Res. **5**(4), 333–340 (2018)
5. Rouan, E., Safi, S., Boumezzough, A.: An automated parking access control system based on RFID technology. In: 2020 15th Design & Technology of Integrated Systems in Nanoscale Era (DTIS), Marrakech, Morocco, pp. 1–2 (2020). https://doi.org/10.1109/DTIS48698.2020.9080913
6. Salah, B.: Design, simulation, and performance evaluation-based validation of a novel RFID-based automatic parking system. J. Simul. **96**(5), 487–497 (2019)
7. Castro, L., Wamba, F.: An inside look at RFID technology. J. Technol. Manage. Innov. **2**(1), 128–141 (2007)
8. Mbacke, A.A., Mitton, N., Rivano, H.: RFID reader anticollision protocols for dense and mobile deployments, MDPI electronics special issue. RFID Syst. Appl. 22 (2016)
9. Mohammed, A.M.: Modelling and optimization of an RFID-based supply chain network, thesis (2018)

Correction to: Game Theoretic Approaches to Mitigate Cloud Security Risks: An Initial Insight

Abdelkarim Ait Temghart, M'hamed Outanoute, and Mbarek Marwan

Correction to:
Chapter "Game Theoretic Approaches to Mitigate Cloud Security Risks: An Initial Insight" in: M. Fakir et al. (Eds.):
***Business Intelligence*, LNBIP 416,**
https://doi.org/10.1007/978-3-030-76508-8_24

In the originally published version of chapter 24, the author Driss Ait Omar was erroneously included in the author list. This has now been corrected.

The updated version of this chapter can be found at
https://doi.org/10.1007/978-3-030-76508-8_24

Author Index

Printed in the United States
by Baker & Taylor Publisher Services